Generative Social Science

Princeton Studies in Complexity

Simon A. Levin (Princeton University)
and Steven H. Strogatz (Cornell University), *Editors*

Lars-Erik Cederman, *Emergent Actors in World Politics:
How States and Nations Develop and Dissolve*

Robert Axelrod, *The Complexity of Cooperation: Agent-Based Models
of Competition and Collaboration*

Peter S. Albin, *Barriers and Bounds to Rationality: Essays on Economic
Complexity and Dynamics in Interactive Systems. Edited and with an
introduction by Duncan K. Foley*

Duncan J. Watts, *Small Worlds: The Dynamics of Networks between
Order and Randomness*

Scott Camazine, Jean-Louis Deneubourg, Nigel R. Franks,
James Sneyd, Guy Theraulaz, Eric Bonabeau,
Self-Organization in Biological Systems

Peter Turchin, *Historical Dynamics: Why States Rise and Fall*

Andreas Wagner, *Robustness and Evolvability in Living Systems*

Mark Newman, Albert-Laszlo Barabasi, and Duncan Watts, eds.,
The Structure and Dynamics of Networks

J. Stephen Lansing, *Perfect Order: Recognizing Complexity in Bali*

Joshua M. Epstein, *Generative Social Science: Studies in Agent-Based
Computational Modeling*

Generative
Social Science

STUDIES IN AGENT-BASED
COMPUTATIONAL MODELING

Joshua M. Epstein

PRINCETON UNIVERSITY PRESS

PRINCETON AND OXFORD

Published by Princeton University Press, 41 William Street,
Princeton, New Jersey 08540
In the United Kingdom: Princeton University Press, 3 Market Place, Woodstock,
Oxfordshire OX20 1SY

Requests for permission to reproduce material from
this work should be sent to Permissions,
Princeton University Press.

Library of Congress Cataloging-in-Publication Data
Epstien, Joshua M., 1951–
Generative social science: studies in agent-based computational
modeling / Joshua M. Epstein
p. cm. — (Princeton studies in complexity)
Includes bibliographical references and index.
ISBN-13: 978-0-691-12547-3 (cloth : alk. paper)
ISBN-10: 0-691-12547-3 (cloth : alk. paper)
1. Social sciences—Computer simulation.   2. Social sciences—
Mathematical models.   I. Title.   II. Series.
H61.3.E67 2007
300.1'13—dc22   2006004238

British Library Cataloging-in-Publication Data is available

This book has been composed in Sabon

Printed on acid-free paper. ∞

pup.princeton.edu

Printed in China

1 3 5 7 9 10 8 6 4 2

For Melissa, Matilda, and Joey

_____

CONTENTS

INTRODUCTION

The introduction to *Growing Artificial Societies* offers the following thought on the future of explanation:

> What constitutes an explanation of an observed social phenomenon? Perhaps one day people will interpret the question, "Can you explain it?" as asking "Can you *grow* it?" Artificial society modeling allows us to "grow" social structures *in silico* demonstrating that certain sets of microspecifications are *sufficient to generate* the macrophenomena of interest.... We can, of course, use statistics to test the match between the true, observed, structures and the ones we grow. But the ability to grow them... is what is new. Indeed, it holds out the prospect of a new, *generative*, kind of social science.[1]

A concluding section of the same work, entitled "Generative Social Science," restates the point even more broadly:

> *In effect, we are proposing a generative program for the social sciences and see the artificial society as its principal scientific instrument.* (177)

This book presents some of the achievements of that, now quite vibrant, program, and illustrates the scope of (at least my own) agent-based computational research since *Growing Artificial Societies*. Indeed, one candidate title for the present volume was *Growing Artificial Societies II*. But that book had its own flavor. While it made a substantial number of concrete claims (some of which will be recalled here), it was more a general "call to arms" than a concerted attack on any particular problem, more methodological than applied, more a laboratory than any particular experiment.

By contrast, the chapters that follow are much more focused studies in particular areas: the history of the Anasazi; the emergence of economic classes; the timing of retirement; the evolution of norms; the dynamics of ethnic conflict; the spread of epidemics, and organizational adaptation among them. While the chapters span the social sciences from archaeology to economics to epidemiology, there is unity to the volume. Indeed, each subsequent chapter illustrates core points made in the overarching methodological statement of chapter 1: "Agent-Based

---

[1] Joshua M. Epstein and Robert Axtell, *Growing Artificial Societies: Social Science from the Bottom Up* (Cambridge: MIT Press, 1996), 20.

Computational Models and Generative Social Science."[2] As such, the book is more than a collection; it makes an *argument*.

The Stakes: Explanation

To me, the core of that argument concerns the notion of a scientific explanation. This is really what is "at stake," if you will, in the advent of agent-based models: *What is to be the accepted standard of explanation in the social sciences?* In this book, I define and argue for a *generative* standard and highlight a tool—the agent-based computational model, or artificial society—that facilitates the construction of scientific models satisfying that standard. The notion of a generative explanation, which was not defined at any length in *Growing Artificial Societies*, is discussed at length in the opening chapter below, but is encapsulated nicely in the motto: "If you didn't grow it, you didn't explain it."[3] Or, under the obvious interpretation of the symbols:

$$\forall x(\neg Gx \supset \neg Ex) \tag{1}$$

Dynamic Attainment versus Static Existence of Equilibrium

This represents a sharp departure from prevailing practice. While there are notable dynamic exceptions, game theory and mathematical economics (the twin pillars of contemporary social science) are over-whelmingly concerned with *equilibria*, Nash equilibrium being the most important example. Indeed, in these quarters, "explaining an observed social pattern" is basically understood to mean "demonstrating that it is the Nash equilibrium (or a distinguished Nash equilibrium) of some game." However, these are mere demonstrations of existence. Per se, they do *not* demonstrate that the configurations of interest—the patterns allegedly explained—are attainable at all, much less attainable on time scales of interest to humans.[4] Moreover, standard equilibrium models impose very stringent demands on the individual's information and

---

[2] Joshua M. Epstein, "Agent-Based Computational Models and Generative Social Science," *Complexity* 4, no. 5 (1999): 41–60.

[3] Epstein, "Agent-Based Computational Models," 43. For an eloquent statement and powerful defense of this same generative explanatory standard in the field of linguistics, see Samuel David Epstein and T. Daniel Seely, eds., *Derivation and Explanation in the Minimalist Program* (Oxford: Blackwell, 2002), 1–18 (introduction) and 65–67.

[4] Debreu is like Proust: more noted than read. Debreu himself was actually very attentive to the problem of attainability, but did not resolve it in his seminal work, Gerard Debreu, *Theory of Value* (New York: Wiley, 1959). Later, Simon and Saari would show that he was right not to try—equilibrium was not attainable given inescapable limits on information and computing power. See Donald G. Saari and Carl P. Simon, "Effective Price Mechanisms," *Econometrica* 46(5) (1978): 1097–1125.

computing (optimizing) power. They often ignore space, assume global (not local) interactions, and involve little if any heterogeneity.

To the generativist, this is unsatisfactory; to explain a pattern, it does *not* suffice to demonstrate that—under this ensemble of strictures—*if* society is placed in that pattern, no (rational) individual would unilaterally depart (which is the Nash equilibrium condition). Rather, one must show how a population of boundedly rational (i.e., cognitively plausible) and heterogeneous agents, interacting locally in some space, could actually arrive at the pattern on time scales of interest—be it a wealth distribution, spatial settlement pattern, or pattern of violence. Hence, to explain macroscopic social patterns, we try to "grow" them in multi-agent models.

Nonequilibrium Systems

The preceding critique applies even when the pattern to be explained is an equilibrium. But what if it isn't? What if the social pattern of interest is itself a *nonequilibrium dynamic*? What if equilibrium exists, but is not attainable on acceptable time scales, or is unattainable outright? I hope the book demonstrates that the agent-based generative approach can be explanatory even in such cases—where "the equilibrium approach," if I may call it that, is either infeasible or is devoid of explanatory significance.

The Computer Is Not the Point

It is certainly true that recent advances in computing permit this agent-based generative social science *programme* to be pursued with unprecedented scope and vigor. The computer is a powerful laboratory in which to conduct experiments concerning the generative sufficiency of agent specifications. That is its contribution. But the essential move is conceptual, not technological. Using a computer to calculate equilibrium does not challenge equilibrium as an explanatory standard. Likewise, building an agent-based computational model in which all agents behave exactly as in neoclassical microeconomic theory does not constitute a departure from that field. *The computer is not the point.* The point is whether one's explanatory standards are generative or not. Indeed, the genre's pioneering models (e.g., Schelling's segregation model) were developed without computers at all. There are other misconceptions worth early rebuttal.

Agent Models Are Expressible as Equations

For one, the oft-claimed distinction between computational models and equation-based models is, in principle, illusory. Every agent-based computational model is, after all, a computer program, typically written in an object-oriented programming language, such as C++ or Java. Any such model is clearly Turing computable—computable by a Turing machine. But, it is a central result in logic and computability that for every Turing Machine, there exists a unique corresponding and equivalent recursive (partial) function. So, in principle, one could cast any agent-based computational model as an explicit set of mathematical formulas (recursive functions). In practice, these mathematical formulas would be gargantuan in size and imposingly complex, but in principle, they exist. One could have called the approach "recursive social science," or "effectively computable social science," "constructive social science," or any number of other equivalent things. The use of "generative" was inspired by Chomsky's usage. In any event, the issue is not whether equivalent equations exist, but which representation (equations or programs) is most illuminating. To all but the most adept practitioners, and perhaps to them as well, the recursive function representation would be utterly unrecognizable as a model of social interaction, while the equivalent agent model is immediately intelligible as such.

Agent Models Deduce

Another misconception is that the equation-based (or "axiom-theorem-proof") approach is *deductive*, whereas the agent-based computational approach is not. This is also incorrect. Every realization of an agent-based model *is* a strict deduction. Indeed, three demonstrations of this point are offered in chapter 2, which seeks to clarify a set of basic foundational issues, among them the following:

1. Generative sufficiency versus explanatory necessity
2. Generative agent-based models versus explicit mathematical models
3. Generative explanation versus deductive explanation
4. Generative explanation versus inductive explanation
5. Incompleteness and uncomputability in mathematical social science
6. Generality of agent models
7. Beauty of agent models

The last of these topics should not be underestimated. G. H. Hardy wrote that, in mathematics, "Beauty is the first test: there is no permanent place in the world for ugly mathematics."[5] And at first blush, agent-based

---

[5]G. H. Hardy, *A Mathematician's Apology* (1940), Canto version (Cambridge: Cambridge University Press, 1992), 85.

computational models may appear to lack the kind of parsimonious elegance toward which theoretical minds are drawn, and toward which the best mathematical social science aspires. However, I argue in chapter 2 and its prelude that agent models can be parsimonious, elegant, minimal, and beautiful in the mathematician's sense, a sense (by the way) that has nothing to do with the colorful visualizations associated with the approach.

The Plan of the Book

The opening chapter, "Agent-Based Computational Models and Generative Social Science," sets the stage for all the chapters that follow. Indeed, they function as supporting evidence for the argument advanced in that overarching Generative statement. Chapter 1 argues in detail that the agent-based computational model—the artificial society—is a new scientific instrument, and that it permits a distinctive approach to social science, for which the term *generative* is appropriate. While the chapter is focused on foundational issues, each of the subsequent chapters is discussed there in some connection, or (if published more recently) reinforces core points made there.[6] For example, the Generative chapter argues that the artificial society permits data-driven empirical research. The Artificial Anasazi research (chapters 4–6) and the Smallpox epidemic model (chapter 12) are examples. The more theoretical Classes model (chapter 8) is discussed in connection with nonequilibrium social dynamics. Some chapters are more exploratory than explanatory, "Growing Adaptive Organizations" (chapter 13) being a case in point. Chapter 2, "Remarks on the Foundations of Agent-Based Generative Social Science," amplifies on, and extends, a number of epistemological points made in the overarching first chapter. It also discusses the sense in which agent models can be beautiful.

Chapters 2 and 13 have not been previously published. The others appeared in far-flung books and journals where their relations to one another are obscure. Here interconnections are emphasized. From very diverse quarters, these studies are assembled here under one roof, and in support of one particular argument.

---

[6]Some of the previously published chapters contain references to works that were forthcoming when the chapters were first published. Many of these have been published since the original chapter's publication. For the reader's convenience, these references have been updated. This explains the occasional appearance of references more recent than the original chapter's publication date.

The Preludes

Preceding each chapter, I offer a brief discourse, or prelude, highlighting those points in the Generative essay, chapter 1, the model is illustrating, and identifying what phenomena it is generating. (Where pertinent or interesting, I also recall how the project got started.) The intended result is a unified statement in which theory (the Generative chapter) and applications (the subsequent chapters) enhance and amplify one another.

Rather than give a précis of each chapter, it may be more interesting to now review some ways in which the models illustrate the distinguishing features of agent-based models, as enumerated in the Generative statement. In addition to autonomous agents (which all the models display), the typical artificial society model involves heterogeneous agents, bounded rationality, explicit space, local interactions, and (often) nonequilibrium dynamics. The way in which the various models handle these things is itself quite heterogeneous!

Heterogeneity

To begin, the agents differ differently in the different models. In the Anasazi chapters they differ by age and fertility. In the Retirement model, they have different decision rules. Some decide by tossing coins; some (very few!) decide like *homo economicus*; and some play a coordination game in their social networks, where the social networks are themselves heterogeneous and dynamic. In the Classes model, agents differ by what they store in their memory (their recent interactions). In the model of Thoughtless conformity, they differ by dynamic search radius. In the Demographic Prisoner's Dilemma, they exhibit diversity in ages and levels of accumulated wealth. Civil Violence agents are heterogeneous by economic hardship, as well as by local information and political grievance, both of which are dynamic. Agents differ by disease stage in the Smallpox model. In the adaptive organizations of chapter 13, managers control different sets of resources, and face different and dynamic local environments.

Bounded Rationality

While all the models assume bounded rationality, the way in which it is bounded varies widely. In the Classes model, agents play best reply to recent sample evidence, where the sample is of a fixed size on the order of 10 percent of the population—bounded. In the model of Thoughtless conformity, the sample size itself is dynamic and can shrink to zero. In the extreme case of the Demographic Prisoner's Dilemma, agents are pegged at zero: they are hardwired to play a given strategy. One's rationality can't be more bounded than that! By contrast, in the Retirement model, a few agents decide when to retire by tossing a coin each year and retiring if it's heads. The preponderance plays a coordination game (retire, work) in their dynamic social network. Neither strategy is economically rational. In the Civil Violence model, agents use only local information and play "rebel" (or not) based on a type of expected utility calculation, but (as in the voting paradox) do *not* consider how their individual choice affects the stability of the ruling regime (the probability of successful revolution).

Space and Local Interactions

The interaction spaces used range from the very concrete to the abstract. In the Anasazi models, the space is real—a reconstructed landscape with estimated values of environmental variables (hydrology, maize potential, topsoil aggradation, etc). In the Retirement model, the space in which agents interact is a dynamic social network. In the Civil Violence and Prisoner's Dilemma models, it is a 2-dimensional lattice (a topological torus). In the Thoughtless conformity model it is a 1-dimensional ring. In the Smallpox epidemic model, it is a realistic two-town county with homes, workplaces, schools, and a hospital (from the pathogen's viewpoint, it is a landscape of hosts moving around this space!). In the model of Adaptive Organizations, the space is a variable geometry graph. In all cases, agents interact locally with others in the space of interest—within a finite vision for Civil Violence, Prisoner's Dilemma, and Thoughtless conformity; within social networks for Retirement and Epidemics; on a subgraph for Organizations.

Nonequilibrium Dynamics

Only one chapter focuses on equilibrium, and it does so in large part to highlight its weakness as an explanatory tool. Chapter 3, entitled "Non-Explanatory Equilibria," presents a simple game most of whose equilibria are unattainable outright. Moreover, for most of the remaining equilibria, convergence times scale exponentially in the number of agents, making them unattainable in any practical sense. The other models included exhibit various types of nonequilibrium dynamics. The Civil Violence and Thoughtless conformity models show punctuated equilibrium, or intermittence, depending on one's definition. The Prisoner's Dilemma and Organization models show cycling. Even where—as in the Classes model—asymptotic equilibrium can be proven, waiting times are astronomical, and the long-lived transients are studied computationally. The Smallpox model is concerned with spatiotemporal epidemic dynamics and "quenching" phenomena, not equilibrium. Since they are calibrated to historical data, the Smallpox and Anasazi models are both examples of empirical nonequilibrium social science.

The book, then, offers a kind of gallery of ways in which agent models can display the distinctive characteristics of the general approach, as identified in the opening chapter.

The CD

The book also includes a CD, which can be used in a variety of modes.[7]

PLAY MOVIES

The CD Movies folder contains roughly forty movies (primarily Quick-Time) published here for the first time. Each movie is an animated realization of an agent-based model presented in one of the chapters. One can simply play the movies corresponding to model runs discussed in the various chapters. The movies are read-only and cannot be modified.

RUN AND MODIFY MODELS PARAMETRICALLY

However, all the models in the book are implemented in the Ascape agent-based modeling environment.[8] One can run selected models as Ascape applets[9] from the CD, adjusting parameters at will, exploring how changes in assumptions affect model outcomes. One does not need to know any computer programming (coding) to run the models in this mode. Simply open the Ascape Applets folder and double-click on the Ascape.exe file. A dialog box then appears from which one can select models from the pull-down menu. In courses on agent-based modeling, one could read a book chapter and play the movies of runs discussed in the text. Then, as projects, students could load the models in Ascape and alter the assumptions, explore sensitivities, or analyze variations.[10] For example, the Demographic Prisoners' Dilemma (PD) Ascape code is completely general. To explore the Demographic Coordination Game,[11] simply load the PD Model, change the payoffs, and re-run. Beyond their pedagogical applications, the same models can obviously be used for research. The variations possible in this mode, while extensive, are largely

---

[7] As discussed on the CD's Read Me file, administrative privileges may be required in some cases.

[8] Ascape was developed at The Brookings Institution by Miles T. Parker.

[9] All applets on this CD download with Ascape, and were implemented at The Brookings Institution by Miles T. Parker.

[10] Full instructions for doing this are given on the CD's Read Me file.

[11] See the Appendix to Chapter 9.

parametric. To change a particular model's agent rules syntactically, or to make up completely new ones, access to the model's source code is required. Subject to the copyright restrictions noted below (and elaborated on the CD's Read Me file), source code for all models is available from the author on request.

THE READ ME FILE

Instructions for using the CD are provided in full detail in the CD's Read Me file. Beyond that, a detailed description of Ascape and its runtime functionality is also available at: http://jasss.soc.surrey.ac.uk/4/1/5.html.[12] In addition to most of the models in this book, a "positive externality" is that certain models *not* in this book also download with Ascape, and are hence included as applets on this CD. These are: The Epstein-Axtell Sugarscape model and Conway's Game of Life. Sugarscape (and Schelling Segregation model) movies are available on the CD accompanying Epstein and Axtell (1996).

COPYRIGHT RESTRICTIONS

Note that the CD's Read Me file contains important copyright restrictions. Ascape is copyright 1998–2000, The Brookings Institution. This software may be used without fee for noncommercial purposes. For more information, please refer to the full copyright statement on the CD.

COLLEAGUES

The argument for, and the terminology of, *generative explanation* elaborated in chapter 1 are mine; each of the subsequent chapters functions here as supporting evidence—as "exhibits A, B, and C"—for that overall generative "case." Neither that case nor the generative terminology is necessarily shared by my coauthors on various chapters. These scholars, of course, are credited fully where those chapters were originally published, and again here. Here, however, the chapters function as supporting evidence, as data, for the generative argument I alone am making. Hence, Princeton University Press considers the book to be an authored (rather than an edited) work and I appear as the book's author. But I wish to emphasize my profound debt to all chapter coauthors, and to underscore that they deserve full credit on the individual models presented. At the same time they are absolved of any responsibility for

---

[12] Miles T. Parker, "What is Ascape and Why Should You Care?" *Journal of Artificial Societies and Social Simulation* 4(1)(2001).

the overarching argument I am making here, and for the use to which those models are being put in this volume.

A second point is also worth making. My assignment was to consolidate work in which I have been directly involved since *Growing Artificial Societies*. It was not to collect what I consider to be the best pieces of agent-based social science and edit the collection. Had that been my charge, the result would certainly include seminal solo agent-based articles by my coauthors in the present volume and doubtless by others with whom I have never directly worked. Whether *that* "greatest agent hits" book would rightfully include all—indeed any of—the present work's chapters, I leave it for the reader to judge.

Acknowledgments

Beyond my coauthors, I wish to especially acknowledge my brother, and comrade in generative arms, Samuel David Epstein. I also thank my colleagues at the Brookings Institution, Santa Fe Institute, and Johns Hopkins University. For thoughtful reviews of the entire manuscript, I thank Brian Skyrms, John D. Steinbruner, Duncan Foley, Leigh Tesfatsion, and Herbert Gintis. For insightful comments on an earlier draft, I thank Samuel Bowles. Chapter coauthors Ross Hammond, Jon Parker, and Shubha Chakravarty did double duty in also wrestling the CD to bed. Danielle Feher was invaluable in preparing the manuscript for publication. At Princeton University Press, I thank Vickie Kearn, Peter Dougherty, Dimitri Karetnikov, Meera Vaidyanathan, and Richard Isomaki.

Tolstoy wrote that "Happy families are all alike..."[13] While our family is distinctive in certain respects, I thank my beloved Melissa, Matilda, and Joey for making it undistinctive in this all-important one. I dedicate this book to them, in the memory of my most distinctive parents, Lucille and Joseph.

---

[13] The complete line, with which Tolstoy begins *Anna Karenina*, is, "Happy families are all alike; every unhappy family is unhappy in its own way."

Generative Social Science

THE GENERATIVIST MANIFESTO

EVERY SO OFTEN, there is a special conference of the Santa Fe Institute External Faculty. Initially called the Integrative Themes Workshops, their laudable purpose has been to explore crosscutting, unifying aspects of all the diverse efforts underway in the far-flung Institute, from physics, to computer science, to evolutionary ecology, to archaeology, to economics, to immunology.

Over the years, these external faculty conferences have continued. Just as one would expect of a high-powered transdisciplinary community like SFI, the areas of disagreement probably dominate those of consensus. And so, with a healthy self-deprecating sense of its own heterogeneity (if not incoherence), the Institute took to calling these external faculty conferences the *Disintegrative* Themes Workshops.

It was for the 1998 workshop that I wrote the first chapter. I had a number of narrower topics in mind, but I became convinced that the time was ripe for a *manifesto* on agent-based social science. Truth be known, not everyone in the SFI community shared either my interpretation of, or my enthusiasm for, the approach. And I wrote the chapter as much to sway Institute colleagues—or at least "clear the air"—as for the external audience.

The chapter is the overarching statement for this entire book. Its function, in relation to the other chapters, is well described in the volume's Introduction above, and its own abstract offers a concise précis of its contents. I will not repeat those here. However, one topic I especially wanted to "clear the air" about in Santa Fe was "emergence," a central term in the complexity community, and one I thought was being used uncritically. Unexpectedly, the "emergence" section of the chapter may prove to be among its more controversial, which is fine with me, so long as the controversy is not founded on misinterpretation. To ensure that it isn't, a few new words on that old topic may be permissible.

Emergence

W. V. Quine insightfully observed: "The less a science is advanced, the more its terminology rests on an uncritical assumption of mutual understanding."[1] It seems to me that in our young field of complexity, the terms "emergence" and "emergent phenomena" are excellent examples of Quine's point. My own gedankenexperiment, having attended a great many complexity conferences, is this: Everyone at the conference is handed a piece of paper with the following five statements on it, and they are asked the check those with which they agree:

1. Emergent phenomena are undeducible.
2. Emergent phenomena are unexplainable.
3. Emergent phenomena are unpredictable.
4. Emergent phenomena are unreducible.
5. Emergent phenomena are none of the above.

Leaving aside the obvious questions of what, exactly, any of these statements asserts and how they may be related (e.g., does 1 imply 4?), my strong suspicion is that the set of responses would be far from uniform. I do not insist that the term "emergent phenomenon" *cannot* be defined clearly. And when so defined (perhaps as in *Growing Artificial Societies*),[2] I certainly do not object to its use; but neither do I see that it denotes anything particularly new. The general questions of interest to all of us—*how ensembles achieve functionalities (or properties) their constituents lack*[3]—are certainly important, deep, and fascinating; and the new techniques will shed fundamentally new light on them, but the questions themselves are not new. Neither, unfortunately, is the terminology of "emergence." However, its intellectual pedigree is dreadful, beginning in the 1920s with antiscientific deists like Lloyd Morgan (who would have checked 1 at the very least), whose works were appropriately savaged by the philosophical likes of Bertrand Russell[4] and Ernest Nagel.[5]

---

[1] W. V. Quine, "Truth by Convention" (1936), in *Philosophy of Mathematics: Selected Readings*, ed. Paul Benacerraf and Hilary Putnam, 2nd ed. (Cambridge: Cambridge University Press, 1983).

[2] Joshua M. Epstein and Robert L. Axtell, *Growing Artificial Societies: Social Science from the Ground Up* (Washington, DC: Brookings Institution Press; Cambridge: MIT Press, 1996).

[3] I am prepared to grant that this is what is meant when people use the vague terminology of a "whole exceeding the sum of its parts." I wonder if the whole also exceeds the *product* of its parts.

[4] Russell's critique is characteristically biting and colorful. Referring to Morgan's 1923 treatise, *Emergent Evolution*, Russell, The Scientific Outlook (New York: W. W. Norton), (1931) writes:

Lloyd Morgan believes that there is a Divine Purpose underlying the course of evolution, more particularly of what he calls "emergent evolution." The definition

While granting that contemporary complexity researchers are *not* all using the term "emergence" exactly as the classical emergentists did, I also felt that a discussion of the latter's many confusions should be of more than antiquarian interest. As Santayana admonished, "Those who cannot remember the past are condemned to repeat it."

All of that said, the overarching Generative chapter that follows is focused on agent-based modeling and its many potential applications in the social sciences, a number of which are offered in the subsequent chapters.

---

of emergent evolution, if I understand it rightly, is as follows:

> *It sometimes happens that a collection of objects arranged in a suitable pattern will have a new property which does not belong to the objects singly, and which cannot, as far as we can see, be deduced from their several properties together with the way in which they are arranged.*

> He [Morgan] considers that there are examples of the same kind of thing even in the inorganic realm. The atom, the molecule, and the crystal will all have properties which, if I understand Lloyd Morgan right, he regards as not deductible from the properties of their constituents. The same holds in a higher degree of living organisms, and most of all with those higher organisms which possess what are called minds... "Emergent evolution," he [Morgan] says, "is from first to last a revelation and manifestation of that which I speak of as Divine Purpose." (129–30; emphasis added)

It seems to me that the central italicized passage is identical to much contemporary usage, to which position Russell responds:

> It would be easier to deal with this view if any reasons were advanced in its favor, but so far as I have been able to discover from professor Lloyd Morgan's pages, he considers that the doctrine is its own recommendation and does not need to be demonstrated by appeals to the mere understanding.

[5]See the discussion in Ernest Nagel, *The Structure of Science* (New York: Harcourt, Brace, and World), 1961.

Chapter 1

AGENT-BASED COMPUTATIONAL MODELS AND

GENERATIVE SOCIAL SCIENCE

Joshua M. Epstein*

THIS article argues that the agent-based computational model permits a distinctive approach to social science for which the term "generative" is suitable. In defending this terminology, features distinguishing the approach from both "inductive" and "deductive" science are given. Then, the following specific contributions to social science are discussed: The agent-based computational model is a new tool for empirical research. It offers a natural environment for the study of connectionist phenomena in social science. Agent-based modeling provides a powerful way to address certain enduring—and especially interdisciplinary—questions. It allows one to subject certain core theories—such as neoclassical microeconomics—to important types of stress (e.g., the effect of evolving preferences). It permits one to study how rules of individual behavior give rise—or "map up"—to macroscopic regularities and organizations. In turn, one can employ laboratory behavioral research findings to select among competing agent-based ("bottom up") models. The agent-based approach may well have the important effect of decoupling individual rationality from macroscopic equilibrium and of separating decision science from social science more generally. Agent-based modeling offers powerful new forms of hybrid theoretical-computational work; these are particularly relevant to the study of non-equilibrium systems. The agent-based approach invites the interpretation of society as a distributed computational device, and in turn the interpretation of social dynamics as a type of computation. This interpretation raises important foundational issues in social science—some related to intractability, and some to undecidability proper. Finally, since "emergence"

*The author is a senior fellow in Economic Studies at The Brookings Institution and a member of the External Faculty of the Santa Fe Institute.

For insightful comments and valuable discussions, the author thanks George Akerlof, Robert Axtell, Bruce Blair, Samuel Bowles, Art DeVany, Malcolm DeBevoise, Steven Durlauf, Samuel David Epstein, Herbert Gintis, Alvin Goldman, Scott Page, Miles Parker, Brian Skyrms, Elliott Sober, Leigh Tesfatsion, Eric Verhoogen, and Peyton Young. For production assistance he thanks David Hines.

This essay was published previously in *Complexity* 4(5): 41–60.

figures prominently in this literature, I take up the connection between agent-based modeling and classical emergentism, criticizing the latter and arguing that the two are incompatible.

GENERATIVE SOCIAL SCIENCE

The agent-based computational model—or artificial society—is a new scientific instrument.[1] It can powerfully advance a distinctive approach to social science, one for which the term "generative" seems appropriate. I will discuss this term more fully below, but in a strong form, the central idea is this: To the generativist, explaining the emergence[2] of macroscopic societal regularities, such as norms or price equilibria, requires that one answer the following question:

The Generativist's Question

＊How could the decentralized local interactions of heterogeneous autonomous agents generate the given regularity?

The agent-based computational model is well-suited to the study of this question since the following features are characteristic:[3]

HETEROGENEITY

Representative agent methods—common in macroeconomics—are not used in agent-based models (see Kirman 1992). Nor are agents

---

[1] A basic exposure to agent-based computational modeling—or artificial societies—is assumed. For an introduction to agent-based modeling and a discussion of its intellectual lineage, see Epstein and Axtell 1996. I use the term "computational" to distinguish artificial societies from various equation-based models in mathematical economics, n-person game theory, and mathematical ecology that (while not computational) can legitimately be called agent-based. These equation-based models typically lack one or more of the characteristic features of computational agent models noted below. Equation based models are often called "analytical" (as distinct from computational), which occasions no confusion so long as one understands that "analytical" does not mean analytically tractable. Indeed, computer simulation is often needed to approximate the behavior of particular solutions. The relationship of agent-based models and equations is discussed further below.

[2] The term "emergence" and its history are discussed at length below. Here, I use the term "emergent" as defined in Epstein and Axtell (1996, 35), to mean simply "arising from the local interaction of agents."

[3] The features noted here are not meant as a rigid definition; not all agent-based models exhibit all these features. Hence, I note that the exposition is in a strong form. The point is that these characteristics are easily arranged in agent-based models.

aggregated into a few homogeneous pools. Rather, agent populations are heterogeneous; individuals may differ in myriad ways—genetically, culturally, by social network, by preferences—all of which may change or adapt endogenously over time.

AUTONOMY

There is no central, or "top-down," control over individual behavior in agent-based models. Of course, there will generally be feedback from macrostructures to microstructures, as where newborn agents are conditioned by social norms or institutions that have taken shape endogenously through earlier agent interactions. In this sense, micro and macro will typically co-evolve. But as a matter of model specification, no central controllers or other higher authorities are posited *ab initio*.

EXPLICIT SPACE

Events typically transpire on an explicit space, which may be a landscape of renewable resources, as in Epstein and Axtell (1996), an n-dimensional lattice, or a dynamic social network. The main *desideratum* is that the notion of "local" be well posed.

LOCAL INTERACTIONS

Typically, agents interact with neighbors in this space (and perhaps with environmental sites in their vicinity). Uniform mixing is generically not the rule.[4] It is worth noting that although this next feature is logically distinct from generativity, many computational agent-based models also assume:

BOUNDED RATIONALITY

There are two components of this: bounded information and bounded computing power. Agents do not have global information, and they do not have infinite computing power. Typically, they make use of simple rules based on local information (see Simon 1982 and Rubinstein 1998).

The agent-based model, then, is especially powerful in representing spatially distributed systems of heterogeneous autonomous actors with bounded information and computing capacity who interact locally.

---

[4] For analytical models of local interactions, see Blume and Durlauf 2001.

The Generativist's Experiment

In turn, given some macroscopic *explanandum*—a regularity to be explained—the canonical agent-based experiment is as follows:

> **Situate an initial population of autonomous heterogeneous agents in a relevant spatial environment; allow them to interact according to simple local rules, and thereby generate—or "grow"—the macroscopic regularity from the bottom up.[5]

Concisely, ** is the way generative social scientists answer *. In fact, this type of experiment is not new[6] and, in principle, it does not necessarily involve computers.[7] However, recent advances in computing, and the advent of large-scale agent-based computational modeling, permit a generative research program to be pursued with unprecedented scope and vigor.

Examples

A range of important social phenomena have been generated in agent-based computational models, including: right-skewed wealth distributions (Epstein and Axtell 1996), right-skewed firm size and growth rate distributions (Axtell 1999), price distributions (Bak *et al.* 1993), spatial settlement patterns (Dean *et al.* 1999), economic classes (Axtell *et al.* 2001), price equilibria in decentralized markets (Albin and Foley 1990; Epstein and Axtell 1996), trade networks (Tesfatsion 1995; Epstein and Axtell 1996), spatial unemployment patterns (Topa 1997), excess volatility in returns to capital (Bullard and Duffy 1998), military tactics (Ilachinski 1997), organizational behaviors (Prietula, Carley, and Gasser

---

[5]We will refer to an initial agent-environment specification as a microspecification. While, subject to outright computational constraints, agent-based modeling *permits* extreme methodological individualism, the "agents" in agent-based computational models are not always individual humans. Thus, the term "microspecification" implies substantial—but not necessarily complete—disaggregation. Agent-based models are naturally implemented in object-oriented programming languages in which agents and environmental sites are objects with fixed and variable internal states (called instance variables), such as location or wealth, and behavioral rules (called methods) governing, for example, movement, trade, or reproduction. For more on software engineering aspects of agent-based modeling, see Epstein and Axtell 1996.

[6]Though he does not use this terminology, Schelling's (1971) segregation model is a pioneering example.

[7]In fact, Schelling did his early experiments without a computer. More to the point, one might argue that, for example, Uzawa's (1962) analytical model of non-equilibrium trade in a population of agents with heterogeneous endowments is generative.

1998), epidemics (Epstein and Axtell 1996), traffic congestion patterns (Nagel and Rasmussen 1994), cultural patterns (Axelrod 1997c; Epstein and Axtell 1996), alliances (Axelrod and Bennett 1993; Cederman 1997), stock market price time series (Arthur *et al.* 1997), voting behaviors (Kollman, Miller, and Page 1992), cooperation in spatial games (Lindgren and Nordahl 1994; Epstein 1998; Huberman and Glance 1993; Nowak and May 1992; Miller 1996), and demographic histories (Dean *et al.* 1999). These examples manifest a wide range of (often implicit) objectives and levels of quantitative testing.

Before discussing specific models, it will be useful to identify certain changes in perspective that this approach may impose on the social sciences. Perhaps the most fundamental of these changes involves explanation itself.

EXPLANATION AND GENERATIVE SUFFICIENCY

Agent-based models provide computational demonstrations that a given microspecification is in fact *sufficient to generate* a macrostructure of interest. Agent-based modelers may use statistics to gauge the generative sufficiency of a given microspecification—to test the agreement between real-world and generated macro structures. (On levels of agreement, see Axtell and Epstein 1994.) A good fit demonstrates that the target macrostructure—the *explanandum*—be it a wealth distribution, segregation pattern, price equilibrium, norm, or some other macrostructure, is effectively attainable under repeated application of agent-interaction rules: It is *effectively computable by agent society.* (The view of society as a distributed computational device is developed more fully below.) Indeed, *this demonstration is taken as a necessary condition for explanation itself.* To the generativist—concerned with *formation dynamics*—it does not suffice to establish that, if deposited in some macroconfiguration, the system will stay there. Rather, the generativist wants an account of the configuration's *attainment by a decentralized system of heterogeneous autonomous agents.* Thus, the motto of *generative social science*, if you will, is: If you didn't grow it, you didn't explain its emergence. Or, in the notation of first-order logic:

$$(\forall x)(\neg Gx \supset \neg Ex) \tag{1}$$

It must be emphasized that the motto applies only to that domain of problems involving the formation or emergence of macroscopic regularities. Proving that some configuration is a Nash equilibrium, for example, arguably *does* explain its persistence, but does not account for its attainment.[8]

Regarding the converse of expression (1), if a microspecification, m, generates a macrostructure of interest, then m is a *candidate* explanation. But it may be a relatively weak candidate; merely generating a macrostructure does not necessarily explain its formation particularly well. Perhaps Barnsley's fern (Barnsley 1988) is a good mathematical example. The limit object indeed looks very much like a black spleenwort fern. But—under iteration of a certain affine function system—it assembles itself in a completely unbiological way, with the tip first, then a few outer branches, eventually a chunk of root, back to the tip, and so forth—not connectedly from the bottom up (now speaking literally).

It may happen that there are distinct microspecifications having equivalent generative power (their generated macrostructures fit the macro-data equally well). Then, as in any other science, one must do more work, figuring out which of the microspecifications is most tenable empirically. In the context of social science, this may dictate that competing microspecifications with equal generative power be adjudicated experimentally—perhaps in the psychology lab.

In summary, if the microspecification m does not generate the macrostructure x, then m is not a candidate explanation. If m does generate x, it is a candidate.[9] If there is more than one candidate, further work is required at the micro-level to determine which m is the most tenable explanation empirically.[10]

---

[8]Likewise, it would be wrong to claim that Arrow-Debreu general equilibrium theory is devoid of explanatory power because it is not generative. It addresses different questions than those of primary concern here.

[9]For expository purposes, I write as though a macrostructure is either generated or not. In practice, it will generally be a question of degree.

[10]Locating this (admittedly informal) usage of "explanation" in the vast and contentious literature on that topic is not simple and requires a separate essay. For a good collection on scientific explanation, see Pitt 1988. See also Salmon 1984, Cartwright 1983, and Hausman 1992. Very briefly, because no general scientific (covering) laws are involved, generative sufficiency would clearly fail one of Hempel and Oppenheim's (1948) classic deductive-nomological requirements. Perhaps surprisingly, however, it meets the deduction requirement itself, as shown by the Theorem below. That being the case, the approach would appear to fall within the hypothetico-deductive framework described in Hausman (1992, 304). A microspecification's failure to generate a macrostructure falsifies the hypothesis of its sufficiency and disqualifies it as an explanatory candidate, consistent with Popper (1959). Of course, sorting out exactly what component of the microspecification— core agent rules or auxiliary conditions—is producing the generative failure is the Duhem problem. Our weak requirements for explanatory candidacy would seem to have much in common with the *constructive empiricism* of van Fraassen (1980). On this antirealist position, truth (assuming it has been acceptably defined) is eschewed as a goal. Rather,

For most of the social sciences, it must be said, the problem of multiple competing generative accounts would be an embarrassment of riches. The immediate agenda is to produce generative accounts *per se*. The principal instrument in this research program is the agent-based computational model. And as the earlier examples suggest, the effort is underway.

This agenda imposes a constructivist (intuitionistic) philosophy on social science.[11] In the air is a foundational debate on the nature of explanation reminiscent of the controversy on foundations of mathematics in the 1920s–30s. Central to that debate was the intuitionists' rejection of nonconstructive existence proofs (see below): their insistence that meaningful "existence in mathematics coincides with constructibility" (Fraenkel and Bar-Hillel 1958, 207). While the specifics are of course different here—and I am not discussing intuitionism in mathematics proper—this is the impulse, the spirit, of the agent-based modelers: If the distributed interactions of heterogeneous agents can't generate it, then we haven't explained its emergence.

Generative versus Inductive and Deductive

From an epistemological standpoint, generative social science, while empirical (see below), is not *inductive*, at least as that term is typically used in the social sciences (e.g., as where one assembles macroeconomic data and estimates aggregate relations econometrically). (For a nice introduction to general problems of induction, beginning with Hume, see Chalmers 1982. On inductive logic, see Skyrms 1986. For Bayesians and their critics, see, respectively, Howson and Urbach 1993 and Glymour 1980.)

The relation of generative social science to *deduction* is more subtle. The connection is of particular interest because there is an intellectual tradition in which we account an observation as explained precisely when we can *deduce the proposition expressing that observation from other, more general, propositions.* For example, we *explain* Galileo's leaning

---

"science aims to give us theories which are empirically adequate; and acceptance of a theory involves as belief only that it is empirically adequate" (von Fraassen 1980, 12). However, faced with competing microspecifications that are *equally adequate* empirically (i.e., do equally well in generating a macro target), one would choose by the criterion of empirical plausibility at the micro level, as determined experimentally. On realism in social science, see Hausman 1998.

[11] Constructivism in this mathematical sense should not be confused with the doctrine of social constructionism sometimes identified with so-called "post-modernism" in other fields.

Tower of Pisa observation (that heavy and light objects dropped from the same height hit the ground simultaneously) by strictly deducing, from Newton's Second Law and the Law of Universal Gravitation, the following proposition: "The acceleration of a freely falling body near the surface of the earth is independent of its mass." In the present connection, we seek to explain macroscopic social phenomena. And we are requiring that they be generated in an agent-based computational model. Surprisingly, in that event, we can legitimately claim that they are strictly deducible. In particular, if one accepts the Church-Turing thesis, then every computation—including every agent-based computation—can be executed by a suitable register machine (Hodel 1995; Jeffrey 1991). It is then a theorem of logic and computability that every program can be simulated by a first-order language. In particular, with N denoting the natural numbers:

Theorem. Let P be a program. There is a first-order language L, and for each $a \in N$ a sentence $C(a)$ of L, such that for all $a \in N$, the P-computation with input a halts $\Leftrightarrow$ the sentence $C(a)$ is logically valid.

This theorem allows one to use the recursive unsolvability of the halting problem to establish the recursive unsolvability of the validity problem in first-order logic (see Kleene 1967). Explicit constructions of the correspondence between register machine programs and the associated logical arguments are laid out in detail by Jeffrey (1991) and Hodel (1995). The point here is that for every computation, there is a corresponding logical deduction. (And this holds even when the computation involves "stochastic" features, since, on a computer, these are produced by *deterministic* pseudo-random number generation (see Knuth 1969). Even if one conducts a statistical analysis over some distribution of runs—using different random seeds—each run is itself a deduction. Indeed, it would be quite legitimate to speak, in that case, of a distribution of theorems.)[12] In any case, from a technical standpoint, *generative implies deductive*, a point that will loom large later, when we argue that *agent-based modeling and classical emergentism are incompatible*.

Importantly, however, the converse does not apply: Not all deductive argument has the constructive character of agent-based modeling. Nonconstructive existence proofs are obvious examples. These work as follows: Suppose we wish to prove the existence of an x with some

---

[12]In such applications, it may be accurate to speak of an inductive statistical (see Salmon 1984) account over many realizations, each one of which is, technically, a deduction (by the Theorem above).

property (e.g., that it is an equilibrium). We take as an axiom the so-called Law of the Excluded Middle that (i) either x exists or x does not exist. Next, we (ii) assume that x does *not* exist, and (iii) derive a contradiction. From this we conclude that (iv) x must exist. But we have failed to exhibit x, or indicate any algorithm that would generate it, patently violating the generative motto (1).[13] The same holds for many nonconstructive proofs in mathematical economics and game theory (e.g., deductions establishing the existence of equilibria using fixed-point theorems). See Lewis 1985. In summary, then, generative implies deductive, but the converse is not true.

Given the differences between agent-based modeling and both inductive and deductive social science, a distinguishing term seems appropriate. The choice of "generative" was inspired by Chomsky's (1965) early usage: Syntactic theory seeks minimal rule systems that are sufficient to generate the structures of interest, grammatical constructions among them.[14] The generated structures of interest here are, of course, social.

Now, at the outset, I claimed that the agent-based computational model was a *scientific* instrument. A fair question, then, is whether agent-based computational modeling offers a powerful new way to do empirical research. I will argue that it does. Interestingly, one of the early efforts involves the seemingly remote fields of archaeology and agent-based computation.

EMPIRICAL AGENT-BASED RESEARCH

The Artificial Anasazi project of Dean, Gumerman, Epstein, Axtell, Swedlund, McCarroll, and Parker aims to grow an actual 500-year spatio-temporal demographic history—the population time series and spatial settlement dynamics of the Anasazi—testing against data. The Artificial Anasazi computational model proper is a hybrid in which the physical environment is "real" (reconstructed from dendroclimatalogical and other data) and the agents are artificial. In particular, we are attempting to model the Kayenta Anasazi of Long House Valley, a small region in northeastern Arizona, over the period 800 to 1300 AD, at which point the Anasazi mysteriously vanished from the Valley. The

---

[13] An agent-based model can be interpreted as furnishing a kind of *constructive* existence proof. See Axelrod 1997.

[14] See Chomsky 1965, 3. The "syntactic component of a generative grammar," he writes, is concerned with "rules that specify the well formed strings of minimal syntactically functioning units . . . ." I thank Samuel David Epstein for many fruitful discussions of this parallel.

enigma of the Anasazi has long been a central question in Southwestern archaeology. One basic issue is whether environmental (i.e., subsistence) factors alone can account for their sudden disappearance. Or do other factors—property rights, clan relationships, conflict, disease—have to be admitted to generate the true history? In bringing agents to bear on this controversy, we have the benefits of (a) a very accurate reconstruction of the physical environment (hydrology, aggradation, maize potential, and drought severity) on a square hectare basis for each year of the study period, and (b) an excellent reconstruction of household numbers and locations.

The logic of the exercise has been, first, to digitize the true history— we can now watch it unfold on a digitized map of Longhouse Valley. This data set (what really happened) is the target—the *explanandum*. The aim is to develop, in collaboration with anthropologists, micro-specifications—ethnographically plausible rules of agent behavior—that will generate the true history. The computational challenge, in other words, is to place artificial Anasazi where the true ones were in 800 AD and see if—under the postulated rules—the simulated evolution matches the true one. Is the microspecification *empirically adequate*, to use van Fraassen's (1980) phrase?[15] From a contemporary social science standpoint, the research also bears on the adequacy of simple "satisficing" rules—rather than elaborate optimizing ones—to account for the observed behavior.

A comprehensive report on Phase 1 (environmental rules only) of this research is given in Dean *et al.* 1999. The full microspecification, includ- ing hypothesized agent rules for choosing residences and farming plots, is elaborated there. The central result is that the purely environmental rules explored thus far account for (retrodict) important features of the Anasazi's demography, including the observed coupling between environ- mental and population fluctuations, as well as important observed spatial dynamics: agglomerations and zonal occupation series. These rules also generate a precipitous decline in population around 1300. However, they do not generate the outright disappearance that occurred. One interpretation of this finding is that subsistence considerations alone do *not* fully explain the Anasazi's departure, and that institutional or other cultural factors were likely involved. This work thus suggests the power

---

[15] More precisely, for each candidate rule (or agent specification), one runs a large population of simulated histories—each with its own random seed. The question then becomes: where, in the population of simulated histories is the true history? Rules that generate distributions with the true history (i.e., its statistic) at the mean enjoy more explanatory power than rules generating distributions with the true history at a tail.

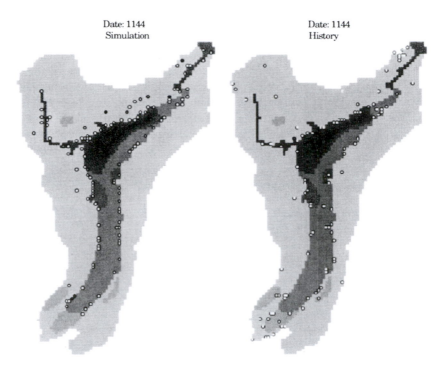

Figure 1.1. Actual and simulated Anasazi compared. (Source: Dean *et al.* 1999, 204.)

and limits of a purely environmental account, a finding that advances the archaeological debate.

Simply to convey the flavor of these simulations—which unfold as animations on the computer—figure 1.1 gives a comparison.[16] Each dot is an Anasazi household. The graphic shows the true situation on the right and a simulation outcome on the left for the year 1144. In both cases, agents are located at the border of the central farming area— associated with a high water table (dark shade)—and the household numbers are interestingly related.

The population time series (see Dean *et al.* 1999) comparing actual and simulated for a typical run is also revealing. The simulated Anasazi curve is qualitatively encouraging, matching the turning points, including a big crash in 1300, but quantitatively inaccurate, generally overestimating population levels, and failing to generate the "extinction" event of interest.

---

[16]The complete animation is included on this book's CD.

As noted earlier, one intriguing interpretation of these results is that the Valley could have supported the Anasazi in fact, so their departure may have been the result of *institutional* factors not captured in the purely environmental account.

The claim is not that the current model has solved—or that the planned extensions will ultimately solve—the mystery of the Anasazi. Rather, the point is that agent-based modeling permits a new kind of empirical research (and, it might be noted, a novel kind of interdisciplinary collaboration).

This is by no means the only example of data-driven empirical research with agents. For example, Axtell (1999) gives an agent-based computational model of firm formation that generates distributions of firm sizes and growth rates close to those observed in the U.S. economy. Specifically, citing the work of Stanley *et al.* (1996, 806), Axtell writes that "there are three important empirical facts that an *accurate* theory of the firm should reproduce: (a) firm sizes must be right-skewed, approximating a power law; (b) firm growth rates must be Laplace distributed; (c) the standard deviation in log growth rates as a function of size must follow a power law with exponent -0.15 ± 0.03." He further requires that the model be written at the level of individual human agents—that it be methodologically individualist. Aside from his own agent-based computational model, Axtell writes, ". . . theories of the firm that satisfy all these requirements are unknown to us" (1999, 88).

Similarly, observed empirical size-frequency distributions for traffic jams are generated in the agent-based model of Nagel and Rasmussen (1994). Bak, Paczuski, and Shubik (1996) present an agent-based trading model that succeeds in generating the relevant statistical distribution of prices.

Axelrod (1993) develops an agent-based model of alliance formation that generates the alignment of seventeen nations in the Second World War with high fidelity. Other exercises in which agent-based models are confronted with data include Kirman and Vriend 1998 and Arthur *et al.* 1997.

As in the case of the Anasazi work, I am not claiming that any of these models permanently resolves the empirical question it addresses. The claim, rather, is that agent-based modeling is a powerful empirical technique. In some of these cases (e.g., Axtell 1999), the agents are individual humans, and in others (Dean *et al.* 1999; Axelrod 1993) they are not. But, in all these cases, the empirical issue is the same: *Does the hypothesized microspecification suffice to generate the observed phenomenon?*—be it a stationary firm size distribution, a pattern of alliances, or a nonequilibrium price time series. The answer may be yes and, crucially, it may be no. Indeed, it is precisely the latter possibility—empirical falsifiability—that qualifies the agent-based computational model as a *scientific* instrument.

In addition to "hard" quantitative empirical targets, agent-based computational models may aim to generate important social phenomena qualitatively. Examples of "stylized facts" generated in such models include: right-skewed wealth distributions (Epstein and Axtell 1996), cultural differentiation (Epstein and Axtell 1996; Axelrod 1997c), multi-polarity in interstate systems (Cederman 1997), new political actors (Axelrod 1997d), epidemics (Epstein and Axtell 1996), economic classes (Axtell, Epstein, and Young 2001), and the dynamics of retirement (Axtell and Epstein 1999) to name a few. This "computational theorizing,"[17] if you will, can offer basic insights of the sort exemplified in Schelling's (1971) pioneering models of racial segregation, and may, of course, evolve into models directly comparable to data. Indeed, they may inspire the collection of data not yet in hand. (Without theory, it is not always clear what data to collect.) Turning from empirical phenomena, the generated phenomenon may be computation itself.

CONNECTIONIST SOCIAL SCIENCE

Certain social systems, such as trade networks (markets), are essentially computational architectures. They are distributed, asynchronous, and decentralized and have endogenous dynamic connection topologies. For example, the CD-ROM version of Epstein and Axtell 1996 presents animations of dynamic endogenous trade networks. (For other work on endogenous trade networks, see Tesfatsion 1995.) There, agents are represented as nodes, and lines joining agents represent trades. The connection pattern—computing architecture—changes as agents move about and interact economically, as shown in figure 1.2.

Whether they realize it or not, when economists say "the market arrives at equilibrium," they are asserting that this type of dynamic "social neural net" has executed a computation—it has computed P*, an equilibrium price vector. No individual has tried to compute this, but the society of agents does so nonetheless. Similarly, convergence to social norms, convergence to strategy distributions (in n-person games), or convergence to stable cultural or even settlement patterns (as in the Anasazi case) are all *social computations* in this sense.

It is clear that the efficiency—indeed the very feasibility—of a social computation may depend on the way in which agents are connected.

---

[17]I thank Robert Axtell for this term.

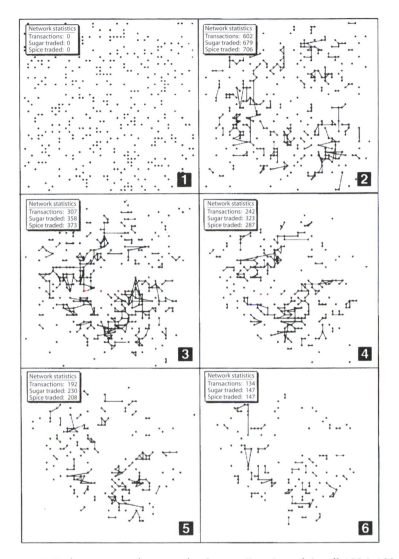

Figure 1.2. Endogenous trade network. (Source: Epstein and Axtell 1996, 132.)

After all, information in society is not manna from heaven; it is collected and processed at the agent level and transmitted through interaction structures that are endogenous. How then does the endogenous connectivity—the topology—of a social network affect its performance as a distributed computational device, one that, for example, computes price equilibria, or converges to (computes) social norms, or converges to spatial settlement patterns such as cities?[18] Agent-based models allow us to pursue such connectionist social science questions in new and systematic ways.

INTERDISCIPLINARY SOCIAL SCIENCE

Many important social processes are not neatly decomposable into separate subprocesses—economic, demographic, cultural, spatial—whose isolated analysis can be somehow "aggregated" to yield an adequate analysis of the process as whole. Yet this is exactly how academic social science is organized—into more or less insular departments and journals of economics, demography, anthropology, and so on. While many social scientists would agree that these divisions are artificial, they would argue that there is no "natural methodology" for studying these processes together, as they interact, though attempts have been made. Social scientists have taken highly aggregated mathematical models— of entire national economies, political systems, and so on—and have "connected" them, yielding "mega-models" that have been attacked on several grounds (see Nordhaus 1992). But attacks on specific models have had the effect of discrediting interdisciplinary inquiry itself, and this is most unfortunate. The line of inquiry remains crucially important. And agent-based modeling offers an alternative, and very natural, technique.

For example, in the agent-based model Sugarscape (Epstein and Axtell 1996), each individual agent has simple local rules governing movement, sexual reproduction, trading behavior, combat, interaction with the environment, and the transmission of cultural attributes and diseases. These rules can all be "active" at once. When an initial population of such agents is released into an artificial environment in which, and with which, they interact, the resulting artificial society unavoidably links demography, economics, cultural adaptation, genetic evolution, combat, environmental effects, and epidemiology. *Because the individual is multidimensional, so is the society.*

Now, obviously, not all social phenomena involve such diverse spheres of life. If one is interested in modeling short-term price dynamics in a local fish market, then human immune learning and epidemic processes may not be relevant. But if one wishes to capture long-term social dynamics of the sort discussed in William McNeill's 1976 book *Plagues and Peoples*, they are essential. Agent-based modelers do not insist that everything be studied all at once. The claim is that the new techniques

---

[18]In a different context, the sensitivity to network topology is studied computationally by Bagley and Farmer (1992).

allow us to transcend certain artificial boundaries that may limit our insight.

Nature-Nurture

For example, Sugarscape agents (Epstein and Axtell 1996) engage in sexual reproduction, transmitting genes for, *inter alia*, vision (the distance they can see in foraging for sugar). An offspring's vision is determined by strictly Mendelian (one locus–two allele) genetics, with equal probability of inheriting the father's or mother's vision. One can easily plot average vision in society over time. Selection will favor agents with relatively high vision—since they'll do better in the competition to find sugar—and, as good Darwinians, we expect to see average vision increase over time, which it does. Now, suppose we wish to study the effect of various *social* conventions on this *biological* evolution. What, for example, is the effect of inheritance—the social convention of passing on accumulated sugar wealth to offspring—on the curve of average vision? Neither traditional economics nor traditional population genetics offer particularly natural ways to study this sort of "nature-nurture" problem. But they are naturally studied in an agent-based artificial society: Just turn inheritance "off" in one run and "on" in another, and compare![19] Figure 1.3 gives a typical realization.

With inheritance, the average vision curve (gray) is lower: Inheritance "dilutes" selection. Because they inherit sugar, the offspring of wealthy agents are buffered from selection pressure. Hence, low-vision genes persist that would be selected out in the absence of this social convention. We do not offer this as a general law, nor are we claiming that agent-based models are the only ones permitting exploration of such topics.[20] The claim is that they offer a new, and particularly natural, methodology for approaching certain interdisciplinary questions, including this one. Some of these questions can be posed in ways that subject dominant theories to stress.

THEORY STRESSING

One can use agent-based models to test the robustness of standard theory. Specifically, one can relax assumptions about individual—micro

---

[19] This is shorthand for the appropriate procedure in which one would generate distributions of outcomes for the two assumptions and test the hypothesis that these are indistinguishable statistically.

[20] For deep work on gene-culture co-evolution generally, using different techniques, see Feldman and Laland 1996.

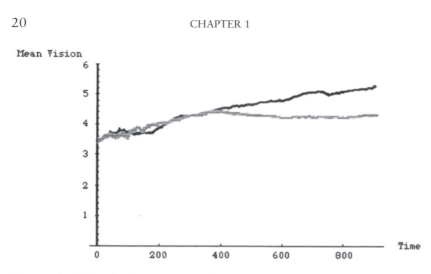

Figure 1.3. Effect of inheritance on selection. (Source: Epstein and Axtell 1996, 68.)

level—behavior and see if standard results on macro behavior collapse. For example, in neoclassical microeconomic theory, individual preferences are assumed to be fixed for the lifetime of the agent. On this assumption (and certain others), individual utility maximization leads to price equilibrium and allocative efficiency (the First Welfare Theorem). But, what if individual preferences are *not* fixed but vary culturally? In Epstein and Axtell 1996, we introduce this assumption into trading agents who are neoclassical in all other respects (e.g., they have Cobb-Douglas utility functions and engage only in Pareto-improving trades with neighbors). The result is far-from-equilibrium markets. The standard theory is not robust to this relaxation in a core assumption about individual behavior. For a review of the literature on this central fixed preferences assumption, see Bowles 1998.

Agents, Behavioral Social Science, and the Micro-Macro Mapping

What can agent-based modeling and behavioral research do for one another? It is hard to pinpoint the dawn of experimental economics, though Simon (1996) credits Katona (1951) with the fundamental studies of expectation formation. In any event, there has been a resurgence of important laboratory and other experimental work on individual decision making over the last two decades. See, for example, Camerer 1997, Rabin 1998, Camerer and Thaler 1995, Tversky and Kahneman 1986, and Kagel and Roth 1995. This body of laboratory social science, if I may call it that, is giving us an ever-clearer picture of how *homo sapiens*—as against *homo economicus*—actually makes decisions. However, a crucial lesson of Schelling's segregation model, and of many subsequent Cellular Automaton models, such as "Life"—not to mention agent-based models themselves—is that *even perfect knowledge of individual decision rules does not always allow us to predict macroscopic structure.* We get macro-surprises despite complete micro-knowledge. Agent-based models allow us to study the micro-to-macro mapping. It is obviously essential to begin with solid foundations regarding individual behavior, and behavioral research is closing in on these. However, we will still need techniques for "projecting up" to the macro level from there (particularly for spatially-distributed systems of heterogeneous individuals). Agent modeling offers behavioral social science a powerful way to do that.

Agent-based models may also furnish laboratory research with counterintuitive hypotheses regarding individual behavior. Some, apparently bizarre, system of individual agent rules may generate macrostructures that mimic the observed ones. Is it possible that those are, in fact, the operative micro-rules? It might be fruitful to design laboratory experiments to test hypotheses arising from the unexpected generative sufficiency of certain rules.

What does behavioral research offer agent-based modeling? Earlier, we noted that different agent-based models might have equal generative (explanatory) power and that, in such cases, further work would be necessary to adjudicate between them. But if two models are doing equally well in generating the macrostructure, preference should go to the one that is best at the micro level. So, if we took the two microspecifications as competing hypotheses about individual behavior, then—apropos of the preceding remark—behavioral experiments might be designed to identify the better hypothesis (microspecification) and, in turn, the better agent model. These, then, are further ways in which agent-based computational modeling can contribute to empirical social science research.

DECOUPLINGS

As noted earlier, to adopt agent-based modeling does not compel one to adopt methodological individualism. However, extreme methodological individualism is certainly possible (indeed common) in agent-based models. And individual-based models may have the important effect of *decoupling individual rationality from macroscopic equilibrium.* For example, in the individual-based retirement model of Axtell and Epstein

(1999), macroscopic equilibrium is attained—through a process of imitation in social networks—even though the vast preponderance of individuals are *not* rational. Hence—as in much evolutionary modeling—micro rationality is not a *necessary* condition for the attainment of macro equilibrium.[21] Now, we also have agent-models in which macroscopic equilibrium is not attained despite orthodox utility maximization at the individual level. The non-equilibrium economy under evolving preferences (Epstein and Axtell 1996) noted earlier is an example. Hence, micro rationality is not a *sufficient* condition for macro equilibrium. But if individual rationality is thus neither necessary nor sufficient for macro equilibrium, the two are logically independent—or decoupled, if you will.

Now, the *fraction* of agents in an imitative system (such as the retirement model) who are rational will definitely affect the *rate* at which any selected equilibrium sets in. But the asymptotic equilibrium behavior *per se* does not depend on the dial of rationality, despite much behavioral research on this latter topic. Perhaps the main issue is not how much rationality there is (at the micro level), but *how little is enough* to generate the macro equilibrium.

In passing, it is worth noting that this is of course a huge issue for policy, where "fad creation" may be far more effective than real education. Often, the aim is not to equip target populations with the data and analytical tools needed to make rational choices; rather, one displays exemplars and then presses for mindless imitation. "Just say no to drugs" not because it's rational—in a calculus of expected lifetime earnings—but because a famous athlete says "no" and it's a norm to imitate him. The manipulation of uncritical imitative impulses may be more effective in getting to a desired macro equilibrium than policies based on individual rationality. The social problem, of course, is that populations of uncritical imitators are also easy fodder for lynch mobs, witch hunts, Nazi parties, and so forth. Agent-based modeling is certainly not the only way to study social contagion (see, for example, Kuran 1989), but it is a particularly powerful way when the phenomenon is spatial and the population in question is heterogeneous.

Relatedly, agent-based approaches may decouple social science from decision science. In the main, individuals do not *decide*—they do not *choose*—in any sensible meaning of that term, to be ethnic Serbs, to be native Spanish speakers, or to consider monkey brain a delicacy. Game theory may do an interesting job explaining the decision of one

---

[21] More precisely, micro rationality is not necessary for *some* equilibrium, but it may a different equilibrium from the one that would occur were agents rational.

ethnic group to attack another at a certain place or time, but it doesn't explain how the ethnic group arises in the first place or how the ethnic divisions are transmitted across the generations. Similarly for economics, what makes monkey brain a delicacy in one society and not in another? Cultural (including preference) patterns, and their nonlinear tippings, are topics of study in their own right with agents.[22] See Axelrod 1997*b* and Epstein and Axtell 1996.

ANALYTICAL-COMPUTATIONAL APPROACH TO NON-EQUILIBRIUM SOCIAL SYSTEMS

For many social systems, it is possible to prove deep theorems about asymptotic equilibria. However, the time required for the system to attain (or closely approximate) such equilibria can be astronomical. The transient, out-of-equilibrium dynamics of the system arc then of fundamental interest. A powerful approach is to combine analytical proofs regarding asymptotic equilibria with agent-based computational analyses of long-lived transient behaviors, the meta-stability of certain attractors, and broken ergodicity in social systems.

One example of this hybrid analytical-computational approach is Axtell, Epstein, and Young 2001. We develop an agent-based model to study the emergence and stability of equity norms in society. (In that article, we explicitly define the term "emergent" to mean simply "arising from decentralized bilateral agent-interactions.") Specifically, agents with finite memory play Best Reply to Recent Sample Evidence (Young 1995, 1998) in a three-strategy Nash Demand Game, and condition on an arbitrary "tag" (e.g., a color) that initially has no social or economic significance—it is simply a distinguishing mark. Expectations are generated endogenously through bilateral interactions. And, over time, these tags acquire socially organizing salience. In particular, tag-based classes arise. (The phenomenon is akin to the evolution of meaning discussed in Skyrms 1998.)

Now, introducing noise, it is possible to cast the entire model as a Markov process and to prove rigorously that it has a unique stationary strategy distribution. When the noise level is positive and sufficiently small, the following asymptotic result can be proved: *The state with the highest long-run probability is the equity norm, both between and within groups.*

---

[22] Again, as a policy application, agent-based modeling might suggest ways to operate on—or "tip"—ethnic animosity itself.

Salutary as this asymptotic result may appear, the transition from inequitable states to these equitable ones can be subject to tremendous inertia. Agent-based models allow us to systematically study long-lived transient behaviors. We know that, beginning in an inequitable regime, the system will ultimately "tip" into the equity norm. But how does the waiting time to this transition depend on the number of agents and on memory length? In this case, the waiting time scales *exponentially in memory length, m, and exponentially in* N, *the number of agents.* Overall, then, the waiting time is immense for $m = 10$ and merely $N = 100$, for example.

Speaking rigorously, the equity norm is stochastically stable (see Young 1998). The agent-based computational model reveals, however, that—depending on the number of agents and their memory lengths—the waiting time to transit from an inequitable regime to the equitable one may be astronomically long.

This combination of formal (asymptotic) and agent-based (non-equilibrium) analysis seems to offer insights unavailable from either approach alone, and to represent a useful hybrid form of analytical-computational study. For sophisticated work relating individual-based models to analytical ones in biology, see Flierl *et al.* 1999.

Foundational Issues

We noted earlier that markets can be seen as massively parallel spatially distributed computational devices with agents as processing nodes. To say that "the market clears" is to say that this device has completed a computation. Similarly, convergence to social norms, convergence to strategy distributions (in *n*-person games), or convergence to stable cultural or settlement patterns, are all social computations in this sense. Minsky's (1985) famous phrase was "the Society *of* Mind." What I'm interested in here is "the Society *as* Mind," society as a computational device. (On that strain of functionalism which would be involved in literally asserting that a society could be a mind, see Sober 1996.)

Now, once we say "computation" we think of Turing machines (or, equivalently, of partial recursive functions). In the context of *n*-person games, for example, the isomorphism with societies is direct: Initial strategies are tallies on a Turing machine's input tape; agent interactions function to update the strategies (tallies) and thus represent the machine's state transition function; an equilibrium is a halting state of the machine; the equilibrium strategy distribution is given by the tape contents in the halting state; and initial strategy distributions that run to equilibrium are languages accepted by the machine. The isomorphism is clear. Now, we know what makes for an intractable, or "hard," computational problem. So, given our isomorphism, is there a computational answer to the question, "What's a hard social problem?"

A Computational Characterization of Hard Social Problems

In the model of tag-based classes discussed earlier (Axtell, Epstein, and Young 2001), we prove rigorously that, asymptotically, the equity norm will set in. However, beginning from any other (meta-stable) equilibrium, the time to transit into the equitable state scales exponentially in the number of agents and exponentially in the agents' memory length. If we adopt the definition that social states are *hard to attain* if they are *not effectively computable by agent society in polynomial time*, then equity is hard. (The point applies to this particular setup; I am emphatically not claiming that there is anything immutable about social inequity.) In a number of models, the analogous point applies to economic equilibria: There are nonconstructive proofs of their existence but computational arguments that their *attainment* requires time that scales exponentially in, for instance, the dimension of the commodity space.[23] On our tentative definition, then, computation of (attainment of) economic equilibria would qualify as another *hard social problem*.

So far we have been concerned with the question, "Does an initial social state run to equilibrium?" or, equivalently, "Does the machine halt given input tape x?" Now, like satisfiability, or truth-table validity in sentential logic, these problems are in principle decidable (that is, the equilibria are effectively computable), but not on time scales of interest to humans. (Here, with Simon [1978], we use the term "time" to denote "the number of elementary computation steps that must be executed to solve the problem.")

Gödelian Limits

But there are social science problems that are undecidable in principle, now in the sense of Gödel or the Halting Problem. Rabin (1957) showed that "there are actual win-lose games which are strictly determined for which there is no effectively computable winning strategy." He continues, "Intuitively, our result means that there are games in which the player who in theory can always win, cannot do so in practice because it is impossible to supply him with effective instructions regarding how

---

[23]See, for example, Hirsch *et al.* 1989.

he should play in order to win." Another nice example, based on the unsolvability of Hilbert's Tenth Problem, is given by Prasad (1997):

> For n-player games with polynomial utility functions and natural number strategy sets the problem of finding an equilibrium is not computable. There does not exist an algorithm which will decide, for any such game, whether it has an equilibrium or not... When the class of games is specified by a finite set of players, whose choice sets are natural numbers, and payoffs are given by polynomial functions, the problem of devising a procedure which computes Nash equilibria is unsolvable.

Other results of comparable strength have been obtained by Lewis (1985, 1992a, and 1992b).[24]

Implications for Rational Choice Theory

Here lies the deepest conceivable critique of rational choice theory. There are strategic settings in which the individually optimizing behavior is uncomputable in principle. A second powerful critique is that, while possible in principle, optimization is computationally intractable. As Duncan Foley summarizes, "The theory of computability and computational complexity suggest that there are two inherent limitations to the rational choice paradigm. One limitation stems from the possibility that the agent's problem is in fact undecidable, so that no computational procedure exists which for all inputs will give her the needed answer in finite time. A second limitation is posed by computational complexity in that even if her problem is decidable, the computational cost of solving it may in many situations be so large as to overwhelm any possible gains from the optimal choice of action" (see Albin 1998, 46). For a fundamental statement, see Simon 1978.

These possibilities are disturbing to many economists. They implicitly believe that if the individual is insufficiently rational it must follow that decentralized behavior is doomed to produce suboptimality at the aggregate level. The invisible hand requires rational fingers, if you will. There are doubtless cases in which this holds. But it is not so in all cases. As noted earlier, in the retirement model of Axtell and Epstein (1999), as well as in much evolutionary modeling, an ensemble of locally interacting agents—none of whom are canonically rational—can nonetheless attain efficiency in the aggregate. Even here, of course, issues of exponential

---

[24] The important Arrow Impossibility Theorem (Arrow 1963) strikes me as different in nature from these sorts of results. It does not turn—as these results do—on the existence of sets that are recursively enumerable but not recursive.

waiting time arise (as in the classes model above). But it is important to sort the issues out.

The agent-based approach forces on us the interpretation of society as a computational device, and this immediately raises foundational specters of computational intractability and undecidability. Much of the economic complexity literature concerns the uncomputability of optimal strategies by *individual rational agents*, surely an important issue. However, our central concern is with the effective computability (attainment) of equilibria *by societies of boundedly rational agents*. In that case, it is irrelevant that equilibrium can be computed by an economist external to the system using the Scarf, or other such, algorithm. The entire issue is whether it can be attained—generated—through decentralized local interactions of heterogeneous boundedly rational actors. And the agent-based computational model is a powerful tool in exploring that central issue. In some settings, it may be the only tool.

EQUATIONS VERSUS AGENT-BASED MODELS

Three questions arise frequently and deserve treatment: Given an agent-based model, are there equivalent equations? Can one "understand" one's computational model without such equations? If one has equations for the macroscopic regularities, why does one need the "bottom-up" agent model?

Regarding the first question—are there equivalent equations for every computational model—the answer is immediate and unequivocal: absolutely. On the Church-Turing Thesis, every computation (and hence every agent-based model) can be implemented by a Turing machine. For every Turing machine there is a unique corresponding and equivalent Partial Recursive Function (see Rogers 1967). Hence, in principle, for any computation there exist equivalent equations (involving recursive functions). Alternatively, any computer model uses some finite set of memory locations, which are updated as the program executes. One can think of each location as a variable in a discrete dynamical system. In principle, there is some—perhaps very high dimensional—set of equations describing those discrete dynamics. Now, could a human write the equations out? Solve them or even find their equilibria (if such exist)? The answer is not clear. If the equations are meant to represent large populations of discrete heterogeneous agents coevolving on a separate space, with which they interact, it is not obvious how to formulate the equations, or how to solve them if formulated. And, for certain classes of problems (e.g., the PSPACE Complete problems), it can be proved

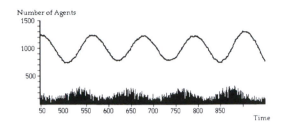

Figure 1.4. Oscillatory population time series. Black vertical spikes represent births. (Source: Epstein and Axtell 1996, 161.)

rigorously that simulation is—in a definite sense—the best one can do in principle (see Buss, Papadimitriou, and Tsitsiklis 1991). But that does not mean—turning to the second question—we have no idea what's going on in the model.

To be sure, a theorem is better than no theorem. And many complex social phenomena may ultimately yield to analytical methods of the sort being pioneered by Young (1998), Durlauf (1997b), and others. But an experimental attitude is also appropriate. Consider biology. No one would fault a "theoremless" laboratory biologist for claiming to understand population dynamics in beetles when he reports a regularity observed over a large number of experiments. But when agent-based modelers show such results—indeed, far more robust ones—there's a demand for equations and proofs. These would be valuable, and we should endeavor to produce them. Meanwhile, one can do perfectly legitimate "laboratory" science with computers, sweeping the parameter space of one's model, and conducting extensive sensitivity analysis, and claiming substantial understanding of the relationships between model inputs and model outputs, just as in any other empirical science for which general laws are not yet in hand.[25]

The third question involves confusion between explanation and description, and might best be addressed through an example. In Epstein and Axtell 1996, spatially distributed local agent interactions generate the oscillatory aggregate population time series shown in figure 1.4.

The question then arises: Could you not get that same curve from some low-dimensional differential equation, and if so, why do you need the agent model? Let us imagine that we can formulate and analytically solve such an equation, and that the population trajectory is exactly $P(t) = A + B\,Sin(Ct)$ for constants A, B, and C. Now, *what is the explanatory significance of that descriptively accurate result?*

---

[25]Here, we are discussing regularities in model output alone, not the relationship of model output to some real-world data set, as in the Anasazi project.

It depends on one's criteria for explanation. If we are generativists, the question is: How could the spatially decentralized interactions of heterogeneous autonomous agents generate that macroscopic regularity? If that is one's question, then the mere formula $P(t) = A + B \operatorname{Sin}(Ct)$ is *devoid of explanatory power despite its descriptive accuracy.* The choice of agents versus equations always hinges on the objectives of the analysis. Given some perfectly legitimate objectives, differential equations are the tool of choice; given others, they're not. If we are explicit as to our objectives, or explanatory criteria, no confusion need arise. And it may be that hybrid models of a second sort are obtainable in which the macrodynamics are well described by an explicit low-dimensional mathematical model, but are also generated from the bottom up in a model population of heterogeneous autonomous agents. That would be a powerful combination. In addition to important opportunities, the field of agent-based modeling, like any young discipline, faces a number of challenges.

CHALLENGES

First, the field lacks standards for model comparison and replication of results; see Axtell *et al.* 1996. Implicit in this is the need for standards in reportage of assumptions and certain procedures. Subtle differences can have momentous consequences. For example, how, exactly, are agents being updated? The Huberman and Glance (1993) critique of Nowak and May (1992) is striking proof that asynchronous updating of agents produces radically different results from synchronous updating. Huberman and Glance show that Nowak and May's main result—the persistence of cooperation in a spatial Prisoner's Dilemma game— depends crucially on synchronous updating. When, *ceteris paribus*, Huberman and Glance introduce asynchronous updating into the Nowak and May model, the result is convergence to pure defection. (For a spatial Prisoner's Dilemma model with asynchronous updating in which cooperation *can* persist, see Epstein 1998.) The same sorts of issues arise in randomizing the agent call order, where various methods—with different effects on output—are possible.

It is also fair to say that solution concepts are weak. Certainly, hitting the "Go" button and watching the screen does not qualify as *solving* anything—any more than an evening at the casino *solves* the Gambler's Ruin Problem from Markov Theory. An individual model run offers a sample path of a (typically) stochastic process, but that is not a general solution—a specific element of some well defined function space (e.g., a Hilbert or Sobolev space). As noted earlier, it is often possible to sweep the parameter space of one's model quite systematically and thereby obtain a statistical portrait of the relationship between inputs and outputs, as in Axelrod 1997c or Epstein 1998. But it is fair to say that this practice has not been institutionalized.

A deeper issue is that sweeping a model's numerical parameter space is easier than exploring the space of possible agent behavioral *rules* (e.g., "If a neighboring agent is bigger than you, run away" or "Always share food with kin agents"). For artificial societies of any complexity (e.g., Sugarscape), we have no efficient method of searching the space of possible individual rules for those that exhibit generative power. One can imagine using evolutionary approaches to this. First, one would define a metric such that, given a microspecification, the distance from model outputs (generated macrostructures) to targets (observed macrostructures) could be computed. The better the match (the smaller this distance) the "fitter" is the microspecification. Second, one would encode the space of candidate micro specifications and turn, say, a Genetic Algorithm (GA) (see Holland 1992; Mitchell 1998) loose on it. The GA might turn up counterintuitive *boundedly* rational rules that are highly "fit" in this sense of generating macrostructures "close" to the targets. (These then become hypotheses for behavioral research, as discussed earlier.)

This strikes me as a far more useful application of GAs than the usual one: finding *hyper-rational* individual strategies, which we now have strong experimental evidence are *not* being employed by humans. The problem is how to encode the vast space of possible individual rules (not to mention the raw computational challenge of searching it once encoded). In some restricted cases, this has been done successfully (Axelrod 1987; Crutchfield and Mitchell 1995), but for high dimensional agents engaged in myriad social interactions—economic, cultural, demographic—it is far from clear how to proceed.

One of the central concepts in dynamics is sensitivity. Sensitivity involves the effect on output (generated macrostructure) of small changes in input (microspecification). To assess sensitivity in agent models, we have to do more than encode the space of rules—we have to *metrize* it. To clarify the issue, consider the following agent rules (*methods* of agent-objects):

> Rule a = Never attack neighbors.
> Rule b = Attack a neighbor if he's green.
> Rule c = Attack a neighbor if he's smaller than you.

Which rule—b or c—represents a "smaller departure from" Rule a? Obviously, the question is ill-posed. And yet we speak of "small changes in the rules" of agent-based models. Some areas (e.g., Cellular Automata) admit binary encodings of rule space for which certain metrics—taxicab or Hamming distance—are natural. But for artificial societies generally, no such simple avenues present themselves. What then constitutes a *small* rule change? Without some metric, we really cannot develop the analogue, for agent-based models, of structural stability—or equivalently, of bifurcation theory—in dynamical systems.

Some challenges are sociological. Generating collective behavior that to the naked eye "looks like flocking" can be extremely valuable, but it is a radically different enterprise from generating, say, a specific distribution of wealth with parameters close to those observed in society. Crude qualitative caricature is a perfectly respectable goal. But if that is one's goal, the fact must be stated explicitly—perhaps using the terminology proposed in Axtell and Epstein 1994. This will avert needless resistance from other fields where "normal science" proceeds under established empirical standards patently not met by cartoon "boid" flocks, however stimulating and pedagogically valuable these may be. On the pedagogical value of agent-based simulation generally, see Resnick 1994.

A number of other challenges include building community and sharing results and are covered in Axelrod 1997a. In addition to foundational, procedural, and other scientific challenges, the field of "complexity" and agent-based modeling faces terminological ones. In particular, the term "emergence" figures very prominently in this literature. It warrants an audit.

"Emergence"

I have always been uncomfortable with the vagueness and occasional mysticism surrounding this word and, accordingly, tried to define it quite narrowly in Epstein and Axtell 1996. There, we defined "emergent phenomena" to be simply "stable macroscopic patterns arising from local interaction of agents."[26] Many researchers define the term in the same straightforward way (e.g., Axelrod 1997a). Since our work's publication, I have researched this term more deeply and find myself questioning its adoption altogether.

"Emergence" has a history, and it is an extremely spotty one, beginning with classical British emergentism in the 1920s and the works of Samuel

---

[26] As we wrote there, "A particularly loose usage of 'emergent' simply equates it with 'surprising,' or 'unexpected,' as when researchers are unprepared for the kind of systematic behavior that emanates from their computers." We continued, "This usage obviously begs the question, 'Surprising to whom?'" (Epstein and Axtell 1996, 35).

Alexander (*Space, Time, and Deity*, 1920), C. D. Broad (*The Mind and Its Place in Nature*, 1925), and C. Lloyd Morgan (*Emergent Evolution*, 1923). The complexity community should be alerted to this history. There is an unmistakably anti-scientific—even deistic—flavor to this movement, which claimed *absolute unexplainability* for emergent phenomena. In the view of these authors, emergent phenomena are unexplainable in principle. "The existence of emergent qualities ... admits no explanation," wrote Alexander (1920).[27] As philosopher Terence Horgan recounts, emergent phenomena were to be "accepted (in Samuel Alexander's striking phrase) 'with natural piety.'"

Striking indeed, this sort of language, and classical emergentism's avowedly vitalist cast (see Morgan 1923) stimulated a vigorous—and to my mind, annihilative—attack by philosophers of science. In particular, Hempel and Oppenheim (1948) wrote, "This version of emergence ... is objectionable not only because it involves and perpetuates certain logical confusions but also because not unlike the ideas of neovitalism, it encourages an attitude of resignation which is stifling to scientific research. No doubt it is this characteristic, together with its theoretical sterility, which accounts for the rejection, by the majority of contemporary scientists, of the classical absolutist doctrine of emergence."

Classical absolute emergentism is encapsulated nicely in the following formalization of Broad's (1925, 61):

> Put in abstract terms the emergent theory asserts that there are certain wholes, composed (say) of constituents A, B, and C in a relation R to each other ... and that the characteristic properties of the whole R(A,B,C) cannot, even in theory, be *deduced* from the most complete knowledge of the properties of A, B, and C in isolation or in other wholes which are not in the form R(A,B,C). (Emphasis in original)

Before explicating the logical confusion noted by Hempel and Oppenheim, we can fruitfully apply a bit of logic ourselves. Notice that we have actually accumulated a number of first-order propositions. For predicates, let C stand for classically emergent, D for deducible, E for explained, and G for generated (in a computational model). Then, if x is

---

[27] Although many contemporary researchers do not use the term in this way, others assume that this is the generally accepted meaning. For example, Jennings, Sycara, and Woolridge (1998) write that "... the very term 'emerges' suggests that the relationship between individual behaviors, environment, and overall behavior is not understandable," which is entirely consistent with the classical usage.

a system property, we have:

(1) $(\forall x)(Cx \supset \neg Dx)$ Broad (emergent implies not deducible)[28]
(2) $(\forall x)(Cx \supset \neg Ex)$ Alexander (emergent implies not explainable)
(3) $(\forall x)(\neg Gx \supset \neg Ex)$ Generativist Motto (not generated
                                           implies not explained)
(4) $(\forall x)(Gx \supset Dx)$ Theorem (generated implies deduced)

Although a number of derivations are possible,[29] the essential point involves (1) and (4). By the earlier Theorem (4), if x is generable, then it is deducible. But, by Broad (1), if x is emergent, it is not deducible. But it then follows that if x is generable, then it cannot be emergent![30] In particular, if x is generated in an agent-based model, it cannot be classically emergent. *Agent-based modeling and classical emergentism are incompatible.* Further incompatibilities between agent-based modeling and classical emergentism will be taken up below.

Logical Confusion

Now, the logical confusion noted earlier is set forth clearly in Hempel and Oppenheim 1948, is discussed at length in Nagel 1961, and is recounted more recently by Hendriks-Jansen 1996. To summarize, like Broad, emergentists typically assert things like, "One cannot deduce higher properties from lower ones; macro properties from micro ones; the properties of the whole from the parts' properties." But, we do not deduce properties. We deduce propositions in formal languages from other propositions in those languages.[31] This is not hair-splitting: If the macro theory contains terms (predicates, variable names) that are not terms of the micro theory, then of course it is impossible to deduce

---

[28]To highlight the chasm between this classical and certain modern usages, while Broad defines emergence as undeducible, Axelrod (1997c, 194) writes that "there are some models ... in which emergent properties can be formally deduced."

[29]For example, note that we can deduce Alexander's Law (2) from the others. By Broad (1), if x is classically emergent then it is not deducible; but then by (4) and *modus tollens*, x is not generable; and then by the Motto (3), x is not explainable. So, by hypothetical syllogism, we obtain Alexander (2). In a punctilious derivation, we would of course invoke universal instantiation first; then rules of an explicit sentential calculus (e.g., Copi 1979), and then use universal generalization.

[30]Filling in for any x, (4) is Gx ⊃ Dx; but Dx = ¬(¬Dx), from which ¬Cx follows from (1) by *modus tollens*.

[31]Formal systems are closed under their rules of inference (e.g., *modus ponens*) in the sense that propositions in a formal system can only be deduced from other propositions of that system.

macro claims involving those terms from propositions of the micro theory. It is logically impossible. So the "higher emergent" property of water, "translucence," is trivially not deducible from the micro theory of oxygen (O) and hydrogen (H) since "translucent" is not a term of the micro theory. Many so called "emergent properties" of "wholes" are not deducible from "parts" for this purely logical reason. So emergence, as nondeducibility, is always *relative* to some theory (some set of well-formed formulae and inference rules); it is not absolute as the classicals would have it.

A *relative* version of emergence due to Hempel and Oppenheim (1948) is formalized in Stephan 1992 as follows. Consider a system with constituents C1,... ,Cn in relation O to one another (analogous to Broad's A, B, C, and R). "This combination is termed a *microstructure* [C1,... Cn;O]. And let T be a theory. Then, a system property P is emergent, *relative to* this microstructure and theory T, if:

(a). There is a law LP which holds: for all x, when x has microstructure [C1.... Cn;O] then x has property P, and
(b). By means of theory T, LP cannot be deduced from laws governing the Cl.... Cn in isolation or in other microstructures than the given."

Stephan continues, "By this formulation the original absolute claim has been changed into a merely relative one which just states that at a certain time according to the available scientific theories we are not able to deduce the so-called emergent laws" (1992, 39).[32] But now, as Hempel and Oppenheim write, "If the assertion that life and mind have an emergent status is interpreted in *this* sense, then its import can be summarized approximately by the statement that no explanation, in terms of microstructure theories, is available *at present* for large classes of phenomena studied in biology and psychology" (emphases added). This quite unglamorous point, they continue, would "appear to represent the rational core of the doctrine of emergence." Not only does this relative formulation strip the term of all higher Gestalt harmonics, but it suggests that, for any given phenomenon, emergent status itself may be fleeting.

---

[32]Contemporary efforts (see Baas 1994) to define a kind of relative (or hierarchical) emergence by way of Gödel's First Theorem (see Smullyan 1992) seem problematic. Relative to a given (consistent and finitely axiomatized) theory, T, Baas calls undecidable sentences "observationally emergent." However, we are presumably interested in generating "emergent phenomena" in computational models. And it is quite unclear from what *computational* process Baas's observationally emergent entities—undecidable sentences of T—would actually emerge since, by Tarski's Theorem, the set of true and undecidable propositions is not recursively enumerable (Hodel 1995, 310, 354).

Scientific Progress

As scientific theories progress, in other words, that which was unexplainable and "emergent" ceases to be. The chemical bond—a favorite of the British emergentists—is an excellent example. Here, Terrence Horgan (1993) is worth quoting at length:

> When Broad wrote, "Nothing that we know about Oxygen by itself or in its combination with anything but Hydrogen would give us the least reason to suppose that it would combine with Hydrogen at all. Nothing that we know about Hydrogen by itself or in its combinations with anything but Oxygen would give us the least reason to expect that it would combine with Oxygen at all" (1925, pp. 62–63), his claim was true. Classical physics could not explain chemical bonding. But the claim didn't stay true for long: by the end of the decade quantum mechanics had come into being, and quantum-mechanical explanations of chemical bonding were in sight.

The chemical bond no longer seemed mysterious and "emergent." Another example was biology, for the classical emergentists a rich source of higher "emergent novelties," putatively unexplainable in physical terms. Horgan continues,

> Within another two decades, James Watson and Francis Crick, drawing upon the work of Linus Pauling and others on chemical bonding, explained the information-coding and self-replicating properties of the DNA molecule, thereby ushering in physical explanations of biological phenomena in general.

As he writes, "These kinds of advances in science itself, rather than any internal conceptual difficulties, were what led to the downfall of British emergentism, as McLaughlin (1992) persuasively argues." Or, as Herbert Simon (1996) writes, "Applied to living systems the strong claim [quoting the "holist" philosopher J. C. Smuts] that 'the putting together of their parts will not produce them or account for their characters and behaviors' implies a vitalism that is wholly antithetical to modern molecular biology."

In its strong classical usage, the term "emergent" simply "baptizes our ignorance," to use Nagel's phrase (1961, 371). And, when de-mystified, it can mean nothing more than "not presently explained." But, this is profoundly different from "not explainable in principle," as Alexander and his emergentist colleagues would have it, which is stifling, not to mention baseless empirically. As Hempel and Oppenheim wrote,

> Emergence is not an ontological trait inherent in some phenomena; rather it is indicative of the scope of our knowledge at a given time; thus it has

no absolute, but a relative character; and what is emergent with respect to the theories available today may lose its emergent status tomorrow. (1948, 263)

Good Questions

Now, all the questions posed by agent-based modelers and complexity scientists in this connection are fine: How do individuals combine to form firms, or cities, or institutions, or ant colonies, or computing devices? These are all excellent questions. The point is that they are posable—indeed most productively posed—without the imprecise and possibly self-mystifying terminology of "emergence," or "supervenience," as Morgan called it. Obviously, "wholes" may have attributes or capabilities that their constituent parts cannot have (e.g., "whole" conscious people can have happy memories of childhood while, presumably, individual neurons cannot). Equally obvious, the parts have to be hooked up right—or interact in specific, and perhaps complicated, ways—for the whole to exhibit those attributes.[33] We *at present* may be able to explain why these specific relationships among parts eventuate in the stated attributes of wholes, and we may not. But, unlike classical emergentists, we do not *preclude* such explanation in principle.

Indeed, by attempting to generate these very phenomena on computers or in mathematical models, we are *denying* that they are unexplainable or undeducible in principle—we're trying to explain them precisely by figuring out microrules that will generate them. In short, we agent-based modelers and complexity researchers actually part company with those, like Alexander and company, whose terminology we have, perhaps unwittingly, adopted. Lax definitions can compound the problem.

Operational Definitions

Typical of classical emergentism would be the claim: *No description of the individual bee can ever explain the emergent phenomenon of the hive.* How would one know that? Is this a falsifiable empirical claim, or something that seems true because of a lax definition of terms? Perhaps the latter. The mischievous piece of the formulation is the phrase "description of the individual bee." What is that? Does "the

---

[33]There is no reason to present these points as if they were notable, as in the following representative example: "...put the parts of an aeroplane together in the correct relationship and you get the emergent property of flying, even though none of the parts can fly" Johnson (1995, 26).

bee's" description not include its rules for interacting with other bees? Certainly, it makes little sense to speak of a Joshua Epstein devoid of all relationships with family, friends, colleagues, and so forth. "Man is a social animal," quoth Aristotle. My "rules of social interaction" are, in part, what make me me. And, likewise, the bee's interaction rules are what make it a bee—and not a lump. When (as a designer of agent objects) you get these rules right—when you get "the individual bee" right—you get the hive, too. Indeed, from an operationist (Hempel 1956) viewpoint, "the bee" might be defined as that x which, when put together with other x's, *makes* the hive (the "emergent entity"). Unless the theoretical (model) bees generate the hive when you put a bunch of them together, you haven't described "the bee" adequately. Thus, contrary to the opening emergentist claim, *it is precisely the adequate description of "the individual bee" that explains the hive.* An admirable modeling effort along precisely such lines is Theraulaz, Bonabeau, and Deneubourg 1998.

Agent-Based Modeling Is Reductionist

Classical emergentism holds that the parts (the microspecification) cannot explain the whole (the macrostructure), while to the agent-based modeler, it is precisely the generative sufficiency of the parts (the microspecification) that constitutes the whole's explanation! In this particular sense, agent-based modeling is reductionist.[34] Classical emergentism seeks to preserve a "mystery gap" between micro and macro; agent-based modeling seeks to demystify this alleged gap by identifying microspecifications that are sufficient to generate—robustly and replicably—the macro (whole). Perhaps the following thoughts of C. S. Peirce (1879) are apposite:

> One singular deception... which often occurs, is to mistake the sensation produced by our own unclearness of thought for a character of the object we are thinking. Instead of perceiving that the obscurity is purely subjective, we fancy that we contemplate a quality of the object which is essentially mysterious; and if our conception be afterward presented to us in a clear form we do not recognize it as the same, owing to the absence of the feeling of unintelligibility.

---

[34] The term "reductionist" admits a number of definitions. We are not speaking here of the reduction of *theories*, as in the reduction of thermodynamics to statistical mechanics. See Nagel 1961, Garfinkel 1991, and Anderson 1972.

Explanation and Prediction

A final point is that classical emergentism traffics on a crucial (and to this day quite common) confusion: between explanation and prediction. It may well be that certain phenomena are unpredictable in principle (e.g., stochastic). But that does not mean—as classical emergentists would have it—that they are unexplainable in principle. Plate tectonics explains earthquakes but does not predict their occurrence; electrostatics explains lightning but does not predict where it will hit; evolutionary theory explains species diversity but does not predict observed phenotypes. In short, one may grant unpredictability without embracing "emergence," as absolute unexplainability, à la Alexander and colleagues.[35] And, of course, it may be that in some cases prediction is a perfectly reasonable goal. (For further distinctions between prediction and explanation, see Scheffler 1960; Suppes 1985; and Newton-Smith 1981.)

In its strong classical usages—connoting absolute nondeducibility and absolute unexplainability—"emergentism" is logically confused and antiscientific. In weak level-headed usages—like "arising from local agent interactions"—a special term hardly seems necessary. For other attempts to grapple with the term "emergent," see Cariani 1992, Baas 1994, Gilbert 1995, and Darley 1996. At the very least, practitioners—and I include myself—should define this term carefully when they use it and distinguish their, perhaps quite sensible, meaning from others with which the term is strongly associated historically. To anyone literate in the philosophy of science, "emergence" has a history, and it is one with which many scientists may—indeed should—wish to part company. Doubtless, my own usage has been far too lax, so this admonition is directed as much at myself as at colleagues.

RECAPITULATION AND CONCLUSION

I am not a soldier in an agent-based methodological crusade. For some explanatory purposes, low dimensional differential equations are perfect. For others, aggregate regression is appropriate. Game theory offers deep insight in numerous contexts and so forth. But agent-based modeling is clearly a powerful tool in the analysis of *spatially distributed systems of heterogeneous autonomous actors with bounded information and computing capacity*. It is the main scientific instrument in a generative approach to social science, and a powerful tool in empirical research. It

---

[35] On fundamental sources of unpredictability, see Gell-Mann 1997.

is well suited to the study of connectionist phenomena in social science. It offers a natural environment for the study of certain interdisciplinary questions. It allows us to test the sensitivity of theories, such as neo-classical microeconomics, to relaxations in core assumptions (e.g., the assumption of fixed preferences). It allows us to trace how individual (micro) rules generate macroscopic regularities. In turn, we can employ laboratory behavioral research to select among competing multiagent models having equal generative power. The agent-based approach may decouple individual rationality from macroscopic equilibrium and sep-arate decision science from social science more generally. It invites a synthesis of analytical and computational perspectives that is particularly relevant to the study of non-equilibrium systems. Agent-based models have significant pedagogical value. Finally, the computational interpre-tation of social dynamics raises foundational issues in social science—some related to intractability, and some to undecidability proper. Despite a number of significant challenges, agent-based computational modeling can make major contributions to the social sciences.

References

Akerlof, G. 1997. Social Distance and Social Decisions. *Econometrica* 65:1005–27.

Albin, P. S. 1998. *Barriers and Bounds to Rationality*. Princeton: Princeton University Press.

Albin, P. S., and D. K. Foley. 1990. Decentralized, Dispersed Exchange without an Auctioneer: A Simulation Study. *Journal of Economic Behavior and Organization* 18(1): 27–51.

———. 1997. The Evolution of Cooperation in the Local-Interaction Multi-person Prisoners' Dilemma. Working paper, Department of Economics, Barnard College, Columbia University.

Alexander, S. 1920. *Space, Time, and Deity: The Gifford Lectures at Glasgow, 1916–1918*. New York: Dover.

Anderson, P. W. 1972. More Is Different. *Science* 177:393–96.

Arrow, K. 1963. *Social Choice and Individual Values*. New York: Wiley.

Arthur, W. B., B. LeBaron, R. Palmer, and P. Tayler. 1997. Asset Pricing under Endogenous Expectations in an Artificial Stock Market. In *The Economy as a Complex Evolving System II*, ed. W. B. Arthur, S. Durlauf, and D. Lane. Menlo Park, CA: Addison-Wesley.

Ashlock, D., D. M. Smucker, E. A. Stanley, and L. Tesfatsion. 1996. Preferential Partner Selection in an Evolutionary Study of Prisoner's Dilemma. *BioSystems* 37:99–125.

Axelrod, R. 1987. The Evolution of Strategies in the Iterated Prisoner's Dilemma. In *Genetic Algorithms and Simulated Annealing*, ed. L. Davis. London: Pitman.

————. 1997a. Advancing the Art of Simulation in the Social Sciences. *Complexity* 3:193–99.

————. 1997b. *The Complexity of Cooperation: Agent-Based Models of Competition and Collaboration*. Princeton: Princeton University Press.

————. 1997c. The Dissemination of Culture: A Model with Local Convergence and Global Polarization. In *The Complexity of Cooperation: Agent-Based Models of Competition and Collaboration*. Princeton: Princeton University Press.

————. 1997d. A Model of the Emergence of New Political Actors. In *The Complexity of Cooperation: Agent-Based Models of Competition and Collaboration*. Princeton: Princeton University Press.

Axelrod, R., and D. S. Bennett. 1993. A Landscape Theory of Aggregation. *British Journal of Political Science* 23:211–33.

Axtell, R. L. 1999. The Emergence of Firms in a Population of Agents: Local Increasing Returns, Unstable Nash Equilibria, and Power Law Size Distributions. Santa Fe Institute Working Paper 99-03-019.

Axtell, R. L., and J. M. Epstein. 1994. Agent-Based Modeling: Understanding Our Creations. *Bulletin of the Santa Fe Institute* 9:28–32.

————. 1999. Coordination in Transient Social Networks: An Agent-Based Computational Model of the Timing of Retirement. In *Behavioral Dimensions of Retirement Economics*, ed. H. Aaron. New York: Russell Sage Foundation.

Axtell, R., R. Axelrod, J. M. Epstein, and M. D. Cohen. 1996. Aligning Simulation Models: A Case Study and Results. *Computational and Mathematical Organization Theory* 1:123–41.

Axtell, R., J. M. Epstein, and H. P. Young. 2001. The Emergence of Economic Classes in an Agent-Based Bargaining Model. In *Social Dynamics*, ed. S. Durlauf and H. P. Young. Cambridge: MIT Press.

Baas, N. A. 1994. Emergence, Hierarchies, and Hyperstructures. In *Artificial Life III*, ed. C. G. Langton. Reading, MA: Addison-Wesley.

Bagley, R. J., and J. D. Farmer. 1992. Spontaneous Emergence of a Metabolism. In *Artificial Life II*, ed. C. G. Langton, C. Taylor, J. D. Farmer, and S. Rasmussen. New York: Addison-Wesley.

Bak, P., K. Chen, J. Scheinkman, and M. Woodford. 1993. Aggregate Fluctuations from Independent Sectoral Shocks: Self-Organized Criticality in a Model of Production and Inventory Dynamics. *Ricerche Economiche* 47:3–30.

Bak, P., M. Paczuski, and M. Shubik. 1996. Price Variations in a Stock Market with Many Agents. Santa Fe Institute Working Paper 96-09075.

Barnsley, M. 1988. *Fractals Everywhere*. San Diego: Academic Press.

Ben-Porath, E. 1990. The Complexity of Computing Best Response Automata in Repeated Games with Mixed Strategies. *Games and Economic Behavior* 2:1–12.

Bicchieri, C., R. Jeffrey, and B. Skyrms, eds. 1997. *The Dynamics of Norms*. New York: Cambridge University Press.

Binmore, K. G. 1987. Modeling Rational Players, I. *Economics and Philosophy* 3:179–214.

————. 1988. Modeling Rational Players, II. *Economics and Philosophy* 4:9–55.

Blume, L. E., and S. Durlauf. 2001. The Interactions-Based Approach to Socioeconomic Behavior. In *Social Dynamics*, ed. S. Durlauf and H. P. Young. Cambridge: MIT Press.

Bowles, S. 1998. Endogenous Preferences: The Cultural Consequences of Markets and Other Institutions. *Journal of Economic Literature* 36:75–111.

Broad, C. D. 1925. *The Mind and Its Place in Nature*. London: Routledge and Kegan Paul.

Bullard, J., and J. Duffy. 1998. Learning and Excess Volatility. Macroeconomics Seminar Paper, Federal Reserve Bank of St. Louis.

Buss, S., C. H. Papadimitriou, and J. N. Tsitsiklis. 1991. On the Predictability of Coupled Automata. *Complex Systems* 2:525–39.

Camerer, C. F. 1997. Progress in Behavioral Game Theory. *Journal of Economic Perspectives* 11:167–88.

Camerer, C. F., and R. Thaler. 1995. Anomalies: Ultimatums, Dictators, and Manners. *Journal of Economic Perspectives* 9(2): 209–19.

Cariani, P. 1992. Emergence and Artificial Life. In *Artificial Life II*, ed. C. G. Langton, C. Taylor, J. D. Farmer, and S. Rasmussen. New York: Addison-Wesley.

Cartwright, N. 1983. *How the Laws of Physics Lie*. New York: Oxford University Press.

Cederman, L.-E. 1997. *Emergent Actors in World Politics: How States and Nations Develop and Dissolve*. Princeton: Princeton University Press.

Chalmers, A. F. 1982. *What Is This Thing Called Science?* Cambridge: Hackett.

Chomsky, N. 1965. *Aspects of the Theory of Syntax*. Cambridge: MIT Press.

Copi, I. 1979. *Symbolic Logic*. New York: Macmillan.

Crutchfield, J. P., and M. Mitchell. 1995. The Evolution of Emergent Computation. *Proceedings of the National Academy of Sciences* 92:10742–46.

Darley, O. 1996. Emergent Phenomena and Complexity. In *Artificial Life V*, ed. C. G. Langton and Katsunori Shimohara. Cambridge: MIT Press.

Dean, J. S., G. J. Gumerman, J. M. Epstein, R. L. Axtell, A. C. Swedlund, M. T. Parker, and S. McCarroll. 2000. Understanding Anasazi Culture Change through Agent-Based Modeling. In *Dynamics in Human and Primate Societies: Agent-Based Modeling of Social and Spatial Processes*, ed. T. Kohler and G. Gumerman, 179–205. New York: Oxford University Press.

Durlauf, S. N. 1997a. Limits to Science or Limits to Epistemology? *Complexity* 2:31–37.

———. 1997b. Statistical Mechanics Approaches to Socioeconomic Behavior. In *The Economy as a Complex Evolving System II*, ed. W. B. Arthur, S. Durlauf, and D. Lane. Menlo Park: Addison-Wesley.

Epstein, J. M. 1997. *Nonlinear Dynamics, Mathematical Biology, and Social Science*. Menlo Park, CA: Addison-Wesley.

———. 1998. Zones of Cooperation in Demographic Prisoner's Dilemma. *Complexity* 4(2): 36–48.

Epstein, J. M., and R. L. Axtell. 1996. *Growing Artificial Societies: Social Science from the Bottom Up*. Washington, DC: Brookings Institution Press; Cambridge: MIT Press.

―――. 1997. Artificial Societies and Generative Social Science. *Artificial Life and Robotics* 1:33–34.

Feldman, M. W., and L. L. Cavalli-Sforza. 1976. Cultural and Biological Evolutionary Processes: Selection for a Trait under Complex Transmission. *Theoretical Population Biology* 9:239–59.

―――. 1977. The Evolution of Continuous Variation II: Complex Transmission and Assortative Mating. *Theoretical Population Biology* 11:161–81.

Feldman, M. W., and K. N. Laland. 1996. Gene-Culture Coevolutionary Theory. *Trends in Ecology and Evolution* 11:453–57.

Flierl, G., D. Grunbaum, S. A. Levin, and D. Olson. 1999. From Individuals to Aggregations: The Interplay between Behavior and Physics. *Journal of Theoretical Biology* 196:397–454.

Fraenkel, A. A., and Y. Bar-Hillel. 1958. *Foundations of Set Theory*. Amsterdam: North-Holland.

Friedman, M. 1953. The Methodology of Positive Economics. In *Essays in Positive Economics*. Chicago: University of Chicago Press.

Garfinkel, A. 1991. Reductionism. In *The Philosophy of Science*, ed. R. Boyd, P. Gasper, and J. D. Trout. Cambridge: MIT Press.

Gell-Mann, M. 1997. Fundamental Sources of Unpredictability. *Complexity* 3(1): 9–13.

Gilbert, N. 1995. Emergence in Social Stimulation. In *Artificial Societies: The Computer Simulation of Social Life*, ed. N. Gilbert and R Conte. London: University College Press.

Gilbert, N., and R. Conte, eds. 1995. *Artificial Societies: The Computer Simulation of Social Life*. London: University College Press.

Gilboa, I. 1988. The Complexity of Computing Best-Response Automata in Repeated Games. *Journal of Economic Theory* 45:342–52.

Glaeser, E., B. Sacerdote, and J. Scheinkman. 1996. Crime and Social Interactions. *Quarterly Journal of Economics* 111:507–48.

Glymour, C. 1980. *Theory and Evidence*. Princeton: Princeton University Press.

Hausman, D. M. 1992. *The Inexact and Separate Science of Economics*. Cambridge: Cambridge University Press.

―――. 1998. Problems with Realism in Economics. *Economics and Philosophy* 14:185–213.

Hempel C. G. 1956. A Logical Appraisal of Operationism. In *The Validation of Scientific Theories*, ed. P. Frank. Boston: Beacon Press.

Hempel, C. G., and P. Oppenheim. 1948. Studies in the Logic of Explanation. *Philosophy of Science* 15:567–79.

Hendriks-Jansen, H. 1996. In Praise of Interactive Emergence, or Why Explanations Don't Have to Wait for Implementation. In *The Philosophy of Artificial Life*, ed. M. A. Boden. New York: Oxford University Press.

Hirsch, M. D., C. H. Papadimitriou, and S. A. Vavasis. 1989. Exponential Lower Bounds for Finding Brouwer Fixed Points. *Journal of Complexity* 5:379–416.

Hodel, R. E. 1995. *An Introduction to Mathematical Logic*. Boston: PWS Publishing.

Holland, J. H. 1992. Genetic Algorithms. *Scientific American* 267:66–72.

Horgan, T. 1993. From Supervenience to Superdupervenience: Meeting the Demands of a Material World. *Mind* 102:555–86.

Howson, C., and P. Urbach. 1993. *Scientific Reasoning: A Bayesian Approach.* Chicago: Open Court.

Huberman, B., and N. Glance. 1993. Evolutionary Games and Computer Simulations. *Proceedings of the National Academy of Sciences* 90:7715–18.

Ijiri, Y., and H. Simon. 1977. *Skew Distributions and the Sizes of Business Firms.* New York: North-Holland.

Ilachinski, A. 1997. Irreducible Semi-Autonomous Adaptive Combat (ISAAC): An Artificial-Life Approach to Land Warfare. Center for Naval Analyses Research Memorandum CRM 97-61.10.

Jeffrey, R. 1991. *Formal Logic: Its Scope and Limits.* 3rd ed. New York: McGraw-Hill.

Jennings, N., K. Sycara, and M. Woolridge. 1998. A Roadmap of Agent Research and Development. *Autonomous Agents and Multi-Agent Systems* 7–38.

Johnson, J. 1995. A Language of Structure in the Science of Complexity. *Complexity* 1(3): 22–29.

Kagel, J. H., and A. E. Roth, eds. 1995. *The Handbook of Experimental Economics.* Princeton: Princeton University Press.

Kalai, E. 1990. Bounded Rationality and Strategic Complexity in Repeated Games. In *Game Theory and Applications,* ed. T. Ichiishi, A. Neyman, and Y. Taumaneds. San Diego: Academic Press.

Katona, G. 1951. *Psychological Analysis of Economic Behavior.* New York: McGraw-Hill.

Kauffman, S. A. 1993. *The Origins of Order: Self-Organization and Selection in Evolution.* New York: Oxford University Press.

Kirman, A. P. 1992. Whom or What Does the Representative Individual Represent? *Journal of Economic Perspectives* 6:117–36.

Kirman, A. P., and N. J. Vriend. 1998. Evolving Market Structure: A Model of Price Dispersion and Loyalty. Working paper, University of London.

Kleene, S. C. 1967. *Mathematical Logic.* New York: Wiley.

Knuth, D. E. 1969. *The Art of Computer Programming.* Vol. 2, *Seminumerical Algorithms.* Reading: Addison-Wesley.

Kollman, K., J. Miller, and S. Page. 1992. Adaptive Parties in Spatial Elections. *American Political Science Review* 86:929–37.

Kuran, T. 1989. Sparks and Prairie Fires: A Theory of Unanticipated Political Revolution. *Public Choice* 61:41–74.

Langton, C. G., C. Taylor, J. D. Farmer, and S. Rasmussen, eds. 1992. *Artificial Life II.* New York: Addison-Wesley.

Lewis, A. A. 1985. On Effectively Computable Realizations of Choice Functions. *Mathematical Social Sciences* 10:43–80.

———. 1992a. Some Aspects of Effectively Constructive Mathematics That Are Relevant to the Foundations of Neoclassical Mathematical Economics and the Theory of Games. *Mathematical Social Sciences* 24(2–3): 209–35.

————. 1992b. On Turing Degrees of Walrasian Models and a General Impossibility Result in the Theory of Decision-Making. *Mathematical Social Sciences* 24(2–3): 141–71.

Lindgren, K., and M. G. Nordahl. 1994. Cooperation and Community Structure in Artificial Ecosystems. *Artificial Life* 1:15–37.

McLaughlin, B. P. 1992. The Rise and Fall of British Emergentism. In *Emergence or Reduction? Essays on the Prospects of Nonreductive Physicalism*, ed. A. Beckermann, H. Flohr, and J. Kim. New York: Walter de Gruyter.

McNeill, W. H. 1976. *Plagues and Peoples*. New York: Anchor Press/Doubleday.

Miller, J. H. 1996. The Coevolution of Automata in the Repeated Prisoner's Dilemma. *Journal of Economic Behavior and Organization* 29:87–112.

Minsky, M. 1985. *The Society of Mind*. New York: Simon and Schuster.

Mitchell, M. 1998. *An Introduction to Genetic Algorithms*. Cambridge: MIT Press.

Morgan, C. L. 1923. *Emergent Evolution*. London: Williams and Norgate.

Nagel, E. 1961. *The Structure of Science*. New York: Harcourt, Brace, and World.

Nagel, K., and S. Rasmussen. 1994. Traffic at the Edge of Chaos. In *Artificial Life IV*, ed. R. Brooks. Cambridge: MIT Press.

Newton-Smith, W. H. 1981. *The Rationality of Science*. New York: Routledge.

Neyman, A. 1985. Bounded Complexity Justifies Cooperation in the Finitely Repeated Prisoners' Dilemma. *Economics Letters* 19:227–29.

Nordhaus, W. D. 1992. Lethal Model 2: The Limits to Growth Revisited. *Brookings Papers on Economic Activity* 2:1–59.

Nowak, M. A., and R. M. May. 1992. Evolutionary Games and Spatial Chaos. *Nature* 359:826–29.

Papadimitriou, C. H. 1993. Computational Complexity as Bounded Rationality. *Mathematics of Operations Research*.

Papadimitriou, C. H., and M. Yannakakis. 1994. On Complexity as Bounded Rationality. *Proceedings of the Twenty-Sixth Annual ACM Symposium on the Theory of Computing*. Montreal, Quebec, Canada, May 23–25, 1994, pp. 726–33.

Peirce, C. S. 1879. How to Make Our Ideas Clear. In *The Process of Philosophy: A Historical Introduction*, ed. Joseph Epstein and Gail Kennedy. New York: Random House. 1967.

Pitt, J. C., ed. 1988. *Theories of Explanation*. New York: Oxford University Press.

Popper, K. R. 1959. *The Logic of Scientific Discovery*. New York: Routledge.

————. 1963. *Conjectures and Refutations: The Growth of Scientific Knowledge*. New York: Routledge.

Prasad, K. 1997. On the Computability of Nash Equilibria. *Journal of Economic Dynamics and Control* 21:943–53.

Prietula, M. J., K. M. Carley, and L. Gasser. 1998. *Simulating Organizations*. Cambridge: MIT Press.

Rabin, M. 1998. Psychology and Economics. *Journal of Economic Literature* 36:11–46.

Rabin, M. O. 1957. Effective Computability of Winning Strategies. *Annals of Mathematical Studies* 39:147–157.

Resnick, M. 1994. *Turtles, Termites, and Traffic Jams: Explorations in Massively Parallel Microworlds*. Cambridge: MIT Press.

Rogers, H. 1967. *Theory of Recursive Functions and Effective Computability*. New York: McGraw-Hill.

Rubinstein, A. 1986. Finite Automata Play the Repeated Prisoners' Dilemma. *Journal of Economic Theory* 39:83–96.

———. 1998. *Modeling Bounded Rationality*. Cambridge: MIT Press.

Salmon, W. C. 1984. *Scientific Explanation and the Causal Structure of the World*. Princeton: Princeton University Press.

Salmon, W. C., with R. C. Jeffrey and J. G. Greeno. 1971. *Statistical Explanation and Statistical Relevance*. Pittsburgh: University of Pittsburgh Press.

Scheffler, L. 1960. Explanation, Prediction, and Abstraction. In *Philosophy of Science*, ed. A. Danto and S. Morgenbesser. New York: Meridian.

Schelling, T. 1971. Dynamic Models of Segregation. *Journal of Mathematical Sociology* 1:143–86.

———. 1978. *Micromotives and Macrobehavior*. New York: W. W. Norton.

Simon. H. A. 1978. On How to Decide What to Do. *Bell Journal of Economics* 9:494–507.

———. 1982. *Models of Bounded Rationality*. Cambridge: MIT Press.

———. 1996. *The Sciences of the Artificial*. 3rd ed. Cambridge: MIT Press.

Skyrms, B. 1986. *Choice and Chance*. 3rd ed. Belmont, CA: Wadsworth.

———. 1998. *Evolution of the Social Contract*. Cambridge: Cambridge University Press.

Smullyan, R. M. 1992. *Gödel's Incompleteness Theorems*. New York: Oxford University Press.

Sober, E. 1996. Learning from Functionalism—Prospects for Strong Artificial Life. In *The Philosophy of Artificial Life*, ed. M. A. Boden. New York: Oxford University Press.

Stanley, M. H. R., L. A. N. Amaral, S. V. Buldyrev, S. Havlin, H. Leschhorn, P. Maass, M. A. Salinger, and H. E. Stanley. 1996. Scaling Behavior in the Growth of Companies. *Nature* 379:804–6.

Stephan, A. 1992. Emergence—a Systematic View on Its Historical Facets. In *Emergence or Reduction? Essays on the Prospects of Nonreductive Physicalism*, ed. A. Beckermann, H. Flohr, and J. Kim. New York: Walter de Gruyter.

Suppes, P. 1985. Explaining the Unpredictable. *Erkenntnis* 22:187–195.

Tesfatsion, L. 1995. A Trade Network Game with Endogenous Partner Selection. Economic Report 36, Department of Economics, Iowa State University.

Theraulaz, G., E. Bonabeau, and J. L. Deneubourg. The Origin of Nest Complexity in Social Insects. *Complexity* 3(6): 15–25.

Topa, G. 1997. Social Interactions, Local Spillovers, and Unemployment. Manuscript, Department of Economics, New York University.

Tversky, A. and D. Kahneman. 1986. Rational Choice and the Framing of Decisions. In *Rational Choice: The Contrast between Economics and Psychology*, ed. R. M. Hogarth and M. W. Reder. Chicago: University of Chicago Press.

Uzawa, H. 1962. On the Stability of Edgeworth's Barter Process. *International Economic Review* 3:218–32.

van Fraassen, B. C. 1980. *The Scientific Image*. New York: Oxford University Press.

Watkins, J. W. N. 1957. Historical Explanation in the Social Sciences. *British Journal for the Philosophy of Science* 8:106.

Young, H. P. 1993. The Evolution of Conventions. *Econometrica* 61:57–84.

———. 1995. The Economics of Convention. *Journal of Economic Perspectives* 10:105–22.

———. 1998. *Individual Strategy and Social Structure: An Evolutionary Theory of Institutions*. Princeton: Princeton University Press.

Young, H. P., and D. Foster. 1991. Cooperation in the Short and in the Long Run. *Games and Economic Behavior* 3:145–56.

Prelude to Chapter 2

CONFESSION OF A WANDERING BARK

MY INTELLECTUAL LIFE began in music composition. I wrote three chamber pieces for small ensemble. Then, at Amherst, I became absorbed in mathematics and later, at MIT, in mathematical modeling. Now, I am immersed in agent-based modeling. It would certainly appear that I have wandered from music to mathematics to agent modeling, within which (judging by the current volume) I have wandered from civil violence to epidemiology to archaeology. But I do not feel that I have ever changed *fields*. What, then, is my field? In the next chapter, I quote Bertrand Russell on the topic of mathematical beauty. He writes that

> In the most beautiful work, a chain of argument is presented in which every
> link is important on its own account, in which there is an air of ease and
> lucidity throughout, and the premises achieve more than would have been
> thought possible, by means which appear natural and inevitable.

That is, and always has been, my "field." I'm not sure what to call it, but perhaps *Generative Minimalism* would be apt.

In the third movement of my Quartet for Violin, Oboe, Clarinet, and Bassoon, the "premise," the theme, is a single three-note motif: a descending half-step, followed by a descending minor third. Essentially, the entire movement is that motif—that simple rule—expanded, contracted, inverted, reversed, and passed among the instruments. Hopefully, the theme, the Russellian premise, achieves "more than would have been thought possible, by means which appear natural and inevitable."

In my book *Nonlinear Dynamics, Mathematical Biology, and Social Science*,[1] the analogous "premise" is the Lotka-Volterra ecosystem model. The book then shows how this single elegant system of equations can be "morphed" into—can generate—simple models of arms races, wars, revolutions, and drug epidemics in humans. The mathematician Edward Beltrami, in a very generous and graceful review, writes, "It is amazing what a wealth of striking analogies this short book manages to present.... It is like a repertory theater with just a few actors who

---

[1] Joshua M. Epstein, *Nonlinear Dynamics, Mathematical Biology, and Social Science* (Reading, MA: Addision-Wesley, 1997).

reappear in different guises to play multiple and unexpected roles."[2] One may disagree that I achieved that, but it was certainly my goal.

In *Growing Artificial Societies*, with Robert Axtell, the premises are a few minimal individual agent rules, which are shown to generate a wide variety of core phenomena, such as migrations, skewed wealth distributions, and others. However, as stated there, "The surprise consists precisely in the emergence of familiar macrostructures *from the bottom up*—from simple local rules that outwardly appear quite remote from the social, or collective, phenomena they generate." The surprise is "the generative sufficiency of the simple local rules."[3]

And finally, in the present volume, I have the same invariant aesthetic goal. The specific "target" varies from chapter to chapter: Here it is to grow civil wars, there it is to grow adaptive organizations, or conformity to norms, or epidemic dynamics. But in every case, the aim (successful or not) is to generate the phenomenon—I will not shrink from saying it— "beautifully" in Russell's particular sense, from premises and by means which appear "natural and inevitable."

Auguste Renoir said, "I don't care about the subject, so long as it reflects light." In this same special sense, I don't care about any of the subjects in this book. Of course, as a human being, I care deeply about oppressive class structures, devastating epidemics, and genocidal civil wars. And each phenomenon is certainly worthy of the most serious scholarly study. But, as a theoretician or artist (same thing),[4] the topics are mere *constraints against which to exercise this fixed impulse to a kind of generative minimalism.*

I have certainly wandered from one *constraint* to another: generate an entire movement from a three-note theme; generate mathematical models of revolutions, arms races, and the spread of drugs from a simple ecosystem model; generate an artificial society from a small set of simple agent rules. The constraints, the "instrumentation" (from oboes and violins, to differential equations, to computer programs), and the generative mechanisms (from musical transforms to mathematical ones) may all have changed. But the fundamental impulse—the "light" in Renoir's sense, the "field," if you prefer—has not.[5] And I feel about it

---

[2] Edward Beltrami, *Bulletin of Mathematical Biology* 60 (1998): 411–16.

[3] Joshua M. Epstein and Robert Axtell, *Growing Aritifical Societies: Social Science from the Bottom Up* (Cambridge, MA: MIT Press, 1996), pp. 51–52.

[4] G. H. Hardy, *A Mathematician's Apology* (1940), Canto version (Cambridge: Cambridge University Press, 1992), 151.

[5] There are, of course, no academic departments of this field—only departments of its epiphenomena.

rather as Shakespeare felt about love. "It is an ever fixed mark...the star to every wandering bark."[6]

The Next Chapter

While discussing further this question of beauty in agent models, the next chapter is focused on a number of foundational and epistemological questions. Some are covered briefly in chapter 1, and some recur in chapter 3. Hence, there is some duplication, or inefficiency, in the first three chapters of this book. But, in a volume as critical of economics as this one is, an overweaning concern for efficiency would be downright hypocritical.

---

[6] Sonnet CXVI.

Chapter 2

REMARKS ON THE FOUNDATIONS OF

AGENT-BASED GENERATIVE SOCIAL SCIENCE

JOSHUA M. EPSTEIN*

GENERATIVE EXPLANATION

THE SCIENTIFIC ENTERPRISE is, first and foremost, *explanatory*. While agent-based modeling can change the social sciences in a variety of ways, in my view its central contribution is to facilitate *generative* explanation (see Epstein 1999). To the generativist, explaining macroscopic social regularities, such as norms, spatial patterns, contagion dynamics, or institutions requires that one answer the following question:

How could the autonomous local interactions of heterogeneous boundedly rational agents generate the given regularity?

Accordingly, to explain macroscopic social patterns, we generate—or "grow"—them in agent models. This represents a departure from prevailing practice. It is fair to say that, overwhelmingly, game theory, mathematical economics, and rational choice political science are concerned with equilibria. In these quarters, "explaining an observed social pattern" is essentially understood to mean "demonstrating that it is the Nash equilibrium (or a distinguished Nash equilibrium) of some game."

By contrast, to the generativist, it does *not* suffice to demonstrate that, if a society of rational (*homo economicus*) agents were placed in the pattern, no individual would unilaterally depart—the Nash equilibrium condition. Rather, to explain a pattern, one must show how a population

---

*Joshua M. Epstein is a Senior Fellow in Economic Studies at The Brookings Institution, a member of The Brookings–Johns Hopkins Center on Social and Economic Dynamics, and an External Faculty member of The Santa Fe Institute.

For thoughtful comments on this chapter, the author thanks Claudio Cioffi-Revilla, Samuel David Epstein, Carol Graham, Ross Hammond, Kislaya Prasad, Brian Skyrms, Leigh Tesfatsion, and Peyton Young. For assistance in preparing the manuscript for publication, he thanks Danielle Feher.

This essay is published simultaneously in *Handbook of Computational Economics, Volume 2: Agent-Based Computational Economics,* ed. L. Tesfatsion and K. Judd. 2006. North-Holland.

of cognitively plausible agents, interacting under plausible rules, could actually arrive at the pattern on time scales of interest. The motto, in short, is (Epstein 1999): *If you didn't grow it, you didn't explain it.* Or, in the notation of first-order logic:

$$\forall x(\neg Gx \supset \neg Ex) \tag{1}$$

To explain a macroscopic regularity x is to furnish a suitable microspecification that suffices to generate it.[1] The core request is hardly outlandish: To explain a macro-x, please show how it could arise in a plausible society. Demonstrate how a set of recognizable—heterogeneous, autonomous, boundedly rational, locally interacting—agents could actually get there in reasonable time. The agent-based computational model is a new, and especially powerful, instrument for constructing such demonstrations of generative sufficiency.

FEATURES OF AGENT-BASED MODELS

As reviewed in Epstein and Axtell 1996 and Epstein 1999, key features of agent-based models typically include the following:[2]

HETEROGENEITY

Representative agent methods—common in macroeconomics—are not used in agent-based models. Nor are agents aggregated into a few homogeneous pools. Rather, every individual is explicitly represented. And these individuals may differ from one another in myriad ways: by wealth, preferences, memories, decision rules, social network, locations, genetics, culture, and so forth, some or all of which may adapt or change endogenously over time.

AUTONOMY

There is no central, or "top down," control over individual behavior in agent-based models. Of course, there will generally be feedback between macrostructures and microstructures, as where newborn agents are conditioned by social norms or institutions that have taken shape

---

[1] In slightly more detail, if we let $M = \{i : i$ is a microspecification$\}$ and let $G(i, x)$ denote the proposition that i generates x, then the proposition Gx can be expressed as $\exists i\, G(i, x)$. Then, longhand, the motto becomes: $\forall x(\neg\exists i\, G(i, x) \supset \neg Ex)$.

[2] I do not claim that every agent-based model exhibits all these features. My point is that the explanatory *desiderata* enumerated (heterogeneity, local interactions, bounded rationality, etc.) are easily arranged in agent-based models.

endogenously through earlier agent interactions. In this sense, micro and macro will, in general, co-evolve. But as a matter of model specification, no central controllers (e.g., Walrasian auctioneers) or higher authorities are posited *ab initio*.

EXPLICIT SPACE

Events typically transpire on an explicit space, which may be a landscape of renewable resources, as in Epstein and Axtell 1996, an *n*-dimensional lattice, a dynamic social network, or any number of other structures. The main *desideratum* is that the notion of "local" be well-posed.

LOCAL INTERACTIONS

Typically, agents interact with neighbors in this space (and perhaps with sites in their vicinity). Uniform mixing (mass action kinetics) is generically not the rule. Relatedly, many agent-based models, following Herbert Simon, also assume:

BOUNDED RATIONALITY

There are two components of this: bounded information and bounded computing power. Agents have neither global information nor infinite computational capacity. Although they are typically purposive, they are not global optimizers; they use simple rules based on local information.

NON-EQUILIBRIUM DYNAMICS

Non-equilibrium dynamics are of central concern to agent modelers, as are large-scale transitions, "tipping phenomena," and the emergence of macroscopic regularity from decentralized local interaction. These are sharply distinguished from equilibrium existence theorems and comparative statics, as is discussed below.

RECENT EXPANSION

The literature of agent-based models has grown to include a number of good collections (e.g., The Sackler Colloquium, *Proceedings of the National Academy of Sciences, 2002*), special issues of scholarly journals (*Computational Economics 2001, The Journal of Economic Dynamics and Control, 2004*), numerous individual articles in academic journals (such as *Computational and Mathematical Organization Theory*), the science journals (*Nature, Science*) and books (e.g., Epstein and Axtell 1996; Axelrod 1997; Cederman 1997). New journals (e.g., *The Journal of Artificial Societies and Social Simulation*) are emerging, computational platforms are competing (e.g., Ascape, Repast, Swarm, Mason).

International societies for agent-based modeling are being formed. Courses on agent-based modeling are being offered at major universities. Conferences in the U.S., Europe, and Asia are frequent, and agent-based modeling is receiving considerable attention in the press. The landscape is very different than it was a decade ago.

EPISTEMOLOGICAL ISSUES

Einstein wrote that, "Science without epistemology is—in so far as it is thinkable at all—primitive and muddled" (Pais 1982). Given the rapid expansion of agent-based modeling, it is an appropriate juncture at which to sort out and address certain epistemological issues surrounding the approach. In particular, and without claiming comprehensiveness, the following issues strike me as fundamentally important, and in need of clarification, both within the agent modeling community and among its detractors.

1. Generative sufficiency vs. explanatory necessity
2. Generative agent-based models vs. explicit mathematical models
3. Generative explanation vs. deductive explanation
4. Generative explanation vs. inductive explanation
5. Generality of agent models

I will attempt to address these and a variety of related issues. At several points, there will be a need to distinguish claims from their converses. The first example of this follows.

Generative Sufficiency

The generativist *motto* (1) cited above was:

$$\forall x(\neg Gx \supset \neg Ex) \tag{2}$$

If you didn't grow it, you didn't explain it. It is important to note that we reject the converse claim. Merely to generate is not necessarily to explain (at least not well). A microspecification might generate a macroscopic regularity of interest in a patently absurd—and hence non-explanatory—way. For instance, it might be that Artificial Anasazi (Axtell *et al.* 2002) arrive in the observed (true Anasazi) settlement pattern stumbling around backward and blindfolded. But one would not adopt that picture of individual behavior as explanatory. In summary, *generative sufficiency is a necessary, but not sufficient condition for explanation.*

Of course, in principle, there may be competing microspecifications with equal generative sufficiency, none of which can be ruled out so easily.

The mapping from the set of microspecifications to the macroscopic *explanandum* might be many-to-one. In that case, further work is required to adjudicate among the competitors.

For example, if the competing models differ in their rules of individual behavior, appropriate laboratory psychology experiments may be in order to determine the more plausible empirically. In my own experience, given a macroscopic *explanandum*, it is challenging to devise *any* rules that suffice to generate it. In principle, however, the search could be mechanized. One would metrize the set of macroscopic patterns, so that the distance from a generated pattern to the target pattern (the pattern to be explained) could be computed. The "fitter" a microspecification, the smaller the distance from its generated macrostructure to the empirical target. Given this definition of fitness, one would then encode the space of permissible micro-rules and search it mechanically—with a genetic algorithm, for example (as in Crutchfield and Mitchell 1995).

In any event, the first point is that the motto (1) is a criterion for explanatory candidacy. There may be multiple candidates and, as in any other science, selection among them will involve further considerations.[3]

The Indictment: No Equations, Not Deductive, Not General

Plato observed that the doctors would make the best murderers. Likewise, in their heart of hearts, leading practitioners of any approach know themselves to be its most capable detractors. I think it is healthy for experienced proponents of any approach to explicitly formulate its most damaging critique and, if possible, address it. In that spirit, it seems to me that among skeptics toward agent modeling, the central indictment is tripartite: First, that in contrast to mathematical "hard" science, there are no equations for agent-based models. Second, that agent models are not deductive;[4] and third, that they are *ad hoc*, not general. I will argue that the first two claims are false and that, at this stage in the field's development, the third is unimportant.

Equations Exist

The oft-claimed distinction between computational agent models, and equation-based models is illusory. Every agent model is, after all, a

---

[3] As noted, empirical plausibility is one such. Theoretical economy is another. In generative linguistics, for example, S. D. Epstein and N. Hornstein (1999, ix–xviii) convincingly argue that minimalism should be central in selecting among competing theories.

[4] Not everyone who asserts that computational agent modeling is non-deductive necessarily regards it as a defect. See, for example, Axelrod 1997.

computer program (typically coded in a structured or object-oriented programming language). As such, each is clearly Turing computable (computable by a Turing machine). But, for every Turing machine, there is a unique corresponding and equivalent partial recursive function (see Hodel 1995).

This is precisely the function class constructible from the zero function, the successor function, and the "pick out" or projection function (the three so-called initial functions) by finite applications of composition (substitution), bounded minimization, and—the really distinctive manipulation—*primitive recursion*. This, as the defining formula below suggests, can be thought of as a kind of generalized induction.

$$h(\vec{x}, 0) = f(\vec{x})$$

$$h(\vec{x}, n + 1) = g(\vec{x}, n, h(\vec{x}, n))$$

(See Hamilton 1988, Boolos and Jeffrey 1989, Epstein and Carnielli 1989, or Hodel 1995 for a technical definition of this class of functions.) So, in principle, one could cast any agent-based computational model as an explicit set of mathematical formulas (recursive functions). In practice, these formulas might be extremely complex and difficult to interpret. But, speaking technically, they surely exist. Indeed, one might have called the approach "recursive social science," "effectively computable social science," "constructive social science," or any number of other equivalent things. The use of "generative" was inspired by Chomsky's usage (Chomsky 1965). In any case, the issue is not whether equivalent equations exist, but which representation (equations or programs) is most illuminating.

To all but the most adept practitioners, the recursive function representation would be quite unrecognizable as a model of social interaction, while the equivalent agent model is immediately intelligible as such. However, at the dawn of the calculus, the same would doubtless have been true of differential equations. It is worth noting that recursive function theory is still very young, having developed only in the 1930s. And, it is virtually unknown in the social sciences. It is the mathematical formalism directly isomorphic (see Jeffrey 1991) to computer programs, and over time, we may come to feel as comfortable with it as we now do with differential equations. Moreover, it is worth noting that various agent-based models have, in fact, been revealingly mathematized using other, more familiar, techniques. (See Dorofeenko and Shorish 2002; Pollicott and Weiss 2001; Young 1998.)

In sum, the first element of the indictment, that agent models are "just simulations" for which no equations exist, is simply false. Moreover, even if equivalent equations are not in hand, computational agent models have the advantage that they can be run thousands of times to produce large quantities of clean data. These can then be analyzed to produce a robust statistical portrait of model performance over the parameter ranges (and rule variations) of interest.

This critique, moreover, betrays a certain naiveté about contemporary equation-based modeling in many areas of applied science, such as climate modeling. The mathematical models of interest are huge systems of nonlinear reaction diffusion equations. In practice, they are not solved analytically, but are approximated computationally. So, the opposition of analytically soluble mathematical models on the one hand, and computational models on the other, while conceptually enticing, is quite artificial in practice.

Agent Models Deduce

Another misconception is that the explicit equation-based approach is deductive, whereas the agent-based computational approach is not. This, too, is incorrect. Every realization of an agent model is a *strict deduction*. There are a number of ways to establish this. Perhaps the most direct is to note that it follows from the previous point.

Every program can be expressed in recursive functions. But recursive functions are computed deterministically from initial values. They are mechanically (effectively) computable—in principle by hand with pencil and paper. Given the nth (including the initial) state of the system, the $(n + 1)$st state is computable in a strictly mechanical and deterministic way by recursion. Since this mechanical procedure is obviously deductive, so is each realization of an agent model.

A more sweeping equivalence can be established, in fact. It can be shown that Turing machines, recursive functions, and first-order logic itself (the system of deduction *par excellence*) are all strictly intertranslatable (see Hodel 1995). So, in a rigorous sense, every state generated in an agent model is literally a theorem. Since, accepting our motto, to explain is to generate (but not conversely), and to generate a state is to deduce it as a theorem, we are led to assert that to explain a pattern is to show it to be theorematic.

A third, slightly less rigorous way to think of it is this. Every agent program begins in some configuration x—a set of initial (agent) states analogous to axioms—and then repeatedly updates by rules of the form; if x then y. But, $\{x, x \supset y\}$ is just *modus ponens*, so the model as a whole is ultimately one massive inference in a Hilbert-type deductive system. To "grow" a pattern p (and to explain a pattern p) is thus to show that it is one of these terminal y's—in effect, that it is theorematic, very much as in the classic hypothetico-deductive picture of scientific explanation.

What about Randomness?

If every run is a strict deduction, what about stochasticity, a common feature of many agent models? Stochastic realizations are also strict deductions. In a computer, random numbers are in fact produced by strictly deterministic pseudo-random number generators. For example, the famous linear congruential method (Knuth 1998) to generate a series of pseudo-random numbers is as follows:

Define: m, the modulus ($m > 0$); a, the multiplier ($0 \leq a \leq m$); c, the increment ($0 \leq c \leq m$), and $x(0)$, the seed, or staring value ($0 \leq x(0) \leq m$). Then, the (recursion) scheme for generating the pseudo-random sequence is, for $n \geq 0$:

$$x(n + 1) = (ax(n) + c) \bmod m$$

This determinism is why, when we save the seed and re-run the program, we get exactly the same run again.

What Types of Propositions are Deduced?

In principle, the only objects we ever technically deduce are *propositions*. When we deduce the Fundamental Theorem of Calculus, we deduce the proposition: "The definite integral of a continuous real-valued function on an interval is equal to the difference of an anti-derivative's values at the interval's endpoints." The result is normally expressed in mathematical notation, but, in principle, it is a proposition statable in English.[5] In turn, we explain an empirical regularity when that regularity is rendered as a proposition and that proposition is deduced from premises we accept. For example, we *explain* Galileo's leaning Tower of Pisa

---

[5] In principle, it can be further broken down into statements about limits of sums, and so forth. As a completely worked out simple example, consider the mathematical equation

$$\lim_{x \to 2} x^2 = 4.$$

It asserts: "The limit of the square of x, as x approaches two, is four." In further detail, it is the following claim:

$$\forall(\varepsilon > 0)\exists(\delta > 0)[0 < |x - 2| < \delta \Rightarrow \left|x^2 - 4\right| < \varepsilon].$$

In English, "For every number epsilon greater than zero, there exists a number delta greater than zero such that if the absolute value of the difference between x and 2 is strictly between zero and delta, then the absolute value of the difference between the square of x and four is less than epsilon." The fact that it is easier to manipulate and compute with mathematical symbols than with words may say something interesting about human psychology, but it does not demonstrate any limit on the precision or expressive power of English.

observation (i.e., that objects of unequal masses dropped from the same
height land simultaneously) by strictly deducing, from Newton's Second
Law and the Law of Universal Gravitation, the following proposition:
"The acceleration of a freely falling body near the surface of the earth is
independent of its mass."

Well, if agent models explain by generating, and thus deducing, and
if, as I have just argued, the only deducible objects are propositions,
the question arises: what sorts of propositions are deduced when agent
models explain? In many important cases, the answer is: a normal form.

Social Science as the Satisfaction of Normal Forms

We explain a pattern when the pattern is expressed as a proposition and
the proposition is deduced from premises we accept. Seen in this light,
many of the macroscopic patterns we, as social scientists, are trying
to explain are expressible as large disjunctive normal forms, DNFs. In
general a DNF, δ has the logical form

$$\delta = \bigvee_{i=1}^{n} \bigwedge_{j=1}^{m} \phi_{ij}$$

where ϕ_{ij} is a statement form (see Hamilton 1988). Clearly, this discussion applies to arbitrarily large, but *finite*, populations.

EXAMPLE 1. DISTRIBUTIONS

Suppose, then, that we are trying to explain a skewed wealth distribution
observed in some finite population of agents. For simplicity's sake,
imagine three agents: A, B, and C. And suppose we observe that 6
indivisible wealth units (the country's GNP) are distributed as 3:2:1. That
is the empirical target; and our model will be deemed a success if it grows
that distribution, *regardless of who has what*. What that means is that
the successful model will generate any one of the six conjunctions in the
following DNF, shown in braces (where A3 means "Agent A has 3 units,"
and so forth):

$$(A1 \wedge B2 \wedge C3) \vee$$

$$(A1 \wedge B3 \wedge C2) \vee$$

$$(A2 \wedge B1 \wedge C3) \vee$$

$$(A2 \wedge B3 \wedge C1) \vee$$

$$(A3 \wedge B1 \wedge C2) \vee$$

$$(A3 \wedge B2 \wedge C1)$$

The model succeeds if it grows any one of these conjuncts, that is, a conjunction whose truth makes the DNF true.

EXAMPLE 2. SPATIAL PATTERNS

Likewise, suppose we are trying to model segregation in a population composed of two white and two black agents (W1, W2, B1, B2) arranged on a line with four positions: 1, 2, 3, 4. The model works if it generates two contiguous agents of the same color, followed by two contiguous agents of the other color. As above, we don't care who is where so long as we get segregation on the line. The truth of any of the eight conjunctions of the following DNF will therefore suffice (here W12 denotes the proposition: "white agent 1 occupies position 2"):

$$(W11 \wedge W22 \wedge B13 \wedge B24) \vee$$

$$(W11 \wedge W22 \wedge B23 \wedge B14) \vee$$

$$(W21 \wedge W12 \wedge B13 \wedge B24) \vee$$

$$(W21 \wedge W12 \wedge B23 \wedge B14) \vee$$

$$(B11 \wedge B22 \wedge W13 \wedge W24) \vee$$

$$(B11 \wedge B22 \wedge W23 \wedge W14) \vee$$

$$(B21 \wedge B12 \wedge W13 \wedge W24) \vee$$

$$(B21 \wedge B12 \wedge W23 \wedge W14).$$

Again, success in generating "segregation" consists in generating any one of these conjunctions. That suffices to make the DNF true. While this exposition has been couched in terms of wealth distributions and distributions of spatial position, it obviously generalizes to distributions of myriad sorts (e.g., size and power), and with straightforward modification, to sequences of patterns over time. A dynamic sequence of patterns would, in fact, be a Conjunctive Normal Form (CNF), each term of which is a DNF of the sort just discussed.[6]

Generative Implies Deductive, but Not Conversely: Nonconstructive Existence

A generative explanation is a deductive one. Generative implies deductive. The converse, however, does not apply. It is possible to deduce

---

[6] The general problem of satisfying an n-term CNF is NP-Complete (Garey and Johnson 1979). Based on this observation, it is tempting to conjecture that nonequilibrium social science—suitably cast as CNF satisfaction—is computationally hard in a rigorous sense.

without generating. Not all deductive argument has the constructive character of agent-based modeling. Nonconstructive existence proofs are clear examples. Often, these take the form of *reductio ad absurdum*[7] arguments, which work as follows.

Suppose we wish to prove the existence of an x with some property (e.g., that it is an equilibrium). We take as an axiom the so-called Law of the Excluded Middle (LEM), implying that either x exists or x does not exist. Symbolically:

$$\exists x \lor \neg \exists x$$

One of those *must* be true. Next, we assume that x does not exist and derive a contradiction. That is, we show that

$$\neg \exists x \supset [p \land \neg p]$$

Since contradictions are always False, this has the form:

$$\neg \exists x \supset F$$

But this implication can be True only if the antecedent, $\neg \exists x$, is False. From this it follows from the LEM that $\exists x$ is True and *voila*: the x in question must exist!

But we have failed to exhibit x, or specify any algorithm that would generate it, patently violating our generative motto (1). We have failed to show that x is generable at all, much less that it is generable on time scales of interest. But, the existence argument is nonetheless deductive.

Now, there are deductive and nonconstructive existence proofs that do not use *reductio ad absurdum*. One of my favorites is the beautiful and startling index theoretic proof that, in regular economies, the number of equilibria must be an odd integer (see Mas-Colell *et al.* 1995; Epstein 1997). This proof gives no clue how to compute the equilibria. Like *reductio*, it fails to show the equilibria to be generable at all, much less on time scales of interest. But, the existence argument is nonetheless deductive.

Hence, if we insist that explanation requires generability, we are led to the position that deductive arguments can be non-explanatory. *Generative explanation is deductive, but deduction is not necessarily explanatory.*

We have addressed the first two points of the indictment: that there are no equations, and that agent modeling is not deductive. The third issue

---

[7] Reduction to an absurdity.

was the generality of agent models. I would like to approach this topic by a seemingly circuitous route, extending the preceding points on existence and generability into the areas of incompleteness and computational complexity.

INCOMPLETENESS (ATTAINABILITY AT ALL) AND COMPLEXITY
(ATTAINABILITY ON TIME SCALES OF INTEREST) IN SOCIAL SCIENCE

As background, in mathematical logic, there is a fundamental distinction between a statement's being *true* and its being *provable*. I believe that in mathematical social science there is an analogous and equally fundamental distinction between a state of the system (e.g., a strategy distribution) being an *equilibrium* and its being *attainable* (*generable*). I would like to discuss, therefore, the parallel between the following two questions: (1) Is every true statement provable? and (2) Is every equilibrium state attainable?

In general, we are interested in the distinction between *satisfaction* of some criterion (like being true, or being an equilibrium) and *generability* (like being provable through repeated application of inference rules, or being attainable through repeated application of agent behavioral rules).

Now, mathematico-logical systems in which every truth is provable are called *complete*.[8] The great mathematician David Hilbert, and most mathematicians at the turn of the Twentieth Century, had assumed that all mathematical systems of interest were complete, that all truths statable in those systems were also provable in them (i.e., were deducible from the system's axioms via the system's inference rules). A major objective of the so-called Hilbert Programme for mathematics was to prove precisely this. It came as a tremendous shock when, in 1931, Kurt Gödel proved precisely the opposite: *all sufficiently rich[9] mathematical systems are incomplete*. In all such systems, there are true statements that are unprovable! Indeed, he showed that there were true statements that were neither provable nor refutable in the relevant systems—they were *undecidable*.[10] (See Gödel 1931; Smullyan 1992; Hamilton 1988.)

Now, truth is a special criterion that a logical formula may *satisfy*. For example, given an arbitrary formula of the sentential calculus, its truth (i.e., its tautologicity) can be evaluated mechanically, using truth tables. Provability, by contrast, is a special type of *generability*. A formula is provable if, beginning with a distinguished set of "starting statements"

---

[8]Sometimes the terms *adequate* or *analytical* are used.

[9]For a punctilious characterization of precisely those formal systems to which the theorem applies, see Smullyan 1992.

[10]Importantly, he did so *constructively*, displaying a (self-referential) true statement that is *undecidable*; that is, neither it nor its negation are theorems of the relevant system.

called *axioms*, it can be ground out—attained, if you will—by repeated application of the system's rule(s) of inference.

Equilibrium (Nash equilibrium, for example) is strictly analogous to truth: it too is a criterion that a state (a strategy distribution) may satisfy. And the Nash "equilibriumness" of a strategy configuration (just like the truth of a sentential calculus formula) can be checked mechanically.

I venture to say that most contemporary social scientists—analogous to the Hilbertians of the 1920s—assume that if a social configuration is a Nash equilibrium, then it must also be attainable. In short, the implicit assumption in contemporary social science is that these systems are *complete*.

However, we are finding that this is not the case. Epstein and Hammond (2002) offer a simple agent-based game almost all of whose equilibria are unattainable outright. This model is presented in the next chapter. More mathematically sophisticated examples of incompleteness include Prasad's result, based on the unsolvability of Hilbert's 10th Problem:

> For n-player games with polynomial utility functions and natural number strategy sets the problem of finding an equilibrium is not computable. There does not exist an algorithm which will decide, for any such game, whether it has an equilibrium or not.... When the class of games is specified by a finite set of players, whose choice sets are natural numbers, and payoffs are given by polynomial functions, the problem of devising a procedure which computes Nash equilibria is unsolvable. (Prasad 1997)

Other examples of uncomputable (existent) equilibria include Foster and Young 2001; Lewis 1985, 1992a, 1992b; and Nachbar 1997. Some equilibria are unattainable outright.

A separate issue in principle, but one of great practical significance, is whether attainable equilibria can be attained on time scales of interest to humans. Here, too, we are finding models in which the waiting time to (attainable) equilibria scales exponentially in some core variable. In the agent-based model of economic classes of Axtell, Epstein, and Young (2001),[11] we find that the waiting time to equilibrium is exponential in both the number of agents and the memory length per agent, and is astronomical when the first exceeds 100 and the latter 10. Likewise, the number of time steps (rounds of play) required to reach the attainable equilibria of the Epstein and Hammond (2002) model was shown to grow exponentially in the number of agents.

---

[11] Chapter 8 of this book.

One wonders how the core concerns and history of economics would have developed if, instead of being inspired by continuum physics and the work of Lagrange and Hamilton (see Mirowski 1989)—blissfully unconcerned as it is with effective computability—it had been founded on Turing. Finitistic issues of computability, learnability, attainment of equilibrium (rather than mere existence), problem complexity, and undecidability, would then have been central from the start. Their foundational importance is only now being recognized. As Duncan Foley summarizes,

> The theory of computability and computational complexity suggests that there are two inherent limitations to the rational choice paradigm. One limitation stems from the possibility that the agent's problem is in fact undecidable, so that no computational procedure exists which for all inputs will give her the needed answer in finite time. A second limitation is posed by computational complexity in that even if her problem is decidable, the computational cost of solving it may in many situations be so large as to overwhelm any possible gains from the optimal choice of action. (See Albin 1998, 46)

For fundamental statements, see Simon 1982, 1987; Hahn 1991; and Arrow 1987. Of course, beyond these *formal* limits on canonical rationality, there is the body of evidence from psychology and laboratory behavioral economics that *homo sapiens* just doesn't behave (in his decision-making) like *homo economicus*.

Now, the mere fact that an idealization (e.g., *homo economicus*) is not accurate in detail is not grounds for its dismissal. To say that a theory should be dismissed because it is "wrong" is vulgar. Theories are idealizations. There are no frictionless planes, ideal gases, or point masses. But these are useful idealizations in physics. However, in social science, it is appropriate to ask whether the idealization of individual rationality in fact illuminates more than it obscures. By empirical lights, that is quite clearly in doubt.

This brings us to the issue of generality. The entire rational choice *project*, if you will, is challenged by (1) incompleteness and outright uncomputability, by (2) computational complexity (even of computable equilibria), and by (3) powerful psychological evidence of framing effects and myriad other systematic human departures from canonical rationality. Yet, the social science theory that enjoys the greatest formal generality[12] (and mathematical elegance) is precisely the rational choice theory.

---

[12] Here, I mean generality in the theory's formal statement, not in its range of successful empirical application.

Generality Is Quantification over Sets

Now, generality has to do with quantification. Universal gravitation says that *for any two masses* whatsoever, the attractive gravitational force is inversely proportional to the square of the separation distance. Mechanics quantifies over the set of all masses. Axiomatic general equilibrium theory quantifies over the set of all consumers in the economy, positing constrained utility maximization *for every agent* in the system. Rational choice theory likewise posits expected utility maximization for all actors.

Clearly, agent modelers do not quantify over sets this big. There is a great deal of experimentation with tags, imitation, evolution, learning, bounded rationality, and zero-intelligence traders, for example. In many cases, however, the experiment is motivated by responsiveness to data. Empirically successful (generatively sufficient) behavioral rules for the Artificial Anasazi of 900 A.D.[13] probably *should not* look much like the agent rules in the Axtell-Epstein (1999) model of U.S retirement norms,[14] which in turn may have little relation to the rules governing agents in Axtell's (1999) model of firms, or the Epstein *et al.* (2004) model of smallpox response,[15] or the zero-intelligence traders of Farmer *et al.* (1993). Yet, despite their diversity, these models are impressive empirically. If reasonable fidelity to data requires us to be *ad hoc* (i.e., to quantify over smaller sets), with different rules for different settings, then that is the price of empirical progress.

Truth and Beauty

All of this said, the real reason some mathematical social scientists don't like computational agent-based modeling is not that the approach is empirically weak (in notable areas, it's empirically stronger than the neoclassical approach). It's that it isn't beautiful. When theorists, such as Frank Hahn, lament the demise of "pure theory" in favor of computer simulation (Hahn 1991), they are grieving the loss of mathematical beauty. I would argue that reports of its death are premature. Let us face this aesthetic issue squarely.

On the topic of mathematical beauty, none have written more eloquently than Bertrand Russell (1957):

> Mathematics, rightly viewed, possesses not only truth, but supreme beauty—a beauty cold and austere, like that of sculpture, without appeal to

---

[13] Chapters 4–6 of this book.
[14] Chapter 7 of this book.
[15] Chapter 12 of this book.

any part of our weaker nature, without the gorgeous trappings of painting or music, yet sublimely pure, and capable of a stern perfection such as only the greatest art can show.

Later, in the same essay, Russell writes:

In the most beautiful work, a chain of argument is presented in which every link is important on its own account, in which there is an air of ease and lucidity throughout, and *the premises achieve more than would have been thought possible, by means which appear natural and inevitable.* (Emphasis added)

Hahn (1991) defines "pure theory" as "the activity of deducing implications from a small number of fundamental axioms." And when he writes that "with surprising frequency this leads to beauty (Arrow's Theorem, The Core, etc)," it is clear that it is Russell's beauty he has in mind.

Generality (mathematical unification) for its own sake satisfies this fine impulse to beauty and has proven to be highly productive scientifically. Physics is highly general, and so is mathematical equilibrium theory. And, as Mirowski (1989) has documented, "physics envy" was quite explicitly central to its development. This is entirely understandable. Any scientist who doesn't have physics envy is an idiot. I am not advocating that we abandon the quest for elegant generality in favor of a case by case narrative (i.e., purely historical) approach. By comparison to a beautiful (Newton-like) generalization, actual history is just this particular apple bobbing down this particular hill. To me, the mathematical theory of evolution is more beautiful than any particular tiger. One of the most miraculous results of our own evolution is that our search for beauty can lead to truth. But there are different kinds of beauty. An analogy to music history may be apposite.

Just as the German classical composers had the dominant 7th and circle of fifths as harmonic propulsion, so the neoclassical economists have utility maximization to propel their analyses. And it is a style of "composition" subscribed to by an entire school of academic thought. We agent modelers are not of this school. We don't have the Germanic dominant 7th of utility maximization to propel every analysis forward— more like the French impressionists, we must in each case be inventive to solve the problem of social motion, devising unique agent rules model by model. If that makes us *ad hoc*, then so was Debussy, and we are in good artistic company.

Schelling's (1971) segregation model is important not because it's right in all details (which it doesn't purport to be), and it's beautiful not because it's visually appealing (which it happens to be). It's important because—even though highly idealized—it offers a powerful and counter-intuitive insight. And it's beautiful because it does so with startling Russellian parsimony. The mathematics of chaos is beautiful not because of all the pretty fractal pictures it generates, useful as these are in stimulating popular interest. What's beautiful in Russell's sense is the startlingly compact yet sweepingly general Li-Yorke (1975) theorem that "period three implies chaos." And when an agent-based model is beautiful in this deep sense, it has nothing to do with the phantasmagorical "eye candy"—Russell's gorgeous trappings—of animated dot worlds. Rather, its beauty resides in the far-reaching generative power of its simple micro-rules, seemingly remote from the elaborate macro patterns they produce. Precisely as Russell would have it: *the premises achieve more than would have been thought possible, by means which appear natural and inevitable.*

The musical parallels are again irresistible. To be sure, Bach's final work, *The Art of the Fugue*, is gorgeous music, but to Bach, the game was to explore the generative power of a single fugue theme. Bach wrote nineteen stunningly diverse fugues based on this single theme, this "premise," if you will.[16] In Bach's hands, it certainly "achieves more than would have been thought possible." While its musical beauty is clear, the *intellectual* beauty lies not in the sound, but in its silent unified structure. Perhaps the best agent models unfold as "social fugues" in which the apparent complexity is in fact generated by a few simple individual rules.

In any case, and whatever one's aesthetic leanings, agent modelers are in good scientific company trading away a certain degree of generality for fidelity to data. The issue of induction arises in this connection.

Induction over Theorem Distributions

As noted earlier, one powerful mode of agent-based modeling is to run large numbers of stochastic realizations (each with its own random seed), collect clean data, and build up a robust statistical portrait of model output. One goal of such exercises is to understand one's model when closed form analytical expressions are not in hand (though these exist in principle, as discussed). A second aim of such exercises is to explain observed statistical regularities, such as the distribution of firm sizes in the U.S. economy (Axtell 1999, 2001). In either case, one builds up a large sample of model realizations. But, as emphasized earlier, *each realization is a strict deduction*. So, while I have no objection to calling such activity inductive, it is *induction over a sample distribution of theorems*, in fact. And it has quite a different flavor from "inductive"

---

[16] Bach died before completing this work, and doubtless could have composed countless further fugues.

survey research, where one collects real-world data and estimates it by techniques of aggregate regression.

Summary

A number of uses of agent-based models have not been touched on here. These include purely exploratory applications and those related to mechanism design, among others (see Epstein 1999). My focus has been on computational agent models as instruments in the generative explanation of macroscopic social structures. In that connection, the main epistemological points treated are as follows:

1. We distinguish the generative motto from its converse. The position is:

$$\forall x(\neg Gx \supset \neg Ex)$$

   If you didn't grow it, you didn't explain it. But not conversely. A microspecification that generates the *explanandum* is a candidate explanation. Generative sufficiency is explanatorily necessary, but not explanatorily sufficient. There may be more than one explanatory candidate, as in any science where theories compete.
2. For every agent model, there exist unique equivalent equations. One can express any Turing machine (and hence any agent model) in partial recursive functions. Many agent models have been revealingly mathematized in other ways, as stochastic dynamical systems, for example.
3. Every realization of an agent model is a strict deduction. So, $(Gx \supset Dx)$, but not conversely, as in non-constructive (*reductio ad absurdum*) existence proofs. One can have $(Dx \wedge \neg Gx)$ and hence, by (1), $(Dx \wedge \neg Ex)$. Not all deduction is explanatory.
4. We often generate, and hence deduce, conjuncts satisfying Disjunctive Normal Forms, as when we grow distributions or spatial settlement patterns in finite agent populations.
5. We carefully distinguish between existence and attainability in principle. And we furthermore carefully distinguish between asymptotic attainability and attainability on time scales of interest. In short, we are attentive to questions of incompleteness (à la Gödel) and of computational complexity (as in problems whose time complexity is exponential in key variables). These considerations, when combined with powerful psychological evidence, cast severe doubt on the rational choice picture as the most productive idealization of human decision-making, and serve only to enforce the bounded rationality picture insisted on by Simon (1982).
6. Generality, while a commendable impulse, is not of paramount concern to agent-based modelers at this point. Responsiveness to data often

requires that we *quantify over smaller sets* than physics or neoclassical economics. If that is *ad hocism*, I readily choose it over what Simon (1987) rightly indicts as an empirically oblivious *a priorism* in economics.

7. Empirical agent-based modeling can be seen as induction over a sample of realizations, each one of which is a strict deduction, or theorem, and comparison of the generated distribution to statistical data. This differs from inductive survey research where we assemble data and fit it by aggregate regression, for example.

CONCLUSION

As to the core indictment that agent models are non-mathematical, non-deductive, and *ad hoc*, the first two are false, and the third, I argue, is unimportant. Generative explanation is mathematical in principle; recursive functions could be provided. *Ipso facto*, generative explanation is deductive. Granted, agent models typically quantify over smaller sets than rational choice models and, as such, are less general. But, in many cases, they are more responsive to data, and in years to come, may achieve greater generality and unification. After all, a fully unified field theory has eluded even that most enviable of fields, physics.

REFERENCES

Albin, Peter S., ed. 1998. *Barriers and Bounds to Rationality: Essays on Economic Complexity and Dynamics in Interactive Systems*. Princeton: Princeton University Press.

Albin, Peter S., and D. K. Foley. 1990. Decentralized, Dispersed Exchange without an Auctioneer: A Simulation Study. *Journal of Economic Behavior and Organization* 18(1): 27–51.

Arrow, Kenneth J. 1987. Rationality of Self and Others in an Economic System. In *Rational Choice: The Contrast between Economics and Psychology*, ed. Robin M. Hogarth and Melvin W. Reder. Chicago: University of Chicago Press.

Axelrod, Robert. 1997. Advancing the Art of Simulation in the Social Sciences. *Complexity* 3:193–99.

Axtell, Robert L. 1999. The Emergence of Firms in a Population of Agents: Local Increasing Returns, Unstable Nash Equilibria, and Power Law Size Distributions. Santa Fe Institute Working Paper 99-03-019.

———. 2001. U.S. Firm Sizes Are Zipf Distributed. *Science* 93:1818–20.

Axtell, Robert L., and Joshua M. Epstein. 1999. Coordination in Transient Social Networks: An Agent-Based Computational Model of the Timing of Retirement. In *Behavioral Dimensions of Retirement Economics*, ed. Henry Aaron. New York: Russell Sage Foundation.

Axtell, Robert L., Joshua M. Epstein, Jeffrey S. Dean, George J. Gumerman, Alan C. Swedlund, Jason Harburger, Shubha Chakravarty, Ross Hammond, Jon Parker, and Miles T. Parker. 2002. Population Growth and Collapse in a Multiagent Model of the Kayenta Anasazi in Long House Valley. *Proceedings of the National Academy of Sciences* 99, suppl. 3: 7275–79.

Axtell, Robert L., Joshua M. Epstein, and H. Peyton Young. 2001. The Emergence of Economic Classes in an Agent-Based Bargaining Model. In *Social Dynamics*, ed. Steven Durlauf and H. Peyton Young. Cambridge: MIT Press.

Berry, Brian J. L., Douglas Kiel, and Evel Elliott, editors. 2002. Adaptive Agents, Intelligence, and Emergent Human Organization: Capturing Complexity Through Agent-Based Modeling. *Proceedings of the National Academy of Sciences*, 99, suppl. 3: Results from the Arthur M. Sackler Colloquium of the National Academy of Sciences held October 4–6, 2001.

Boolos, George S., and Richard C. Jeffrey. 1989. *Computability and Logic*. 3rd ed. Cambridge: Cambridge University Press.

Cederman, Lars-Erik. 1997. *Emergent Actors in World Politics: How States and Nations Develop and Dissolve*. Princeton: Princeton University Press.

Chomsky, Noam. 1965. *Aspects of the Theory of Syntax*. Cambridge: MIT Press.

Crutchfield, J. P., and M. Mitchell. 1995. The Evolution of Emergent Computation. *Proceedings of the National Academy of Sciences* 92:10740–46.

Dorofeenko, Victor, and Jamsheed Shorish. 2002. "Dynamical Modeling of the Demographic Prisoner's Dilemma," Economic Series 124, Institute for Advanced Studies, Vienna, November.

Epstein, Joshua M. 1997. *Nonlinear Dynamics, Mathematical Biology, and Social Science*. Menlo Park, CA: Addison-Wesley.

———. 1999. Agent-Based Computational Models and Generative Social Science. *Complexity* 4(5): 41–60.

Epstein, Joshua M., and Robert L. Axtell. 1996. *Growing Artificial Societies: Social Science from the Bottom Up*. Washington, DC: Brookings Institution Press; Cambridge: MIT Press.

Epstein, Joshua M., Derek A. T. Cummings, Shubha Chakravarty, Ramesh M. Singha, and Donald S. Burke. 2004. *Toward a Containment Strategy for Smallpox Bioterror: An Individual-Based Computational Approach*. Washington, D.C.: Brookings Institution.

Epstein, Joshua M., and Ross A. Hammond. 2002. Non-Explanatory Equilibria: An Extremely Simple Game with (Mostly) Unattainable Fixed Points. *Complexity* 7(4): 18–22.

Epstein, Richard L., and Walter A. Carnielli. 1989. *Computability: Computable Functions, Logic, and the Foundations of Mathematics*. Belmont, CA: Wadsworth and Brooks.

Epstein, Samuel David, and Norbert Hornstein, eds. 1999. *Working Minimalism*. Cambridge: MIT Press.

Farmer, J. Doyne, P. Patelli, and I. I. Zovko. 2003. The Predictive Power of Zero Intelligence in Financial Markets. Santa Fe Institute Working Paper 03-09-051.

Foster, Dean P., and H. Peyton Young. 2001. On the Impossibility of Predicting the Behavior of Rational Agents. *Proceedings of the National Academy of Sciences* 98(22): 12848–53.

Garey, Michael R., and David S. Johnson. 1979. Computers and Intractability: A Guide to the Theory of NP-Completeness. New York: W. H. Freeman.

Gödel, Kurt. 1931. Uber formal unentscheidbare Satze der Principia Mathematica und verwadter Systeme I. *Monatshefte fur Mathematik und Physik* 38:173–98. For the English version: On Formally Undecidable Propositions of the *Principia Mathematica* and Related Systems, in *Collected Works*, vol. 1, ed. Solomon Feferman (Oxford: Oxford University Press, 1986).

Hahn, Frank. 1991. The Next Hundred Years. *Economic Journal* 101:47–50.

Hamilton, A. G. 1988. *Logic for Mathematicians*. Cambridge: Cambridge University Press.

Hodel, R. E. 1995. *An Introduction to Mathematical Logic*. Boston: PWS Publishing.

Jeffrey, Richard. 1991. *Formal Logic: Its Scope and Limits*. 3rd ed. New York: McGraw-Hill.

Knuth, Donald E. 1998. *The Art of Computer Programming*. 3rd ed. Vol. 2, *Seminumerical Algorithms*. Boston: Addison-Wesley.

Krishnamurthy, E. V. 1985. *Introductory Theory of Computer Science*. New York: Springer-Verlag.

Lewis, A. A. 1985. On Effectively Computable Realizations of Choice Functions. *Mathematical Social Sciences* 10:43–80.

———. 1992a. Some Aspects of Effectively Constructive Mathematics That Are Relevant to the Foundations of Neoclassical Mathematical Economics and the Theory of Games. *Mathematical Social Sciences* 24:209–35.

———. 1992b. On Turing Degrees of Walrasian Models and a General Impossibility Result in the Theory of Decision Making. *Mathematical Social Sciences* 24: 141–71.

Li, T. Y., and J. A. Yorke. 1975. Period Three Implies Chaos. *American Mathematical Monthly* 82:985.

Mas-Colell, Andreu, Michael D. Whinston, and Jerry R. Green. 1995. *Microeconomic Theory*. Oxford: Oxford University Press.

Mirowski, Philip. 1989. *More Heat Than Light: Economics as Social Physics, Physics as Nature's Economics*. Cambridge: Cambridge University Press.

Nachbar, J. H. 1997. Prediction, Optimization, and Learning in Games. *Econometrica* 65:75–309.

Pais, Abraham. 1982. *Subtle is the Lord: The Science and the Life of Albert Einstein*. Oxford: Oxford University Press.

Pollicot, Mark, and Howard Weiss. 2001. The Dynamics of Schelling-Type Segregation Models and a Nonlinear Graph Laplacian Variational Problem. *Advances in Applied Mathematics* 27:17–40.

Prasad, Kislaya. 1997. On the Compatibility of Nash Equilibria. *Journal of Economic Dynamics and Control* 21:943–53.

Russell, Bertrand. 1957. *Mysticism and Logic*. New York: Doubleday Anchor.

Schelling, Thomas C. 1971. Dynamic Models of Segregation. *Journal of Mathematical Sociology* 1:143–86.

Simon, Herbert A. 1982. *Models of Bounded Rationality*. Cambridge: MIT Press.
———. 1987. Rationality in Psychology and Economics. In *Rational Choice: The Contrast between Economics and Psychology*, ed. Robin M. Hogarth and Melvin W. Reder. Chicago: University of Chicago Press.
Smullyan, Raymond M. 1992. *Godel's Incompleteness Theorems*. Oxford: Oxford University Press.
Tufillaro, Nicholas B., Tyler Abbott, and Jeremiah Reilly. 1992. *An Experimental Approach to Nonlinear Dynamics and Chaos*. Boston: Addison-Wesley.
Young, H. Peyton. 1998. *Individual Strategy and Social Structure: An Evolutionary Theory of Institutions*. Princeton: Princeton University Press.
Young, H. Peyton, and Dean Foster. 1991. Cooperation in the Short and in the Long Run. *Games and Economic Behavior* 3:145–56.

Prelude to Chapter 3

EQUILIBRIUM, EXPLANATION,

AND GAUSS'S TOMBSTONE

COUNTLESS ARTICLES purport to *explain* social phenomena by furnishing a Game in which the phenomenon is proved to be an equilibrium, typically a Nash equilibrium or some refinement thereof. To be sure, there are many cases in which the attempt to demonstrate that an important social regularity is Nash is deeply revealing. But there are at least three cases[1] where it won't be:

Case 1. The phenomenon of interest is a *nonequilibrium* dynamic.
Case 2. Equilibrium is attainable in principle, but not on acceptable time scales.
Case 3. Equilibrium exists but is unattainable outright.

Every model in this volume falls into one of those three categories. I hope the book demonstrates that, nevertheless, agent-based modeling can be explanatory where "the equilibrium approach," if you will, is either infeasible (Case 1) or lacking in explanatory significance (Cases 2 and 3).[2]

Case 1

The Anasazi and Smallpox models are designed and empirically calibrated to reproduce observed spatiotemporal *dynamics*, not to arrive at an equilibrium. I see them as examples of empirical nonequilibrium social science. The Retirement model, too, is concerned with explaining the *dynamics* of a shock-induced long-term transition from one distribution of retirement ages to another. The Norms and Civil Violence models both exhibit classic features of complex systems:[3] local conformity, global

---

[1] The possibility of multiple equilibria raises further issues.

[2] Here, I agree entirely with Brian Skyrms: "The explanatory significance of equilibrium depends on the underlying dynamics." Brian Skyrms, "Stability and Explanatory Significance of Some Simple Evolutionary Models," *Philosophy of Science* 67 (March 2000): 94.

[3] H. Peyton Young, *Individual Strategy and Social Structure: An Evolutionary Theory of Institutions* (Princeton: Princeton University Press, 1998).

diversity, and *punctuated* (not stable) equilibrium. Similarly, the Demographic Prisoner's Dilemma model exhibits persistent nonequilibrium
cycling. The Organization model, while more exploratory than explanatory, is concerned with structural adaptation in dynamic environments,
and finds that, in the main case discussed, optimal adaptation involves a
time-varying, rather than an equilibrium, structure.

Case 2

The Classes model is of the second type. There is indeed an asymptotic
equilibrium—a kind of fairness norm. But it is subject to astronomical
waiting times; these are shown to scale *exponentially* in a number of core
variables (including the number of agents). The equilibrium—the equity
norm—is not attainable on practical time scales, and the system spends
most of its time far from the equitable equilibrium.

Case 3

As discussed in the preceding two chapters, there are a number of deep
papers demonstrating that equilibria can be uncomputable in a strict
mathematical sense. However, this literature, while extremely important,
is also quite sophisticated technically. The next chapter, entitled, "Non-
Explanatory Equilibria," offers an extremely simple game most of whose
equilibria are unattainable in principle (Case 3) and in which the time
to attain those equilibria that are attainable grows exponentially in the
number of players (Case 2). It demonstrates the essential distinction
between existence and attainability (both in principle and in practice)
in a very accessible way. And in so doing, I hope it provokes further and
deeper thought regarding the explanatory significance of equilibrium.

Incompleteness

Also as noted in the preceding chapter, the model can be seen as an
example of *incompleteness* in mathematical social science. If "being
true" is taken as the analog of "being an equilibrium" and "being
provable" is the analogue of "being attainable," then the mapping is
clear: Propositions that are true but unprovable from the axioms (à la
Gödel) correspond to model states that are equilibria but are unattainable
from any permissible initial conditions.

Now, even incompleteness is a special case of the general distinction between *satisfaction* of some mechanically checkable condition
(like being a tautology or being a Nash equilibrium) and *generability* (like
being deducible from axioms under rules of inference or being attainable

from initial conditions under rules of agent interaction). And this brings us to Gauss's tombstone.

Gauss's Tombstone

Why is the tombstone of Carl Friedrich Gauss (1777–1855) engraved with a regular 17-gon? Because he was worried about the very same distinction. But, the story actually begins in Greece in roughly 500 BC. The Greeks knew perfectly well what it would mean for a geometrical shape to *satisfy* the following criterion: to be a *regular n-gon*, a polygon with *n* sides of equal length. And they also knew what it would mean to be constructible—that is, *generable*—by a sequence of specific operations with only a straightedge ruler and a compass. In essence, the Greeks asked: *For what values of n is the regular n-gon constructible?* Good question!

It took roughly twenty-two *centuries* to arrive at the answer. A complete (and very beautiful) theory of constructibility was finally achieved by Evariste Galois (1811–32). En route, Gauss proved that the regular 17-gon is, in fact, constructible. This singular achievement had a profound effect on the young Gauss, convincing him to pursue mathematics over philology. And, indeed, it accounts for his unusual tombstone. Now, Gauss actually went a good deal farther, correctly stating that *The regular n-gon is constructible if and only if n equals some nonnegative power of 2, multiplied by a product of distinct Fermat primes;*[4] these are prime numbers of the form $2^{2^k} + 1$. Notice that $2^0(2^{2^2} + 1) = 17$.

Now, one might presume that there is no upper limit on the number of regular *n*-gons that can be constructed. From Gauss's result, it is clear that this set will indeed be infinite if the set of distinct Fermat primes is itself infinite. So, is it? That, too, is a good question. To this day, the answer is not known!

What is known is that, for roughly 2,500 years, worrying about the distinction between satisfaction and generability has been extremely fruitful in logic and mathematics, stimulating profound researches of Gödel, Gauss, Galois, and many others. I think it will be fruitful in the social sciences as well.

---

[4]Gauss proved sufficiency and asserted that he had a proof of necessity. On the history and mathematics of Galois Theory, see Charles R. Hadlock, *Field Theory and Its Classical Problems* (Washington, DC: Mathematical Association of America, 1978); and I. Stewart, *Galois Theory* (London: Chapman and Hall, 1973).

Chapter 3

NON-EXPLANATORY EQUILIBRIA: AN EXTREMELY SIMPLE GAME WITH (MOSTLY) UNATTAINABLE FIXED POINTS

JOSHUA M. EPSTEIN AND ROSS A. HAMMOND*

EQUILIBRIUM ANALYSIS PERVADES mathematical social science. This paper calls into question the explanatory significance of equilibrium by offering an extremely simple game, most of whose equilibria are unattainable in principle from any of its initial conditions. Moreover, the number of computation steps required to reach those (few) equilibria that are attainable is shown to grow exponentially with the number of players—making long-run equilibrium a poor predictor of the game's observed state. The paper also poses a number of combinatorially challenging problems raised by the game.

Much of game theory and mathematical economics is concerned with equilibria (see Kreps 1990, 405). Nash equilibrium is an important example. Indeed, in many quarters, "explaining an observed social pattern" is understood to mean "demonstrating that it is the Nash equilibrium of some game." But, there is no explanatory significance to an equilibrium that is unattainable in principle. And there is debatable significance to equilibria that are attainable only on astronomical time scales. Yet, in a great many instances, the social pattern to be explained is simply shown to be an equilibrium. The questions, "Is the equilibrium attainable?" and "On what time scale is it attainable?" are not raised.

There is a literature on unattainability—or uncomputability—of equilibria, undecidability in games, and related topics. But it is quite

*Joshua M. Epstein: Economic Studies Program, The Brookings Institution, Washington, DC and External Faculty, Santa Fe Institute, Santa Fe, New Mexico; Ross A. Hammond: Department of Political Science, University of Michigan, Ann Arbor, Michigan.

For insightful comments, the authors thank Robert Axelrod, Robert Axtell, Jim Crutchfield, William Dickens, Samuel David Epstein, John Miller, Scott Page, Duncan Watts, and Peyton Young. This research was conducted at the Brookings Institution—Johns Hopkins University Center on Social and Economic Dynamics.

This essay was previously published in *Complexity* 7(4): 18–22.

technical.[1] This article contributes an extremely simple game—easily played by school children—that drives home the core distinction between attainable and unattainable equilibria. Indeed, the overwhelming preponderance of this game's equilibria are unattainable from any initial configuration of the game.

We hope this arresting example stimulates skepticism about the explanatory significance of equilibrium.[2] As we will show, the game—despite its surface simplicity—also raises a number of combinatorially very challenging questions.

DESCRIPTION OF THE GAME

The game's ingredients are few and simple:

1. Events transpire on a linear array of sites, extending from an origin (the leftmost site) to the right.
2. Agents are numbered consecutively from 1 to n. These numbers do not change in the course of the game.
3. Initially, we require that agents be arrayed in a contiguous row, beginning at the origin, in some arbitrary order. Figure 3.1 gives one such admissible initial configuration for three agents. Each agent is represented as a number, and each empty site is represented as an asterisk.

An Initial 3-Agent Configuration

3 2 1 * * *

Figure 3.1. An initial three-agent configuration.

4. The agents' only rule of behavior is as follows:

> *AGENT RULE: If there is a lower-numbered agent anywhere to your right, go to the head of the line (the site immediately to the right of the rightmost agent). Otherwise, remain in place.*

The rule is reminiscent of the Schelling segregation model (1971, 1978) and the variant of Young (1998).[3] In each case, agents have some

---

[1] See, for example, Foster and Young 2001; Saari and Simon 1978; Prasad 1997; Jordan 1993; and Nachbar 1997.

[2] For an insightful discussion of this issue in the context of chaos and evolutionary games, see Skyrms 1997.

[3] Note, however, that the model is not a cellular automaton, because it involves a nonlocal operation (agents go to the head of the line and are queried in sequence order). We thank

preference for immediate neighbors. In our case, agents hate living anywhere with a lower numbered agent to their right. And (with bounded rationality) they move to the one site that is certain to remove the problem, at least in the immediate term—the front site.

5. In any given round, agents are queried in descending order from the highest number.[4] As we shall see below, not all agents may wish to move. The first agent who does wish to move does so, resulting in a new configuration. That ends the round. Play continues until *equilibrium* is reached, where:

6. An *equilibrium is* a configuration from which no agent would move further under the rule. It is a fixed point. An equilibrium is termed *attainable* if there is some *initial* configuration (see under item 3) from which it can be attained. An *unattainable* equilibrium is an equilibrium for which no such *initial* configuration exists.

That is the complete model specification.

CHILD'S PLAY

One can imagine the model as a children's game, played on a linear sequence of hopscotch squares. Assume the kids differ by height. They form a line extending out from the school wall into the playground, one in front of the next, in some random order by height. Then they move, as specified under item 5 above, each according to the simple rule: *If there's a shorter kid anywhere in front of you, jump to the very head of the line (the square immediately in front of the front kid). Otherwise, remain in place.*

The game ends when equilibrium is attained—when no kid would move further under the rule.[5] (This equilibrium notion is Nash-like: no agent has any incentive to unilaterally depart under the rule.)

A NUMERICAL EXAMPLE

As a simple illustration of how the configurations progress, let us walk the game forward from the figure 3.1 configuration (see table 3.1).

Jim Crutchfield for this observation. On cellular automata, see Wolfram 1986; and Toffoli and Margolus 1987.

[4]This sentence is revised from the original 2002 version, which may have suggested, wrongly, that all agents are queried in each round. See also the appendix to this chapter.

[5]In effect, the kids have invented a type of decentralized (albeit highly inefficient) sorting algorithm.

TABLE 3.1
A Complete Game

Configuration Number	Configuration
1	3 2 1 * * *
2	* 2 1 3 * *
3	* * 1 3 2 *
4	* * 1 * 2 3

Starting in Configuration 1, agent "3" (the highest numbered) is queried first. Since there is a lower numbered agent to her right, she jumps to the head of the line, leaving a space in her former position— yielding Configuration 2. That ends the round. So, we begin a new round. As before, we query agent "3" (the highest numbered) first. This time, she declines to change position. So, we query the next highest numbered agent: "2". Since there is a lower numbered agent to her right, she now jumps to the head of the line, leaving a space in her former spot—yielding Configuration 3. This, of course, "upsets" agent 3, who moves when queried at the beginning of the next round, generating Configuration 4. In Configuration 4, agent 3 does not wish to move, so agent 2 is queried. She declines, so agent 1 is (at last) queried, but declines as well (as the lowest numbered agent always does). Configuration 4 is therefore an equilibrium. It is obviously attainable. Notice that it requires 6 spaces in total.

SPACE AND TIME REQUIREMENTS FOR ATTAINABLE EQUILIBRIA

For n agents, how many spaces are required to ensure enough space for all attainable equilibria? Perhaps surprisingly, the answer is

$$s_{\max}(n) = n + \sum_{i=0}^{n-2} 2^i = (n-1) + 2^{(n-1)} \qquad (1)$$

This space requirement grows exponentially in n. Values of $s_{\max}(n)$, for various n values are given in table 3.2.

Regarding time (i.e., number of computation steps), the equilibrium of table 3.1 required 3 rounds to compute, from the initial configuration 321***. In general, equilibria occupying $s_{\max}(n)$ (as in Equation 1) spaces will be obtainable in $s_{\max}(n) - n$ rounds, which, quite notably, is also exponential in n. Daunting numerical examples are left to the reader.

Table 3.2
Maximum Space Requirements for All Attainable Equilibria, Various n

n	Sites
3	6
4	11
5	20
20	524,307
25	16,777,240
30	5.37×10^8
50	5.63×10^{14}
100	6.34×10^{29}

As prosaic examples with kids, assume each hopscotch square is 2 feet deep and that the games begin on a playground in Cambridge, Massachusetts. Then, for 20 kids (an average kindergarten class), there are initial line-ups such that, when (after 524,287 moves) equilibrium is attained, the tallest kid is standing in Central Park. For 25 kids, there are initial line-ups such that, when (after about 17 million moves) equilibrium is attained the tallest kid is standing in Tokyo. For 30 kids, there are initial line-ups such that, when (after more than 500 million moves) equilibrium is attained, the tallest kid has circumnavigated the earth 10 times. For 50 players, there are attainable equilibria extending over roughly 563 trillion sites. And for games involving 100 agents—a standard population size in the literature of n-person games and agent-based models—even the set of *attainable* equilibria is uncomputable on all practical time scales. And, in fact, most equilibria are unattainable in principle.

ATTAINABLE EQUILIBRIA

A full treatment of the $n = 3$ case will be instructive. There are 3! acceptable initial configurations, and 5 distinct attainable equilibria, as shown in table 3.3. Notice that the equilibrium **1*23 is attainable from the initial configurations: 231*** and 321***. In general, a given attainable equilibrium may be attainable from multiple initial configurations.[6]

---

[6]In this connection, the reader might find it interesting to consider the following general problem: Give a formula, $f(n)$, for the number of *distinct* equilibria attainable from the n! distinct initial configurations of the n-agent game.

TABLE 3.3
The 5 Attainable Equilibria for $n = 3$

Initial Configuration	Resulting Equilibrium
1 2 3 * * *	1 2 3 * * *
1 3 2 * * *	1 * 2 3 * *
2 3 1 * * *	* * 1 * 2 3
2 1 3 * * *	* 1 * 2 3 *
3 1 2 * * *	* 1 2 3 * *
3 2 1 * * *	* * 1 * 2 3

UNATTAINABLE EQUILIBRIA

While (as shown in table 3.3) there are 5 distinct attainable equilibria for the $n = 3$ case, there are 20 equilibria in total (see eq. 2). *Ipso facto,* there are 15 unattainable equilibria! They are listed in table 3.4. In each of these configurations, every agent is happy with her immediate neighborhood, but none of these configurations are attainable from any initial configuration.

For $n = 3$, then, unattainability is the norm among equilibria. This pattern only gets more dramatic as n increases. Indeed, the ratio of attainable to unattainable equilibria approaches zero very quickly. For $n = 4$, there are 330 equilibria, of which 12 are attainable, a mere 4%.

TABLE 3.4
The 15 Unattainable Equilibria for $n = 3$

1.	1 * 2 * * 3
2.	* 1 2 * * 3
3.	1 * * 2 * 3
4.	* 1 * 2 * 3
5.	* * 1 2 * 3
6.	1 * * * 2 3
7.	* 1 * * 2 3
8.	* * * 1 2 3
9.	1 2 * 3 * *
10.	1 2 * * 3 *
11.	1 * 2 * 3 *
12.	* 1 2 * 3 *
13.	1 * * 2 3 *
14.	1 2 * * * 3
15.	* * 1 2 3 *

For $n = 5$, there are 15,504 equilibria, of which 41 are attainable, or 0.2%. For $n > 5$, the attainable percentage is effectively zero.

The formula for the total number of equilibria, $T(n)$, even for the $n = 4$ case, turns out to be quite complex:

$$T(4) = (\beta_4 + 1) + \sum_{i=1}^{\beta_4} i + \sum_{i=1}^{\beta_4} \sum_{j=1}^{i} j + \sum_{i=1}^{\beta_4} \sum_{j=1}^{i} \sum_{k=1}^{j} k,$$

where $\beta_4 = s_{\max}(4) - 4 = \sum_{i=0}^{2} 2^i = 7$.[7] For n agents, the appropriate generalization is as follows. First, the index variables will run from v_1 to v_{n-1}. Then,

$$T(n) = (\beta_n + 1) + \sum_{v_1=1}^{\beta_n} v_1 + \sum_{v_1=1}^{\beta_n} \sum_{v_2=1}^{v_1} v_2 + \cdots + \sum_{v_1=1}^{\beta_n} \sum_{v_2=1}^{v_1} \cdots \sum_{v_{n-1}=1}^{v_{n-2}} v_{n-1}, \quad (2)$$

where $\beta_n = s_{\max}(n) - n = \sum_{i=0}^{n-2} 2^i$, as before.

Now, for n agents, the number of distinct initial configurations is $n!$, but the number of attainable equilibria is less than $n!$ (as illustrated in table 3.4). Hence, the fraction of attainable to total is bounded above by $n!/[T(n)]$. Since $n!/[T(n)] \to 0$ extremely fast, so does the fraction of attainables. Hence the generic equilibrium is, in fact, unattainable from any initial conditions.[8]

Clearly, restricting the space of permissible initial configurations is important to this result. While at first glance, such restrictions may seem artificial, they are the norm in games and contests generally. Chess, checkers, and many other board games possess required initial set-ups.

---

[7] A closed form representation of the result would obscure the iterative nature of the solution. Hence, the iterated summations shown.

[8] Whether or not an equilibrium can be easily *diagnosed as* unattainable is beside the point we are making here. But, to discuss this briefly, some cases are clear on inspection. For example, the equilibrium ***123 is unattainable, because the digit "1" never moves (as noted earlier) and appears too far to the right to be permissible initially. Similarly, the equilibrium *1*2*3 can be easily identified as a Garden of Eden configuration (see below) and is therefore not attainable. However, some cases are not so obvious: **123* is unattainable. Now, in *principle*, one can classify equilibria as unattainable by brute force. For each of the $n!$ initial conditions, one simply grinds out the attainable equilibria. Then, for any candidate equilibrium, one "simply" checks—by bitwise comparison—whether it is in the list of attainables or not. However, the number of required comparisons grows exponentially in n. Mechanical "space counting" tests for unattainability, although more direct, nonetheless require inspection of β_n sites, and will be computationally prohibitive in practice for agent populations of any significance.

Straight pool, 9-Ball, and 8-Ball (stripes and solids) each begin with the billiard balls "racked" in a specified way. In racquet sports, such as tennis, squash, and ping-pong, players are not permitted to serve (i.e., begin a point) from "just anywhere." Football prohibits certain line-ups and allows others. Jousts and pistol duels had highly stylized initial positions, as do fencing matches. Further examples will come readily to mind. Indeed, on reflection, some restriction on initial configurations would seem to be the rule across formalized contests, rather than the exception. In this light, our restriction seems natural enough.

EQUILIBRIUM AND EXPLANATION

Here, then, is an extremely simple playground game that admits a huge number of equilibria, virtually all of which are not attainable from any initial configuration, once there are 5 or more players. So, returning to the central issue of explanatory significance, imagine being a theoretical playgroundologist. Your colleagues, the empirical playgroundologists, have documented a powerful regularity: They observe kids all over the world lined up from shortest to tallest on playgrounds; they are spaced in all sorts of bizarre ways, but they're lined up in order by height. What is the *explanation?* This is the central empirical puzzle of playgroundology.

Now, given an analogous empirical regularity, the standard and ostensibly *explanatory* practice in the formal social sciences is as follows: *Provide a game for which the observed regularity is an equilibrium.*

But, this is easily done for playgroundology—the game we've just been exploring fits the bill. Any line-up from shortest to tallest observed by our empiricists in the field will, indeed, be an equilibrium of this game. As we have shown, however, it will almost certainly *not* be attainable: kids could not have arrived there from any initial line-up. Clearly, then, the rules of this particular game are supremely unlikely to be those followed on real playgrounds.

Nonetheless, under the standard practice above, these rules would be regarded as explanatory! This seems unsatisfactory for playgroundology because the generic equilibrium of the game is not attainable even in principle, much less on time scales of any plausibility. So why, absent demonstrations of attainability, should the same practice be accepted as explanatory in social science? We believe it should not be.

An acceptable notion of "explanation" should include attainability. A candidate is the generative notion advanced in Epstein 1999, in which a set of individual rules, a microspecification, is regarded as explanatory only if it suffices to generate the observed regularity—incorporating the requirement of attainability.

Beyond its explanatory shortcomings, equilibrium may be a bad *predictor* of observed configurations.[9] Obviously, *unattainable* equilibria (because they will *never* be observed) are not predictive of the game's state on any time scale. However, even *attainable* equilibria (given the exponential time complexity of the process) are, in almost all cases, poor predictors on time scales of any interest to humans.

CONCLUSION

For the social sciences more broadly, there would appear to be two lessons of this simple exercise. First, implicit claims that equilibrium analysis is explanatory or predictive should be challenged and require the most careful defense. Second, a successful defense of any such claims must include a demonstration of attainability, on time scales of interest, by agents employing plausible rules.[10]

APPENDIX: FURTHER COMBINATORIAL QUESTIONS

Although the playground game was contrived as a stark illustration of these points, it happens to raise a number of interesting combinatorial questions.

Garden of Eden Configurations

First, by way of definition, if there exists *no previous configuration* from which a given configuration can be attained, then the latter is termed a Garden of Eden (GE) configuration.[11] For example, the following configuration is GE:

$$*1*2*3$$

If 3 had been located anywhere to the left of 2 (or 1), it would have jumped to the site immediately to 2's right, *not* to the position shown. This is both an equilibrium and a GE state.

We know that there are unattainable equilibria (i.e., unattainable from any admissible *initial* configurations). Now, for many of these,

---

[9] Explanation and prediction are different matters: plate tectonics *explains* earthquakes but does not predict when they will occur. Similarly, electrostatics *explains* lightning, but does not predict where it will strike.

[10] By plausible rules, we have in mind those involving bounded information and bounded individual computing capacity. See Simon 1982.

[11] According to E. F. Moore (1962), this term was first suggested by John Tukey.

there *are prior* configurations. So, beginning with such an unattainable equilibrium, if we back calculate, we must stop short of the origin (i.e., the set of permissible initial configurations) since otherwise the equilibrium would have been attainable. Where we stop must therefore be a Garden of Eden configuration! Hence, we have the following proposition:

> *Proposition:* For every non-GE unattainable equilibrium, there exists (at least one) GE non-equilibrium preceding configuration.

For example, consider the string: 1**23*. It is an equilibrium, but it is not attainable from any permitted initial condition. The non-equilibrium configurations from which it is derivable, however, are the following: 1*32** and 13*2**, both of which are GE, since neither one has a predecessor that could occur initially.

The set of GE configurations from which a given configuration is attainable shall be referred to as its *basin of attraction*. Naturally, this suggests the following (evidently hard) question: For any equilibrium configuration *not* attainable from an initial configuration, determine its basin of attraction, or since all initial conditions are themselves GE, the general problem is simply:

> *Problem 1.* For any equilibrium (attainable or not), determine its basin of attraction.

In pondering the computational complexity of this general problem, bear in mind that even for $n = 50$ players there are many unattainable equilibria consuming 563 trillion sites—in general, $s_{max}(n)$ sites.

For the sake of completeness, it would be of further interest to solve the following:

> *Problem 2.* From each "point" of a given equilibrium's basin, how many computation steps are required to attain the equilibrium?

REFERENCES

Epstein, Joshua M. 1999. Agent-Based Computational Models and Generative Social Science. *Complexity* 4(5): 41–60.

Foster, Dean P., and H. Peyton Young. 2001. On the Impossibility of Predicting the Behavior of Rational Agents. *Proceedings of the National Academy of Sciences* 98(22): 12848–53.

Jordan, J. S. 1993. Three Problems in Learning Mixed-Strategy Equilibria. *Games and Economic Behavior* 5:368–86.

Kreps, David M. 1990. *A Course in Microeconomic Theory.* Princeton: Princeton University Press.

Moore, E. F. 1962. Machine Models of Self-Reproduction. In *Mathematical Problems in the Biological Sciences*. Proceedings of Symposia in Applied Mathematics, 14. Providence, RI: American Mathematical Society.

Nachbar, J. H. 1997. Prediction, Optimization, and Learning in Games. *Econometrica* 65:275–309.

Prasad, Kislaya. 1997. On the Computability of Nash Equilibria. *Journal of Economic Dynamics and Control* 21:943–53.

Saari, Donald G., and Carl P. Simon. 1978. Effective Price Mechanisms. *Econometrica* 50:1097–1125.

Schelling, Thomas C. 1971. Dynamic Models of Segregation. *Journal of Mathematical Sociology* 1:143–86.

———. 1978. *Micromotives and Macrobehavior*. New York: Norton.

Simon, Herbert A. 1982. *Models of Bounded Rationality*. Vol. 1. Cambridge: MIT Press.

Skyrms, Brian. 1997. Chaos and the Explanatory Significance of Equilibrium: Strange Attractors in Evolutionary Game Dynamics. In *The Dynamics of Norms*, ed. Cristina Bicchieri, Richard Jeffrey, and Brian Skyrms. New York: Cambridge University Press.

Toffoli, Tommaso, and Norman Margolus. 1987. *Cellular Automata Machines: A New Environment for Modeling*. Cambridge: MIT Press.

Wolfram, Stephen, ed. 1986. *Theory and Applications of Cellular Automata*. Singapore: World Scientific.

Young, H. Peyton. 1998. *Individual Strategy and Social Structure: An Evolutionary Theory of Institutions*. Princeton: Princeton University Press.

LARGE EFFECT OF A SUBTLE RULE CHANGE

ONE ISSUE RAISED by the model is unattainability proper, which (as noted in the prelude) is analogous to unprovability. A second issue is the time complexity of attainable equilibria, which is analogous to the length of proofs (or, equivalently, to Turing machine halting times). In the opening Generative chapter, I discussed the issue of structural stability in agent models, and in particular, the difficulty of metrizing the space of local rules, the aim being to make concrete the notion of a "small change" in rules. The present model furnishes a nice example of how a rather subtle rule change produces notably different dynamics.

Recall the rule used in the chapter: Each round, agents are queried, beginning with the highest. The minute *any agent* moves, the round ends. Then, in the next round, agents are again queried, starting with the highest numbered. Here's how it unfolds for the $n = 5$ case. Notice that it takes 8 rounds before "2" is even queried, by which point "5" has already moved 4 times. In total, the "5" agent moves 8 times.

54321 [Initial Configuration]

*43215

**32154

**321*45

***21*453

***21*4*35

21354

213*45

****1***3*452

****1***3*4*25

****1***3***254

****1***3***2*45

****1*******2*453

****1*******2*4*35

****1*******2***354

****1*******2***3*45

If, by contrast, each agent is queried *once per cycle*, in descending order, equilibrium is attained far more quickly, as shown below. Here, the "5" agent moves but 4 times.

```
54321
*43215
**32154
***21543
****15432
****1*4325
****1**3254
****1***2543
****1***2*435
****1***2**354
****1***2**3*45
```

This variant, in which *each agent is queried in each round*, produces very different dynamics. In fact, for the attainable equilibria, this process has polynomial (worst-case) complexity, where the published process—in which not all agents are queried each round—has complexity $O(2^n)$. So this somewhat subtle rule change is the difference between a process convergent in polynomial time and one that is exponential. In agent models, details of the call order—subtle changes in rules—can be crucial.

GENERATING CIVILIZATIONS: THE 1050 PROJECT

AND THE ARTIFICIAL ANASAZI MODEL

THE ARTIFICIAL ANASAZI model grew out of the 2050 Project. This was a multiyear collaboration of the Brookings Institution, the Santa Fe Institute, and the World Resources Institute, funded by the MacArthur Foundation. Although the project's overall aim was to identify the conditions for sustainable development on a global scale to the year 2050, there were many Working Groups. Murray Gell-Mann asked if I would direct the Theoretical one. Having secured from Murray a guarantee that it would not involve any administration ("ordering tuna sandwiches for everybody"), I agreed, but in truth, the responsibilities of my directorship quickly reduced to working on the Sugarscape Model with Rob Axtell, and attending CLAW meetings. Since Murray kept admonishing the project to take "a Crude Look At the Whole," we established a CLAW Group, with Murray—whom some of us dubbed "Dr. CLAW"—as head.

At one of these CLAW meetings at SFI, I showed a Sugarscape run that Rob and I had named the Proto-History. It was a highly idealized, or "toy," history of civilization (and is presented in *Growing Artificial Societies*). Then I asked the audience—a typically unlikely SFI gathering of physicists, archaeologists, biologists, neuroscientists, and refugees from the various social sciences—"does this remind anyone of anything real?" A hand shot up and a voice rang out, "It reminds me of the Anasazi." That was my introduction to George Gumerman and to the Anasazi.

I'm sure I said something like, "The Ana-*what*?" The Anasazi, George explained, were a vibrant civilization that flourished in what is now the southwestern United States, from around AD 800 to AD 1350, at which point they enigmatically vanished. Why? That struck me as a fascinating question. The issue was whether it could be studied computationally. Thinking the answer had to be no, I asked—at that point almost rhetorically—"Is there any data?" To my (continuing) amazement, the answer was yes. Enter dendrochronologist Jeff Dean, of the Tree Ring Laboratory at Arizona.

Well, I returned to Brookings and excitedly told my colleagues Rob and our then assistant Steve McCarroll. The immediate issue was what form the data was in. It was then in a variety of files accumulated over a period of decades by the SARG (Southwestern Anthropological Research Group), detailing the environmental and demographic history of the Longhouse Valley Anasazi for over a millennium. Our first step was to digitize this history. That then became the "target"—the *explanandum*—of the Artificial Anasazi Project. Could we "grow"—generate—that true history in an agent-based model? We were off and running.

There was a delicious irony in the thought that the 2050 Project would end up pioneering the field of agent-based computational archaeology. And as a play on its futuristic origins, we called the effort "The 1050 Project."

I present three papers on the Anasazi. The first was published in the Oxford volume *Dynamics in Human and Primate Societies* and assumes no heterogeneity in fertility ages specifically. The second, published in the *Proceedings of the National Academy of Sciences*, does assume heterogeneity, and the result is a vastly better fit to the data.

Generative Explanation

A core point of the opening Generative chapter is that one can do legitimate, rigorous empirical research with agent-based computational models. Chapters 4–6 demonstrate that point. Indeed, Jared Diamond, writing in *Nature*,[1] said of this work that it "sets a new standard in archaeological research." I probably don't know enough archaeology to competently evaluate that claim, but find it extremely gratifying nonetheless.

This Anasazi work explicitly seeks to satisfy the *generative explanatory standard* set forth in the Generative chapter. Indeed, the *Proceedings* paper ends with precisely this point: "To *explain* an observed spatio-temporal history is to specify agents that generate—or grow—this history. By this criterion, our . . . account of the evolution of this society goes a long way toward *explaining* this history."

A third Anasazi article, published in *Artificial Life*, is focused on spatial settlement clustering and also broaches the question of social hierarchy, itself a focus of chapter 13 on adaptive organizations.

---

[1] Jared M. Diamond, "Life with the Artificial Anasazi," *Nature* 419:567–69.

Chapter 4

UNDERSTANDING ANASAZI CULTURE CHANGE

THROUGH AGENT-BASED MODELING

JEFFREY S. DEAN, GEORGE J. GUMERMAN,
JOSHUA M. EPSTEIN, ROBERT L. AXTELL,
ALAN C. SWEDLUND, MILES T. PARKER,
AND STEPHEN MCCARROLL*

INTRODUCTION

TRADITIONAL narrative explanations of prehistory have become increasingly difficult to operationalize as models and to test against archaeological data. As such models become more sophisticated and complex, they also become less amenable to objective evaluation with anthropological data. Nor is it possible to experiment with living or prehistoric human beings or societies. Agent-based modeling offers intriguing possibilities for overcoming the experimental limitations of archaeology by representing the behavior of culturally relevant agents on landscapes. Manipulating the behavior of artificial agents on such landscapes allows us to, as it were, "rewind the tape" of sociocultural history and to experimentally examine the relative contributions of internal and external factors to sociocultural evolution (Gumerman and Kohler in press).

Agent-based modeling allows the creation of variable resource (or other) landscapes that can be wholly imaginary or that can capture important aspects of real-world situations. These landscapes are populated with heterogeneous agents. Each agent is endowed with various

*We thank David Z. C. Hines of the Brookings Institution and Carrie Dean of the Laboratory of Tree-Ring Research for valuable assistance. The following organizations provided funding and institutional assistance: The Brookings Institution, The National Science Foundation, The John D. and Catherine T. MacArthur Foundation, The Alex C. Walker Educational and Charitable Foundation, the Santa Fe Institute, the Arizona State Museum, and the Laboratory of Tree-Ring Research.

This essay was previously published in *Dynamics in Human and Primate Societies: Agent-Based Modeling of Social and Spatial Processes*, edited by Timothy A. Kohler and George G. Gumerman. 2000. New York: Oxford University Press. Used by permission of Oxford University Press, Inc.

attributes (e.g., life span, vision, movement capabilities, nutritional requirements, consumption and storage capacities) in order to replicate important features of individuals or relevant social units such as households, lineages, clans, and villages. A set of anthropologically plausible rules defines the ways in which agents interact with the environment and with one another. Altering the agents' attributes, their interaction rules, and features of the landscape allows experimental examination of behavioral responses to different initial conditions, relationships, and spatial and temporal parameters. The agents' repeated interactions with their social and physical landscapes reveal ways in which they respond to changing environmental and social conditions. As we will see, even relatively simple models may illuminate complex sociocultural realities.

While potentially powerful, agent-based models in archaeology remain unverified until they are evaluated against actual cases. The degree of fit between a model and real-world situations allows the model's validity to be assessed. A close fit between all or part of a model and the test data indicates that the model, albeit highly simplified, has explanatory power. Lack of fit implies that the model is in some way inadequate. Such "failures" are likely to be as informative as successes because they illuminate deficiencies of explanation and indicate potentially fruitful new research approaches. Departures of real human behavior from the expectations of a model identify potential causal variables not included in the model or specify new evidence to be sought in the archaeological record of human activities.

THE ARTIFICIAL ANASAZI PROJECT

The Artificial Anasazi Project is an agent-based modeling study based on the Sugarscape model created by Joshua M. Epstein and Robert Axtell (1996), both of The Brookings Institution and the Santa Fe Institute. The project was created to provide an empirical, "real-world" evaluation of the principles and procedures embodied in the Sugarscape model and to explore the ways in which bottom-up, agent-based computer simulations can illuminate human behavior in a real-world setting. In this case, the actual "test bed" is prehistoric Long House Valley in northeastern Arizona, which, between roughly 1800 B.C. and A.D. 1300, was occupied by the Kayenta Anasazi, a regionally distinct prehistoric precursor of the modern Pueblo cultures of the Colorado Plateau (figure 4.1). Archaeological information on the Kayenta Anasazi provides an empirical data set against which simulations of human behavior in Long House Valley can be evaluated. The actual spatiotemporal history is the "target" we attempt to recreate and, hence, explain with

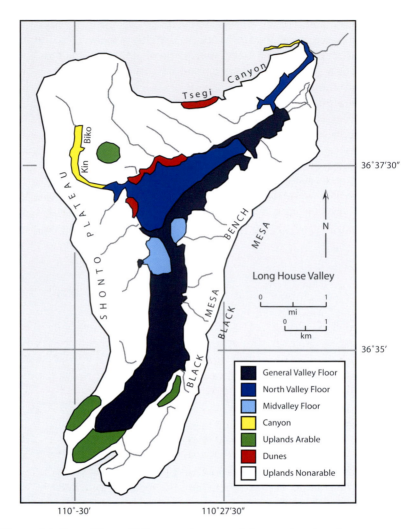

Figure 4.1. Long House Valley, northeastern Arizona, showing the seven potential production zones.

an agent-based model. Ultimately, this target is constructed from the
research of the Long House Valley Project, a multiyear research effort
of the Museum of Northern Arizona and the Laboratory of Tree-Ring
Research at The University of Arizona, which primarily involved a
100-percent survey of the valley (Dean *et al.* 1978). Directly, however,
the data were extracted from the Long House Valley database in the
computer files of the Southwestern Anthropological Research Group
(SARG), an effort at large-scale data accumulation and management and

cooperative research (Gumerman 1971; Euler and Gumerman 1978). These data were downloaded from the SARG master file, modified through the elimination of many categories of data deemed extraneous for our purposes, and then imported into the Artificial Anasazi software. These locational and site data serve as the referents against which the simulations are evaluated.

The simulations take place on a landscape (analogous to Epstein and Axtell's Sugarscape) of annual variations in potential maize production values based on empirical reconstructions of low- and high-frequency paleoenvironmental variability in the area. The production values represent as closely as possible the actual production potential of various segments of the Long House Valley environment over the last 1,600 years. On this empirical landscape, the agents of the Artificial Anasazi model play out their lives, adapting to changes in their physical and social environments.

CHARACTERISTICS OF THE STUDY AREA

Long House Valley (figure 4.1), a topographically discrete, 96 km² land form on the Navajo Indian Reservation in northeastern Arizona, provides an intensively surveyed archaeological case study for the agent-based modeling of settlement and economic behavior among subsistence-level agricultural societies in marginal habitats. This area is well suited for such a test for four reasons. First, it is a topographically bounded, self-contained landscape that can be conveniently reproduced in a computer. Second, a rich paleoenvironmental record, based on alluvial geomorphology, palynology, and dendroclimatology (Gumerman 1988), permits the accurate quantitative reconstruction of annual fluctuations in potential agricultural production (in kilograms of maize per hectare). Combined, these factors permit the creation in the computer of a dynamic resource landscape that replicates conditions in the valley. On this landscape, our artificial agents move about, bring new sites under cultivation, form new households, and so on. Third, detailed regional ethnographies provide an empirical basis for generating plausible behavioral rules for the agents. Fourth, intensive archaeological research, involving a 100-percent survey of the area supplemented by limited excavations, creates a database on human behavior during the last 2,000 years that constitutes the real-world target for the modeling outcomes (Dean *et al.* 1978; Gumerman and Dean 1989). Between roughly 7000 and 1800 B.C., the valley was sparsely occupied by, first, Paleoindian big game hunters and, second, Archaic hunters and gatherers. The introduction of maize around 1800 B.C. initiated a long transition to a

food-producing economy and the beginning of the Anasazi cultural tradition, which persisted until the abandonment of the area about A.D. 1300.
Long House Valley provides archaeological data on economic, settlement, social, and religious conditions among a localized western Anasazi
population. The archaeological record of Anasazi farming groups from
A.D. 200–1300 provides information on a millennium of sociocultural
stasis, variability, change, and adaptation to which the model can be compared. The valley's geologic history has produced seven different environmental zones (figure 4.1) with vastly different productive potentials
for domesticated crops and various degrees of suitability for residential
occupation. One of these habitats, the Uplands Nonarable zone, consists
of exposed bedrock and steep, forested colluvial slopes with no farming
potential. Different soil and water characteristics impart different agricultural potentials to the remaining habitats, in order of increasing potential
productivity the Uplands Arable, General Valley Floor, Midvalley Floor,
North Valley Floor and Canyon, and Sand Dunes zones.

Because the local environment is not temporally stable, modern conditions, which include three soil types, heterogeneous bedrock and surficial
geology, sand dunes, arroyos, seeps, springs, and varied topography,
are only imperfect indicators of the past environmental circumstances
that influenced how and where the Anasazi lived and farmed. Accurate
representations of these circumstances, however, can be achieved through
paleoenvironmental reconstruction. Low- and high-frequency variations
in alluvial hydrologic and depositional conditions, effective moisture, and
climate have been reconstructed in unprecedented detail using surficial
geomorphology, palynology, dendroclimatology, and archaeology. High-
frequency climatic variability is represented by annual Palmer Drought
Severity Indices (PDSI), which reflect the effects of meteorological
drought (moisture and temperature) on crop production (Palmer 1965).
Low-frequency environmental variability is characterized primarily by
the rise and fall of alluvial groundwater and the deposition and erosion of
floodplain sediments. Based on relationships among these variables provided by Van West (1994), these measures of environmental variability
are used to create a dynamic landscape of annual potential maize
production, in kilograms, for each hectare in the study area for the period
A.D. 382 to 1400.

CONSTRUCTING THE PRODUCTION LANDSCAPE

Because there are no crop yield data for any nearby or comparable areas,
maize production in Long House Valley (LHV) cannot be reconstructed
directly from tree growth or from dendroclimatically reconstructed PDSI

values, as was done by Burns (1983) and Van West (1994) for southwestern Colorado, which possesses the only reliable dry-farming crop yield data in the entire Southwest. Rather, the integration of information from several different sources was necessary to extrapolate the likely production record. The sources utilized include Burns' (1983) and Van West's (1994) dendroclimatic research and the Dolores Archaeological Project's soils work (Becker and Petersen 1987; Leonhardy and Clay 1985) in southwestern Colorado, E. and T. Karlstrom's (Karlstrom 1983, 1985; Karlstrom and Karlstrom 1986; Karlstrom 1988) soil and geomorphological studies and Lebo's (1991) dendroagricultural research on nearby Black Mesa, Bradfield's (1969, 1971) Hopi farming studies, and Soil Conservation Service (SCS) soils surveys in Apache (Miller and Larsen 1975) and Coconino (Taylor 1983) counties in Arizona. LHV crop yields were estimated by using relationships between PDSI values and maize production worked out for southwestern Colorado by Van West. In order to employ these relationships, the existing PDSI reconstruction for the LHV area (the Tsegi Canyon reconstruction produced by the Tree-Ring Laboratory's Southwest Paleoclimate Project) had to be related to one (or more) of Van West's 113 PDSI reconstructions. Because PDSI is calculated using specific water-holding attributes of the soils involved, LHV soils had to be matched as closely as possible to one (or more) possible southwestern Colorado equivalents.

The first step in matching LHV and Colorado soils involved characterizing the former so that attributes comparable to the latter might be identified. Because there are no soils data from LHV, one or more LHV soils had to be classified in order to acquire the necessary attributes. Soils research by the Black Mesa Archaeological Project identified possible analogs to one LHV soil, that of the area defined as the General Valley Floor environmental zone, hereafter referred to as LHV gensoil. This soil and several Black Mesa soils are clayey units derived principally from the Mancos shale. Furthermore, the Black Mesa soils were equated with T. Karlstrom's *x* and *y* chronostratigraphic units, which are coeval with the prehistoric LHV soils of interest here. Using these criteria, it was possible to identify six of E. Karlstrom's profiles that contained units potentially equivalent to LHV gensoil: Profiles 3, 4, and 9 (Karlstrom 1983), and 3, 4, and 5 (Karlstrom 1985) in Moenkopi Wash and Reed Valley, respectively.

Although E. Karlstrom provides considerable information on his soil units, he does not include the critical water-holding data necessary to derive PDSIs. Therefore, we had to identify analogs to his soils that had the requisite water capacity data. This was done by using SCS surveys of Apache and Coconino counties to find shale-derived soils that fell into the same typological classes as the Black Mesa soils: soil families fine, loamy, mixed mesic Typic Camborthids, Typic Haplargids, and Ustollic Haplargids. Potential analogs with adequate water capacity data included the Clovis Soil (Ustollic Haplargid) from Apache County and the Epikom Soil (Lithic Camborthid) from Coconino County. These preliminary identifications were checked against Bradfield's data for soils along Oraibi Wash that should share most characteristics with Black Mesa soils farther up the drainage. These procedures led to the recognition of E. Karlstrom's Ustollic Haplargid x/y alluvial soils from Profiles 3 and 4 (Karlstrom 1983) and 4 and 5 (Karlstrom 1985) as satisfactory analogs for LHV gensoil.

At this point, we intended to use the typological and water capacity characteristics inferred for LHVgensoil to identify one or more analogs among the 113 soils Van West used for PDSI calculations. Two problems arose in this regard. First, the Tsegi PDSI values had been calculated using NOAA's (the National Oceanic and Atmospheric Administration) generic soil moisture values of 1" in the first six inches of soil and 5" in the rest of the column (the 1"/5" standard). These values clearly did not mimic the 1"/10+" attributes inferred for LHVgensoil. Two options were open: (1) recalculate PDSI using more realistic water capacity values or (2) find a Colorado analog for the 1"/5" default PDSIs. Lacking resources to do the former, we opted for the latter.

Finding a Colorado analog for the default PDSIs involved identifying a soil (or soils) with attributes that mimicked those of the postulated LHV gensoil. Potential analogs had to have the following characteristics: (1) they had to duplicate the LHV soil families, (2) they had to represent the same elevational range as the floor of LHV, roughly 6,000 to 7,000 feet, (3) they had to have a comparable silt-loam-sand composition and shale derivation, and (4) they had to exhibit the default 1"/5" water capacity used in calculating the Tsegi PDSIs. The first criterion was rejected because, although Van West gives the series names for the 113 soils she used, she does not give their family assignments. Luckily, the Dolores Archaeological Project provided both series and family names and allowed us to assign family designations to Van West's series names. With this information, it was a simple matter to identify soils that exhibited the four characteristics listed above. Two soils came closest to meeting the criteria: Sharps-Pulpit Loam (R7C) and Pulpit Loam (ROHC), both fine, loamy, mixed mesic Ustollic Haplargids that occur between 6,000 and 7,000 feet in elevation. Fortunately, Van West had chosen each of these soils to represent one of eleven soil moisture classes. The ROHC class came closest to LHV gensoil and was chosen as the Colorado analog for that taxon.

The selection of ROHC as the working analog for LHV gensoil permitted the use of PDSI to estimate annual maize crop yields in LHV.

Through a series of statistical operations, Van West calculated the yield of maize in pounds per acre or kilograms per hectare for each representative soil type, including ROHC, under five different growing season conditions: Favorable, Favorable-to-Normal, Normal, Normal-to-Unfavorable, and Unfavorable. She also assigned each yield category a range of PDSI values: Favorable (PDSI $\geq$ 3.000), Favorable-to-Normal (1.000 to 2.999), Normal (-0.999 to 0.999), Normal-to-Unfavorable (-2.999 to -1.000), and Unfavorable (≤-3.000). These concordances allow crop yields to be estimated for each PDSI category. It then becomes a relatively "simple" matter to convert the Tsegi PDSI values to LHV maize crop yields.

Before conversion could begin, some way of integrating the LHV PDSI, Hydrologic Curve (HC), and Aggradation Curve (AC) representations of past environmental variability into a single measure useful for estimating crop yield had to be devised. This was necessary because, during periods of rising and stable high water tables, groundwater basically supports crop production and overrides climatic fluctuations. Therefore, there are long periods when PDSI does not adequately represent environmental potential for farming. We handled this issue by generating Adjusted PDSI values that incorporate HC and AC effects on crop production. This was done by assigning arbitrary PDSI values corresponding to Favorable or Favorable-to-Normal conditions during periods of deposition and rising or stable high water tables. At other times, climate is the primary control on crop yield, and straight PDSI values express environmental input. The new series of Adjusted PDSI values reflects this operation. But, this procedure applies only to the General Valley Floor zone of LHV and not to the five other farmable environments in the valley, the North Valley Floor, Midvalley Floor, Canyon, Uplands Arable, and Sand Dunes zones (figure 4.1). Because the HC and AC are different for each of the environmental zones, a set of Adjusted PDSI values was created for each of five groups of zones: (1) General Valley Floor, (2) North Valley Floor East and West and Canyons, (3) Midvalley Floor East and West, (4) Shonto Plateau–Black Mesa Uplands Arable, and (5) the Sand Dunes along the northeastern margins of the valley. Each set of Adjusted PDSI values is used for its corresponding environmental zone. The four series of Adjusted PDSI values are then converted to maize crop yields for each hectare in each zone.

The conversion takes place by equating specific crop yields in kg/ha with specific PDSI ranges as indicated in table 4.1. Thus, for example, on the General Valley Floor and Midvalley Floor East, a PDSI between 1.000 and 2.999 equals a yield of 824 kg/ha of shelled corn, a PDSI greater than or equal to 3.000 equals a yield of 961 kg/ha, and so forth. This transformation applies only to General Valley Floor and Midvalley Floor

TABLE 4.1
TABLE 4.1

Factors for Converting Long House Valley Adjusted PDSI Values to Maize Crop
Yields in Kilograms/Hectare

	Maize Yield (kilograms/hectare)			
Adjusted PDSI	General Valley Floor[a]	North Valley Floor/Can[b]	Upland Areas[c]	Sand Dune Areas[d]
3.00 to ∞	961	1,153	769	1,201
1.00 to 2.99	824	988	659	1,030
−0.99 to 0.99	684	821	547	855
−2.99 to −1.00	599	719	479	749
−∞ to −3.00	514	617	411	642

[a]Used with General Valley Floor Adjusted PDSIs to estimate crop yields for the General Valley Floor and Midvalley Floor East environmental zones.

[b]Used with North Valley Floor and Canyons Adjusted PDSIs to estimate crop yields for the North Valley Floor East and West, Canyon, and Midvalley Floor West environmental zones.

[c]Used with Upland Adjusted PDSIs to estimate crop yields for the Shonto Plateau–Black Mesa Uplands Arable environmental zones.

[d]Used with North Valley Floor East and West and Canyons Adjusted PDSIs to estimate crop yields for the dune areas in the North Valley Floor and Midvalley Floor West environmental zones.

East, however, because the other environmental zones have different productivities. For example, the North Valley Floor, Midvalley Floor West, and Canyon zones produce higher yields under identical climatic, hydrologic, and aggradational conditions, while the Arable Uplands produce less. These differences are expressed by increasing the yield for the North Valley Floor–Midvalley Floor West–Canyons zones by 20 percent and decreasing the yield of the Uplands zones by 20 percent relative to the General Valley Floor yield as shown in table 4.1. Thus, a PDSI between 1.000 and 2.999 equals crop yields of 988 (North Valley Floor) and 659 (Uplands) kg/ha, and a PDSI ≥ 3.000 produces yields of 1153 and 769 kg/ha, respectively. Yields for the particularly favorable dune areas in the North Valley and Midvalley Floor West zones are calculated by increasing the General Valley Floor yields by 25 percent as shown in table 4.1. Here, a PDSI between 1.000 and 2.999 equals a crop yield of 1030 kg/ha, and a PDSI ≥ 3.000 equals a yield of 1201 kg/ha. Carrying these conversions of Adjusted PDSI values through for each of the environmental zones produces four series of annual crop yield estimates in kg/ha. Multiplying these by the hectarage of each zone produces estimates of total potential crop yield if every bit of land is farmed.

AGENT (HOUSEHOLD) ATTRIBUTES

The constructed physical and resource landscape of Long House Valley is the changing environment on which the agents described here act. Artificial agents representing individual households, the smallest social unit consistently definable in the archaeological record, populate the landscape. These household agents have independent characteristics such as age, location, and grain stocks, and shared characteristics such as death age and nutritional need.

Distinctions between independent and shared characteristics are not always certain. For example, in the current model, nutritional need is the same for all agents but, in other models, nutritional need might be varied stochastically across agents. Agent demographics, nutritional requirements, and marriage characteristics were derived from ethnographic and biological anthropological studies of historic Pueblo groups and other subsistence agriculturalists throughout the world (Hassan 1981; Nelson *et al.* 1994; Swedlund 1994; Weiss 1973; Wood 1994).

In the archaeological view of Long House Valley, five surface rooms or one pithouse is considered to represent a single household, which, based on numerous archaeological and ethnographic analyses, is assumed to comprise five individuals. In our Artificial Anasazi model, household size is fixed at this number for all households at all times. Each simulated household is conceived to be both matrilineal and matrilocal, and so assumptions governing household formation and movement center on females. Males are included in maize consumption calculations.

Every year, household agents harvest the grain that is available at the location they have chosen to farm, as determined by environmental data and modified by stochastic factors. These factors are intended to grossly approximate location-to-location soil quality variation, as well as year-to-year fluctuations caused by weather, blight, and other factors not available in the data.

The agents then consume their nutritional requirements, 800 kg of maize per year, based on an approximation of individual consumption of 160 kg (560 kcal) per year. Households can store any remaining grain for later consumption, but grain that is not consumed within two years of harvest is lost. At this point households may cease to exist, either because they do not have enough grain to satisfy their nutritional needs or because they have aged beyond a certain maximum, 30 years in the current model. Note that a household "death" is not imagined to represent the literal death of all household members. Instead, it represents that a given household no longer exists as a single unit in the valley. Members might die, but they also might be absorbed by other households or simply migrate out of the valley altogether.

Next, household agents estimate the amount of grain that will be available the following year, based on current year harvest and grain stores. If this amount will not satisfy minimum requirements for a given household, the household moves. Determining how, and thus where, a household moves is a critical factor in designing a model that has a meaningful relationship to the historical record. First, the agent finds a new location to farm. In the current model, agents simply search for the most productive land that is available and within 1,600 m of a water source. Household farmlands each occupy one cell in the model, with each cell comprising one hectare. Household residential locations, or settlements, also occupy one cell. To be considered available, land must be unfarmed and unsettled.

Second, the agent looks for a settlement location. The agent finds and settles on the location nearest the farmland that contains a water source. In the current model, if the closest water source is located in a flood plain, the agent instead occupies the closest location to the water source that is on the border of or outside of the floodplain area.

Note that the requirement that a farmland site be within 1,600 m of a water source is not dictated by an overriding need to farm near a water source; water sources in the context of the model provide potable water suitable for household consumption, and they are not important to agriculture. Rather, the farmland must be near water because proximity to water sources is a critical factor in choosing residence locations and because farmplots must be located within reasonable distances of residences. In fact, farm and residence siting searches are really inseparable parts of single decisions on residence and farm locations. This is one reason households are not initially assigned historical settlement locations. As historical farmland locations are not known, they cannot be supplied as initial conditions for running the model and have to be selected by the agents according to the rules of the model. To attempt to constrain farming location choice by using contextually meaningless and predetermined residence locations would be arbitrary and inconsistent.

Finally, household agents may fission. If a household is older than a specified fission age (16 years), it has a defined probability (0.125) of triggering the formation of a new household through the "marriage" of a female child. This summary value is derived from the combined demographic inputs. The use of a minimum fission age combined with fission probability is designed to approximate the probability a household would have daughters, the time it would take such daughters to reach maturity, and the chances of their finding a mate, conceiving a child, and forming a new household. As discussed for household deaths above, the fission process is not meant to be a strict measure of new births within a household. For instance, fission might partially represent immigration, as new arrivals to the valley combine with existing households.

The above completes the specification of agent attributes. Artificial Anasazi household agents are endowed with behavioral rules governing consumption, reproduction ("fissioning"), movement, the selection of farm and residential sites, and ultimately decisions to abandon LHV, which the actual Anasazi did around 1300. Can we explain all or part of local Anasazi history—including the departure—with agents that recognize no social institutions or property rights (rules of land inheritance), or must such factors be built into the model? At present, our agents do not invoke such considerations; they respond purely to environmental stimuli. These are the simplest plausible rules that we could devise. Both the strengths and weaknesses of these rules will prove revealing.

RUNNING THE MODEL

Although the LHV production landscape has been reconstructed for the period A.D. 382 to 1450, our study period runs from A.D. 800 to 1350. The initial agent configuration for each run uses the historically known number of agents but, to be consistent with the agent design, does not use historical settlement locations. Each household executes its full behavioral repertoire (e.g., moving, consuming, reproducing, storing food, and, if need be, leaving) each year. The program tracks household fissions, deaths, grain stocks, and internal demographics. If felicitous decisions are made, the household produces enough food to get through another year; if not, the household runs out of food and is removed from the simulation as a case of either death or emigration.

While a single simulation run may produce plausible and interesting outcomes, many iterations involving altered initial conditions, parameters, and random number generators must be performed in order to assess the model's robustness. Some model outputs (e.g., total population) can be characterized statistically across runs and can be compared to LHV data. Other outputs (e.g., spatial distributions of agents) are not easily characterized statistically, but can be visually compared to real-world patterns.

COMPARING THE SIMULATION WITH THE ARCHAEOLOGICAL DATA

Graphical output of the model includes a map for each year of simulated household residence and field locations, which runs simultaneously with a map of the corresponding archaeological and environmental data. These paired maps facilitate comparison of historical and simulated population dynamics and residence locations. Simultaneously, "real time" histograms and time series plots illustrate annual simulated and historical population numbers, zonal aggregation of population, location and size of residences by environmental zone, the simulated amounts of maize stored and harvested, the zonal distribution of simulated field locations, and the number of simulated households that fission, die out, or leave the valley. Figures 4.2 through 4.5 illustrate representative results for many simulations, all using the parameter values listed in table 4.2. Unless otherwise indicated, the graphs represent mean values for 35 runs, a procedure that characterizes general trends across a number of iterations rather than the idiosyncrasies of individual runs.

Population size curves representing iterations of the model and archaeological estimates are shown in figure 4.2(a) and 4.2(b), respectively, at different scales to facilitate comparison. The stepped appearance of the archaeological population graph is an artifact of the estimation procedure in which ceramic dates for sites begin and end on full, half, or quarter centuries (e.g., 1000, 1150, or 1275). Simulated population typically tracks the archaeological population trajectory; that is, both exhibit similar relative variation. If it were smoothed, the archaeological curve would even more closely resemble the simulated graph. Each shows an increase up to about 900, a leveling off in the tenth century, a major growth surge between 1000 and 1050, another leveling from 1050 to 1150, a drop in the middle 1100s, resurgence in the late 1100s to a peak in the thirteenth century, and a major crash in the late 1200s. The simulated and archaeological curves also exhibit important qualitative differences including a greater and more prolonged simulated population decline in the twelfth century, a more immediate, more gradual, and relatively higher post-1150 recovery in the archaeological population, a slightly earlier thirteenth century decline in the simulated curve, and the failure of the simulated curve to drop to zero at 1300. While there is general qualitative agreement between these two curves, there are significant quantitative disparities in the household numbers and settlement sizes. Both total population (figure 4.2) and individual settlement sizes (figure 4.3) are much larger in the typical simulation than what we infer to have been the actual case. Population aggregation occurs earlier and with greater frequency in the typical simulation than in the historical record (figure 4.3).

Although simulated Long House Valley population aggregation (figure 4.3) departs quantitatively from the archaeological situation, it is nonetheless quite revealing about settlement dynamics. The simulation's tendency to generate aggregation of greater magnitude than that of the

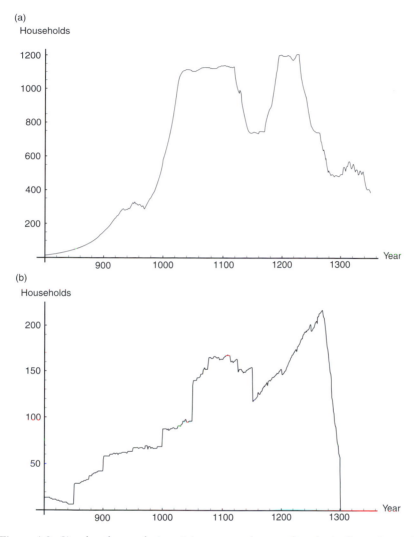

Figure 4.2. Simulated population (a) compared to archaeologically estimated population (b) in numbers of households through time, A.D. 800 to 1360. Numbers of households are graphed at different scales to allow easier comparison of the time series.

study area is evident in the large number of households distributed across settlements larger than 40 households beginning in the early 1100s. This pattern varies considerably from the real situation in which a few large sites appeared only after 1200. The peak at 1180 means that nearly 800 simulated households were living in fewer than 20 sites of 40 or

TABLE 4.2
Long House Valley Model Parameter Summary

Parameter	Value
Random seed	Varies
Simulation begin year	A.D. 800
Simulation termination year	A.D. 1350
Minimum nutritional need	800 kg
Maximum nutritional need	800 kg
Maximum length of grain storage	2 yr
Harvest adjustment	1.00
Harvest variance, year-to-year	0.10
Harvest variance, location-to-location	0.10
Minimum household fission age	16 yr
Maximum household age (death age)	30 yr
Fertility (chance of fission)	0.125
Grain store given to child household	0.33
Maximum farm-to-residence distance	1,600 m
Minimum initial corn stocks	2,000 kg
Maximum initial corn stocks	2,400 kg
Minimum initial agent age	0 yr
Maximum initial agent age	29 yr

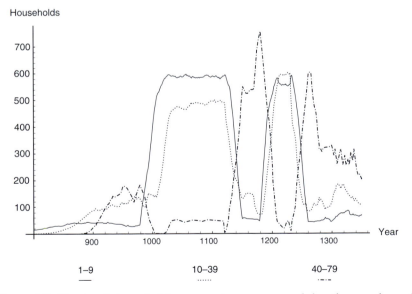

Figure 4.3. Simulated household aggregation represented by the number of households grouped into settlements of 1 to 9 rooms, 10 to 39 rooms, and 40 or more rooms.

more households, when during that period in the real valley there were no sites of the requisite size (200 rooms). The peak around 1260, with 600 households in fewer than 15 sites of more than 40 households, conforms more closely to archaeological reality. Although only one or two individual sites had as many as 200 rooms during the 1250–1300 period (Long House had more than 300), there were at least four clusters of sites each of whose total room count equaled or exceeded that number. Clearly, the simulation packs more households into single residential loci than did the real Anasazi who tended to distribute members of a residential unit across a number of discrete but spatially clustered habitation sites (Dean *et al.* 1978).

On a larger scale, the qualitative aspects of the aggregation graphs (figure 4.3) replicate important aspects of the settlement history of Long House Valley and the surrounding region with uncanny accuracy. Fluctuations in the numbers of simulated households in the two smaller site categories (1 to 9 and 10 to 39 households) parallel one another and, together, exhibit a strong reciprocal relationship with the largest sites (40 and more households). When the simulated population is concentrated in large aggregated sites, there are few small-to-medium-sized sites and, when most of the population is distributed among small-to-medium sites, there are few large sites. Thus, from 900 to 1000, the population was concentrated in large and medium sites and from 1150 to 1200 and 1260 to 1300 it was concentrated in large sites. Conversely, from 1000 to 1150 and 1200 to 1260 it was distributed among small and medium sites.

These patterns of aggregation and dispersal virtually duplicate the settlement history of the eastern Kayenta Anasazi area, which includes Long House Valley (Dean 1996). During Pueblo I times (850–1000), the real population was aggregated into medium-to-large pithouse villages. Large villages disappeared abruptly after 1000, and, in the Pueblo II period (1000–1150), the population dispersed widely across the landscape, living in small-to-medium-sized unit pueblos that rarely comprised more than 30 rooms. During the Transition period (1150–1250), settlements once again exhibited a tendency toward aggregation, although not nearly as strong as that produced by the simulation between 1150 and 1200. After 1150, people began moving out of upland and outlying areas and concentrating in lowland localities like Long House Valley. Although they did not yet aggregate into extremely large sites or unified clusters of sites, site size tended to be larger than that of the Pueblo II period. The magnitude of the simulated return to a dispersed small-to-medium site distribution from 1200 to 1260 also far exceeds the archaeological situation during the second half of the Transition period in which there were minor settlement adjustments toward a more dispersed pattern in the valley (Effland 1979). The simulated shift to residence in large sites between 1260 and 1300 duplicates the Tsegi phase (1250–1300) pattern of aggregation into fewer but larger sites and into organized site clusters throughout Kayenta Anasazi territory.

Eastern Kayenta Anasazi settlement shifts between 800 and 1300 are related to low- and high-frequency environmental fluctuations (Dean et al. 1985, figure 1). During periods of depressed alluvial water tables and channel incision (750–925, 1130–1180, 1250–1450) populations tended to aggregate in the few localities where intensive flood plain agriculture was possible under these conditions, a process of compaction that produced large sites or site clusters. During intervals of high groundwater levels and flood plain deposition (925–1130, 1180–1250), farming was possible nearly everywhere, and people were not constrained to live in a few favored localities. In the 1000–1130 period, a combination of salubrious flood plain circumstances and unusually favorable high-frequency climatic conditions allowed the population to disperse widely across the landscape. Given the simulation outcomes illustrated in figure 4.3, it seems clear that the Artificial Anasazi Project has successfully captured the dynamic relationship between settlement aggregation-dispersal and low- and high-frequency environmental variability in the study area.

After 1250, the area was afflicted with simultaneous low and high frequency environmental degradations. Falling alluvial water tables, rapid arroyo cutting, the Great Drought of 1276–1299, and a breakdown in the spatial coherence of seasonal precipitation (Dean and Funkhouser 1995) combined to create the most severe subsistence crisis of the nearly 2000 years of paleoenvironmental record, an event that was accompanied by the abandonment of the entire Kayenta region. As was the case with population, simulated aggregation does not duplicate the Anasazi abandonment of the valley after 1300. Nonetheless, the behavior of artificial aggregation after 1250 is extremely instructive about the possibilities of human occupancy of the area during intervals of environmental deterioration and high population densities. The number of large settlements (more than 40 households) drops precipitously, but they do not disappear entirely. Conversely, the numbers of small and medium settlements continue unchanged through the period of greatest stress and increase noticeably after 1300. These results, coupled with paleoenvironmental evidence that the valley environment could have supported a reduced population, clearly indicate that many Anasazi could have remained in the area had they disaggregated into smaller communities dispersed into favorable habitats, especially the North Valley Floor. Thus, the model supports extant ideas that environmental factors account for only part of the exodus from the study area and that the total abandonment must be attributed to a combination of environmental and nonenvironmental causes (Dean 1966, 1969). The delicate balance between environmental "push" factors and nonenvironmental (cultural) "pull" factors suggested by the artificial Long House Valley results is compatible with long-standing, archaeologically untestable hypotheses (Dean 1966; Lipe 1995) about the real Anasazi world. The failure of this aspect of the simulation to quantitatively replicate the case study results provides valuable insights into what humans might have done in the real Long House Valley but did not.

The similarities between the simulated and real Long House Valley settlement patterns far outweigh the differences. Figure 4.4(a)–(c) gives side-by-side comparisons of simulated and archaeological site distributions for three years (1000, 1144, and 1261) selected to illustrate relationships between the simulated and known distributions of sites against the backdrop of increasing hydrologic potential represented by progressively darker shades of gray. Figure 4.4(a) shows the paired situations at 1000 when there was considerable hydrologic variability coupled with high corn production potential across the landscape. While the number of simulated sites far exceeds the actual numbers, the simulation accurately reflects the distribution of real sites along the periphery of the flood plain throughout the entire valley. Although crowded, the simulated distribution is what would be expected given the relatively uniform productive potentials across all farmable zones. The large simulated sites along the northeastern margin of the valley represent population aggregates held over from the Pueblo I interval (850–1000) of low alluvial water tables and flood plain erosion. Apart from these similarities, however, the model performs only moderately well for this period.

Figure 4.4(b) represents a year (1144) in which both hydrologic conditions and potential crop production varied across the landscape. The simulation mimics the spread of sites throughout the valley, particularly along the margins of the flood plain, and captures the initial shift in settlement density toward the north end of the valley that occurred during the environmental degradation of the middle twelfth century. The simulation replicates the twelfth century clustering of settlements into five groups, one at the southwestern extremity of the valley, one in each of the Midvalley Floor localities, one at the mouth of Kin Biko on the northwestern margin of the valley floor, and one at the northeastern corner of the valley. In the northeastern corner of the valley, a real group of eight sites is matched in the simulation by two aggregated sites. In addition, the simulation reproduces the scatter of sites in the nonagricultural uplands on the western and northern sides of the valley. In the northeastern corner of the valley, a real group of eight sites is matched in the simulation by two aggregated sites. In addition, the simulation reproduces the scatter of sites in the nonagricultural uplands

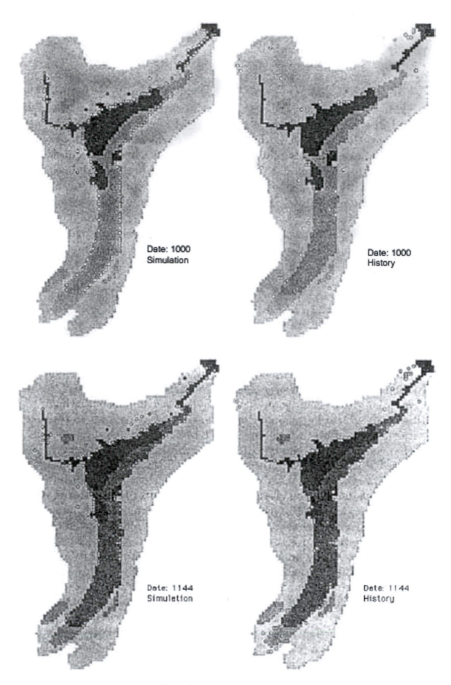

Figure 4.4. Long House Valley showing simulated site size and spatial distributions at three selected years: (a) *top*, A.D. 1000, (b) *bottom*, A.D. 1144, (c) *next page*, A.D. 1261. Sites are represented by circles; the darker the circle, the greater the number of households in the settlement.

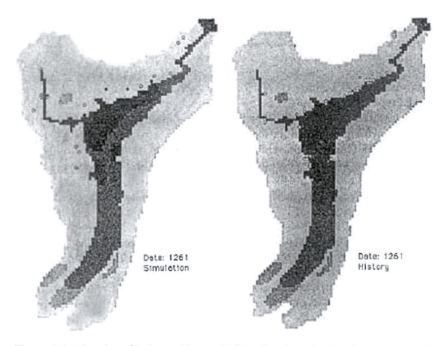

Figure 4.4. (Continued.) Long House Valley showing simulated site size and spatial distributions at three selected years: (c) *above*, A.D. 1261. Sites are represented by circles; the darker the circle, the greater the number of households in the settlement.

on the western and northern sides of the valley. Finally, the simulation accurately locates settlements in the appropriate environmental zones, with the heaviest concentrations in the Midvalley Floor and the North Valley Floor. Major differences between the simulated and real situations are the greater size and number of simulated settlements in the north-central uplands and upper Kin Biko.

Figure 4.4(c) depicts a year (1261) near the beginning of the period of severe environmental stress that began about 1250. This year was characterized by extremely high spatial differentials in hydrologic conditions and crop production potential. The model spectacularly duplicates the abandonment of the southern half of the valley as a place of residence and the concentration of the population along the northwestern edge of the flood plain near the remaining patches of productive farmland. The simulation also reproduces four of the five settlement clusters that characterized Tsegi phase settlement. An upland cluster of four sites at the northeastern end of the valley is represented in the simulation by

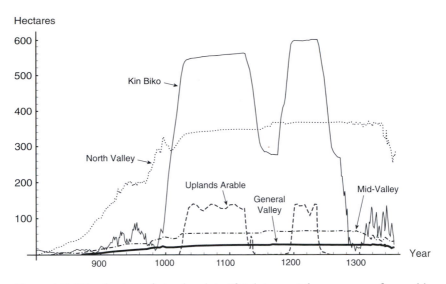

Figure 4.5. Distribution of simulated Artificial Anasazi farms among five arable environmental zones expressed as the number of hectares under cultivation in each zone from A.D. 800 to 1360.

two aggregated sites. A larger cluster of large and small sites along the northern margin of the flood plain is matched by a single, very large site and a row of small to large sites. The Long House cluster at the mouth of Kin Biko is represented in the simulation by a couple of large sites. The simulated Midvalley Floor West settlement group, consisting of two aggregated and two small sites, is displaced into the uplands compared to the actual cluster of nine sites, which adjoins the farmland. In three of the clusters, the simulation reproduces the site size distribution of the actual situation in which the cluster comprises one or two large pueblos and a few smaller sites. In addition, the model located a large site in exactly the same positions as the Anasazi situated Long House in the cluster at the mouth of Kin Biko and Tower House in the cluster along the northern margin of the valley. Significant discrepancies between the artificial and real site distributions are the absence of a Midvalley Floor East group from the simulation and the model's placement of too many settlements in Kin Biko.

Another aspect of the simulation reveals much about general patterns of subsistence farming in the valley, even though real-world information on the utilization of farmland for detailed testing of the simulation is unattainable. The order in which different environmental zones are exploited by the Artificial Anasazi (figure 4.5) is in exact accordance with expectations of the real world (Dean *et al.* 1978). The simulation begins during a period of low-frequency environmental stress, and most fields are located in the zones that are productive during episodes of depressed alluvial water tables and arroyo cutting: the North Valley Floor, Kin Biko, and, to a lesser degree, the Midvalley Floor. Fields saturate the North Valley Floor and Kin Biko by 1000, and farming of these areas fluctuates only slightly thereafter. The Midvalley Floor reaches capacity by 1030. During the 800–1000 period, use of the Uplands Arable and General Valley Floor zones remains low because neither area has groundwater available to support crop production, which would have depended solely on the unreliable and generally deficient rainfall. This dependency is indicated by the high variability in farmland use on the General Valley Floor before 1000. By 1000, changes in flood plain processes that begin early in the tenth century enhance productivity in the General Valley Floor and impact the agricultural decisions made by the agents. Rising alluvial water tables and flood plain deposition provide a stable water supply for crops in the General Valley Floor and replace precipitation as the primary control on production. The result is a major "land rush" to establish fields in the newly productive General Valley Floor that begins around 980 and approaches the zone's carrying capacity within 50 years. Again as would be expected, large-scale use of the Upland Arable Zone, where production is controlled primarily by rainfall, does not begin until after 1020 when all the other zones have achieved virtual saturation. The salubrious agricultural conditions that began around 1000 supported the huge population growth and settlement expansions of the 1000–1120 period in both the artificial and the real Long House Valleys.

After about 1030, figure 4.5 reflects varying use of different farming environments by a population of agents that approaches the carrying capacity of the simulated area. The North Valley Floor, General Valley Floor, Midvalley Floor, and Kin Biko were fully occupied by fields, and because crop production was controlled by stable flood plain conditions, use of these areas exhibits minimal variability. In contrast, field use in the Upland zone varies considerably because production there depends on precipitation rather than hydrologic conditions. Fields in both the General Valley Floor and Upland zones are severely reduced by the secondary fluvial degradation and drought of the middle twelfth century. The greater sensitivity of the Upland farms to environmental perturbations is indicated by the facts that farming in these areas begins to decline before that on the General Valley Floor and that Upland farmland is abandoned during the depth of the crisis. Upward blips in Midvalley Floor and North Valley Floor field numbers reflect the establishment of fields in these more productive areas by a small number of agents forced out of the General Valley Floor and Uplands. Displaced agents that cannot be accommodated in the favorable areas are removed from the simulation (through death or emigration), which accounts for the decline in simulated population during this interval (figure 4.2(a)). When favorable groundwater conditions return, more fields are once again established in the General Valley Floor and Uplands. Once again, the inferior quality of Upland farmland is indicated by the fact that it was not reoccupied until all other zones had been filled to capacity. The Uplands' continued sensitivity to environmental variation is shown by the fluctuations in use between 1190 and 1230 and its abandonment before the primary effects of the next environmental degradation are felt in the other zones. The major low- and high-frequency environmental crises of the last half of the thirteenth century have major repercussions for Artificial Anasazi field selection. Groundwater depletion and arroyo cutting virtually destroy the farming potential of the General Valley Floor and make production there totally dependent on precipitation, which itself is depressed by the Great Drought of 1279–1299. Upland fields are abandoned by 1240, the number of fields on the Midvalley Floor decreases after 1270, and the General Valley Floor is virtually abandoned as a farming area by 1280. Only the North Valley Floor and Kin Biko, where local topographic and depositional factors mitigate the effects of fluvial degradation, retain their farming potential. These areas, already completely filled, remain fully utilized but lack the capacity to absorb agents displaced from the General Valley Floor, Midvalley Floor, and Uplands zones. The disappearance of these agents from the simulation accounts for the major population decline of this period (figure 4.2(a)). Unlike the real Anasazi, simulated agents continue to locate fields in Long House Valley after 1300 but under vastly altered environmental conditions. Far fewer fields are located on the General Valley Floor and, as shown by the rapid fluctuations after 1280, this area is far less productive and far more vulnerable to high-frequency productivity fluctuations caused by the greater control exercised by precipitation. The North Valley Floor continues to be highly productive but even there, after about 1280, high-frequency climatic variability becomes a more important factor in productivity. The decline in field locations on the North Valley Floor after 1320 is due to the depopulation of the virtual valley; for the first time in about 200 years, the population of agents falls below what can be supported on the landscape.

As was the case with simulated population (figure 4.2(a)) and aggregation (figure 4.3), the simulated distribution of farmland clearly shows that the Anasazi need not have totally abandoned Long House Valley as a result of environmental deterioration comparable to that presently built into the model. Instead, a substantial fraction of the population could have stayed behind in small settlements dispersed across suitable farming habitats located in areas (primarily in the North Valley Floor zone) still suitable for agriculture given the detrimental environmental conditions of the post-1250 period. The fact that in the real Long House Valley, that fraction of the population chose not to stay behind but to participate in the exodus from the area, supports the assertion that sociocultural "pull" factors were drawing them away from their homeland.

All runs described above use the parameter estimates of table 4.2. These were deemed the most plausible values. But they are not the only possibilities. In research to be presented elsewhere, these values are systematically varied over a range in which the table 4.2 values are intermediate. It is striking that over that entire range of plausible environments—including some severely degraded ones—we have not observed the complete abandonment of the real Long House Valley that occurred after A.D. 1300. This outcome strongly reinforces the idea that the valley could have supported a reduced, but still viable, population. Thus, the comparison of the results of the simulation with the real world helps differentiate external (environmental) from internal (cultural) causation in cultural variation and change and even provides a clue, in the form of the proportion of the population that could have stayed but elected to go, as to the relative magnitude of these factors. This finding highlights the utility of agent-based modeling in archaeology by demonstrating a predicted response (Dean 1966) that never could be tested with archaeological data. Because the purely environmental rules explored thus far do not fully account for the Anasazi's disappearance from LHV, it could be argued that predominantly nonenvironmental sociological and ideological factors were responsible for the complete abandonment of an area still capable of supporting a substantial population.

CONCLUSIONS

How has agent-based modeling improved understanding of culture change in Anasazi country? First, it allows us to test hypotheses about the past for which we have only indirect evidence. For example, the simulations support predictions about the use of different kinds of farmland under different low- and high-frequency environmental conditions. Second, it illuminates the relative importance of and interactions among various demographic and environmental factors in the processes of sociocultural stability, variation, and change. Third, the generation of similar macroscale results from different microscale specifications elucidates the role of equifinality in sociocultural processes and archaeological analysis. Fourth, progressively augmenting agent specifications allows the experimental manipulation of behavioral modes and assessment of their incremental effects on agent responses to environmental variability.

Finally, agent-based modeling encourages the consideration of previously unspecified, ignored, or discounted factors as consequential mechanisms of cultural adaptation and change. In this regard, Stephen Jay Gould (1989) among others has emphasized that a problem with historical sciences—such as astronomy, geology, paleontology, and archaeology—is that we cannot rewind and rerun the tape of history. While this may be literally true, with agent-based modeling we can execute numerous simulations to investigate alternative outcomes of sociocultural processes under different initial conditions and operational procedures. We can systematically alter the quantitative parameters of a model or make qualitative changes that introduce completely new, and even unlikely, elements into the artificial world of the simulation. Thus, in terms of the Artificial Anasazi model, we could change agent attributes, such as fecundity or food consumption, or introduce new elements, such as mobile raiders, environmental catastrophes, or epidemics.

Ultimately, "to explain" the settlement and farming dynamics of Anasazi society in Long House Valley is to identify rules of agent behavior that account for those dynamics; that is, generate the target spatiotemporal history. Agent-based models are laboratories where competing explanations—hypotheses about Anasazi behavior—can be tested and judged in a disciplined empirical way. The simple agents posited here explain important aspects of Anasazi history while leaving other important aspects unaccounted for. Our future research will attempt to extend and improve the modeling, and we invite colleagues to posit alternative rules, suggest different system parameters, or recommend operational improvements. Agent-based models may never fully explain the real history—these, after all, are simple instruments—but they enable us to make scientific progress in a replicable, cumulative way that does not seem possible with other modeling techniques or through narrative methods alone, crucial as these are in formulating the principles, hypotheses, and experiments that can carry us forward.

References

Bradfield, Maitland. 1969. Soils of the Oraibi Valley, Arizona, in Relation to Plant Cover. *Plateau* 41:133–40.

———. 1971. The Changing Pattern of Hopi Agriculture. Royal Anthropological Institute Occasional Paper No. 30. London: Royal Anthropological Institute of Great Britain and Ireland.

Burns, Barney Tillman. 1983. Simulated Anasazi Storage Behavior Using Crop Yields Reconstructed from Tree Rings: A.D. 652–1968. Ph.D. diss., Department of Anthropology, University of Arizona.

Dean, Jeffrey S. 1966. The Pueblo Abandonment of Tsegi Canyon, Northeastern Arizona. Paper presented at the 31st Annual Meeting of the Society for American Archaeology, Reno, NV.
———. 1969. *Chronological Analysis of Tsegi Phase Sites in Northeastern Arizona.* Tucson: University of Arizona Press.
———. 1996. Kayenta Anasazi Settlement Transformations in Northeastern Arizona, A.D. 1150 to 1350. In *The Prehistoric Pueblo World, A.D. 1150–1350,* ed. Michael A. Adler. Tucson: University of Arizona Press.
Dean, Jeffrey S., Robert C. Euler, George J. Gumerman, Fred Plog, Richard H. Hevly, and Thor N. V. Karlstrom. 1985. Human Behavior, Demography, and Paleoenvironment on the Colorado Plateaus. *American Antiquity* 50:537–54.
Dean, Jeffrey S., and Gary S. Funkhouser. 1995. Dendroclimatic Reconstructions for the Southern Colorado Plateau. In *Climate Change in the Four Corners and Adjacent Regions: Implications for Environmental Restoration and Land-Use Planning,* ed. W. J. Waugh. Grand Junction, CO: U.S. Department of Energy, Grand Junction Projects Office.
Dean, Jeffrey S., Alexander J. Lindsay, Jr., and William J. Robinson. 1978. Prehistoric Settlement in Long House Valley, Northeastern Arizona. In *Investigations of the Southwestern Anthropological Research Group: An Experiment in Archaeological Cooperation,* ed. Robert C. Euler and George J. Gumerman. Flagstaff: Museum of Northern Arizona.
Effland, Richard Wayne, Jr. 1979. A Study of Prehistoric Spatial Behavior: Long House Valley, Northeastern Arizona. Ph.D. diss., Department of Anthropology, Arizona State University.
Epstein, Joshua M., and Robert L. Axtell. 1996. *Growing Artificial Societies: Social Science from the Bottom Up.* Washington, DC: Brookings Institution Press; Cambridge: MIT Press.
Euler, Robert C., and George J. Gumerman, eds. 1978. *Investigations of the Southwestern Anthropological Research Group: An Experiment in Archaeological Cooperation.* Flagstaff: Museum of Northern Arizona.
Gould, Stephen J. 1989. *Wonderful Life: The Burgess Shale and the Nature of History.* New York: Norton.
Gumerman, George J., ed. 1971. *The Distribution of Prehistoric Population Aggregates: Proceedings of the Southwestern Anthropological Research Group.* Prescott College Studies in Anthropology, No. 1. Prescott, AZ: Prescott College Press.
———, ed. 1988. *The Anasazi in a Changing Environment.* Cambridge: Cambridge University Press.
Gumerman, George J., and Jeffrey S. Dean. 1989. Prehistoric Cooperation and Competition in the Western Anasazi Area. In *Dynamics of Southwest Prehistory,* ed. Linda S. Cordell and George J. Gumerman. Washington, DC: Smithsonian Institution Press.
Gumerman, George J., and Timothy A. Kohler. In press. *Creating Alternative Culture Histories in the Prehistoric Southwest: Agent-Based Modeling in Archaeology.* Durango, CO: Fort Lewis College Press.
Hassan, Fekri A. 1981. *Demographic Archaeology.* New York: Academic Press.

116 CHAPTER 4

Karlstrom, Eric. 1983. Soils and Geomorphology of Northern Black Mesa. In *Excavations on Black Mesa, 1981: A Descriptive Report*, ed. F. E. Smiley, Deborah L. Nichols, and Peter P. Andrews. Carbondale: Center for Archaeological Investigations, Southern Illinois University at Carbondale.

————. 1985. Soils and Geomorphology of Excavated Sites. In *Excavations on Black Mesa, 1983: A Descriptive Report*, ed. Andrew L. Christenson and William J. Perry. Carbondale: Center for Archaeological Investigations, Southern Illinois University at Carbondale.

Karlstrom, Eric, and Thor N. V. Karlstrom. 1986. Late Quaternary Alluvial Stratigraphy and Soils of the Black Mesa–Little Colorado River Areas, Northern Arizona. In *Geology of Central and Northern Arizona*, ed. J. D. Nations, C. M. Conway, and G. A. Swann. Flagstaff, AZ: Geological Society of America.

Karlstrom, Thor N. V. 1988. Alluvial Chronology and Hydrologic Change of Black Mesa and Nearby Regions. In *The Anasazi in a Changing Environment*, ed. George J. Gumerman. Cambridge: Cambridge University Press.

Lebo, Cathy J. 1991. Anasazi Harvests: Agroclimate, Harvest Variability, and Agricultural Strategies on Prehistoric Black Mesa, Northeastern Arizona. Ph.D. diss., Department of Anthropology, Indiana University.

Leonhardy, Frank C., and Vickie L. Clay. 1985. Soils. In *Dolores Archaeological Program: Studies in Environmental Archaeology*, comp. Kenneth Lee Petersen, Vickie L. Clay, Meredith H. Matthews, and Sarah W. Neusius. Denver: USDI Bureau of Reclamation.

Lipe, William D. 1995. The Depopulation of the Northern San Juan: Conditions in the Turbulent 1200s. *Journal of Anthropological Archaeology* 14:143–169.

Miller, Mack L., and Kermit Larsen. 1975. *Soil Survey of Apache County, Arizona, Central Part*. Washington, DC: USDA Soil Conservation Service.

Nelson, Ben A., Timothy A. Kohler, and Keith W. Kintigh. 1994. Demographic Alternatives: Consequences for Current Models of Southwestern Prehistory. In *Understanding Complexity in the Prehistoric Southwest*, ed. George J. Gumerman and Murray Gell-Mann. Reading, MA: Addison-Wesley.

Palmer, William C. 1965. Meteorological Drought. Research Paper No. 45, USDC Office of Climatology, U.S. Weather Bureau, Washington, DC.

Swedlund, Alan C. 1994. Issues in Demography and Health. In *Understanding Complexity in the Prehistoric Southwest*, ed. George J. Gumerman and Murray Gell-Mann. Reading, MA: Addison-Wesley.

Taylor, Don R. 1983. *Soil Survey of Coconino County Area, Arizona, Central Part*. Washington, DC: USDA Soil Conservation Service.

Van West, Carla R. 1994. *Modeling Prehistoric Agricultural Productivity in Southwestern Colorado: A GIS Approach*. Pullman: Washington State University, Department of Anthropology.

Weiss, Kenneth M. 1973. *Demographic Models for Anthropology*. Washington, DC: Society for American Archaeology.

Wood, James W. 1994. *Dynamics of Human Reproduction: Biometry, Biology, Demography*. New York: Hawthorne.

Chapter 5

POPULATION GROWTH AND COLLAPSE IN A

MULTIAGENT MODEL OF THE KAYENTA ANASAZI

IN LONG HOUSE VALLEY

ROBERT L. AXTELL, JOSHUA M. EPSTEIN,
JEFFREY S. DEAN, GEORGE J. GUMERMAN,
ALAN C. SWEDLUND, JASON HARBURGER,
SHUBHA CHAKRAVARTY, ROSS HAMMOND,
JON PARKER, AND MILES PARKER*

LONG HOUSE VALLEY in the Black Mesa area of northeastern Arizona (U.S.) was inhabited by the Kayenta Anasazi from about 1800 before Christ to about *anno Domini* 1300. These people were prehistoric ancestors of the modern Pueblo cultures of the Colorado Plateau. Paleoenvironmental research based on alluvial geomorphology, palynology, and dendroclimatology permits accurate quantitative reconstruction of annual fluctuations in potential agricultural production (kg of maize per hectare). The archaeological record of Anasazi farming groups from *anno Domini* 200–1300 provides information on a millennium of sociocultural stasis, variability, change, and adaptation. We report on a multiagent computational model of this society that closely reproduces the main features of its actual history, including population ebb and flow, changing spatial settlement patterns, and eventual rapid decline. The agents in the model are monoagriculturalists, who decide both where to situate their fields as well as the location of their settlements. Nutritional needs constrain fertility. Agent heterogeneity, difficult to model mathematically, is demonstrated to be crucial to the high fidelity of the model.

---

*The authors' affiliations are as follows: Robert L. Axtell: Center on Social and Economic Dynamics, The Brookings Institution, to whom reprint requests should be addressed, e-mail: raxtell@brookings.edu; Joshua M. Epstein: Center on Social and Economic Dynamics, The Brookings Institution, and External Faculty Member, Santa Fe Institute; Jeffrey S. Dean: Laboratory of Tree-Ring Research and Department of Anthropology, University of Arizona, and Arizona State Museum; George J. Gumerman: Department of Anthropology, University of Arizona, and Arizona State Museum; Alan C. Swedlund: Department of Anthropology, University of Massachusetts; Jason Harburger: Center on Social and Economic Dynamics, The Brookings Institution; present address, Departments of

As the only social science that has access to data of sufficient duration to reveal long-term changes in patterned human behavior, archaeology traditionally has been concerned with describing and explaining how societies adapt and evolve in response to changing conditions. A major impediment to rigorous investigation in archaeology—the inability to conduct reproducible experiments—is one shared with certain other sciences, such as astronomy, geophysics, and paleontology. Computational modeling is providing a way around these difficulties.[1]

Within anthropology and archaeology there has been a rapidly growing interest in so-called agent-based computational models (Gilbert and Doran 1994; Gilbert and Conte 1995; Kohler and Gumerman 2000). Such models consist of populations of artificial, autonomous "agents" who live on spatial landscapes (Dean, Lindsay, and Robinson 1978). Each agent is an indivisible social unit—an individual, a household, a clan—endowed with specific attributes (e.g., life span, nutritional requirements, movement capabilities, family ties). A set of anthropologically plausible rules of behavior defines the ways in which agents interact with their physical environment and with one another. Social histories unfold in such models by "turning on" each agent periodically

---

Computer Science, Economics, and Mathematical Sciences, The Johns Hopkins University; Shubha Chakravarty: Center on Social and Economic Dynamics, The Brookings Institution; Ross Hammond: Center on Social and Economic Dynamics, The Brookings Institution; present address, Department of Political Science, University of Michigan; Jon Parker: Center on Social and Economic Dynamics, The Brookings Institution; present address, Departments of Computer Science, Economics, and Mathematical Sciences, The Johns Hopkins University; Miles Parker: Center on Social and Economic Dynamics, The Brookings Institution; present address, Bios Group, Inc.

This paper results from the Arthur M. Sackler Colloquium of the National Academy of Sciences, "Adaptive Agents, Intelligence, and Emergent Human Organization: Capturing Complexity through Agent-Based Modeling," held October 4–6, 2001, at the Arnold and Mabel Beckman Center of the National Academies of Science and Engineering in Irvine, CA.

We thank Samuel Bowles and Colloquium participants for valuable suggestions. We gratefully acknowledge financial support from the National Science Foundation (IIS-9820872), the John D. and Catherine T. MacArthur Foundation, the Alex C. Walker Foundation, the National Park Service, and the Advanced Research Projects Agency, as well as additional support from the Brookings Institution, the Santa Fe Institute, and the University of Arizona.

This essay was published previously in *Proceedings of the National Academy of Sciences, Colloquium* 99(3): 7275–7279.

[1]For example, because large-scale experiments on the Earth's tectonic structure (e.g., mantle and core) are impossible, numerical models play a crucial role in geophysics (Glatzmaier *et al.* 1999). An essentially identical situation exists in planetology, where progress on the origin of the moon, for instance, is achieved numerically (Canup and Asphaug 2001). Computational models are increasingly common in paleontology (Bak and Sneppen 1993).

Figure 5.1. Long House Valley, looking to the south.

and permitting it to interact. Agent models offer intriguing possibilities for overcoming the experimental limitations of archaeology through systematic analyses of alternative histories. Changing the agents' attributes, their rules, and features of the landscape yields alternative behavioral responses to initial conditions, social relationships, and environmental forcing.

Long House Valley, a topographically discrete, 96-km^2 land form (fig. 5.1) on the Navajo Indian Reservation in northeastern Arizona (Dean, Lindsay, and Robinson 1978), provides a realistic archaeological test of the ability of agent-based computational models to explain settlement patterns and demographic behavior among subsistence-level agricultural societies in marginal habitats. Between roughly 7000 and 1000 years before Christ (B.C.), the valley was sparsely occupied, first by Paleo-Indian big game hunters and second by Archaic hunters and gatherers. The introduction of maize around 1800 B.C. initiated a long transition to a food producing economy and began the Anasazi cultural tradition (Epstein and Axtell 1996), which persisted until the abandonment of the region around *anno Domini* (A.D.) 1300 (Gumerman 1984). Anasazi is the term applied to a distinctive archaeological pattern and sequence that is confined to the southern Colorado Plateau and that has given rise to the cultural configurations that characterize the modern Pueblo people of the Southwest. The Anasazi pattern is defined by an emphasis on black-on-white painted ceramics, plain and textured gray cooking pottery, the development from pithouses to stone masonry and adobe pueblos, and the kiva as the principal ceremonial structure. Considerable spatial variability within the general pattern has led to the recognition of several geographic variants of Anasazi. Long House Valley falls within one of the western Anasazi configurations.

Long House Valley is well suited for application of multiagent modeling for a variety of reasons (Dean *et al.* 2000). Its bounded topography combined with the rich paleoenvironmental record permits the creation in the computer of a dynamic resource landscape that accurately replicates actual conditions in the valley from A.D. 200 to 1500. Low- and high-frequency variations in alluvial hydrologic and depositional conditions, effective moisture, and climate have been reconstructed in unprecedented detail with dendroclimatology, surficial geomorphology, palynology, and archaeology (Dean *et al.* 1985; Gumerman 1988). High-frequency climatic variability is represented by annual June Palmer Drought Severity Indices (PDSI), which reflect the effects of meteorological drought (moisture and temperature) on crop production (Palmer 1965). Low-frequency environmental variability is characterized primarily by the rise and fall of alluvial groundwater and the deposition and erosion of flood plain sediments. Based on statistical relationships between PDSI and annual crop yields in southwestern Colorado provided by Van West (1994), these measures of environmental variability are used to create a dynamic landscape of annual potential maize production, in kilograms, for each hectare in the study area for the period A.D. 400–1400. Intensive archaeological research provides a database on human settlement in the valley (Dean, Lindsay, and Robinson 1978; Gumerman and Dean 1989).[2] Finally, detailed regional ethnographies provide an empirical basis for generating plausible behavioral rules for the agents (Forde 1931; Hack 1942; Levy 1992).

The multiagent model is created by instantiating the landscape, reconstructed from paleoenvironmental variables, and then populating it with artificial agents that represent individual families, or households, the smallest social unit consistently definable in the archaeological record

---

[2] The archaeological survey data were generated by the Long House Valley Project, a joint venture of the Museum of Northern Arizona and the Laboratory of Tree-Ring Research at the University of Arizona (Dean, Lindsay, and Robinson 1978). The availability of the Long House Valley data in the Southwestern Anthropological Research Group (Euler and Gumerman 1978) automated database greatly facilitated the development of the model.

TABLE 5.1
Household (agent) Attributes

1. Five surface rooms or one pithouse is considered to represent a single household.
2. Each household that is both matrilineal and matrilocal consists of 5 individuals. Only female marriage and residence location are tracked, although males are included in maize-consumption calculations.
3. Each household consumes 160 kg of maize per year per individual.
4. Each household can store a maximum of 2 years' total corn consumption (1,600 kg), i.e., if at harvest 800 kg of corn remains in storage an additional 800 kg can be added to that from the current crop.
5. Households use only 64% of the total potential maize yield. (The unutilized production is attributed to fallow, loss to rodents, insects, and mildew, and seed for the next planting.)

(Dean 1969; Rohn 1965).[3] Each household agent is initialized based on demographic characteristics and nutritional requirements derived from ethnographic studies of historic Pueblo groups and from other subsistence agriculturists.[4] Each family agent is defined by certain attributes (table 5.1), including its age, size, composition, and amount of maize storage. Similarly, each agent has specific rules of behavior (table 5.2). These rules determine how the households select their planting and dwelling locations.

Once all agents are initialized, the model proceeds according to internal clocks (table 5.3). Essentially, all agents engage in agricultural activity during each period (1 calendar year) and move their plots or dwellings or both based on their success in meeting nutritional needs. Simulated household and field locations, as well as the size of each community (the number of households at each site), are updated annually. A map of annual simulated field locations and household residence locations and sizes runs simultaneously with a map of the

---

[3]The model is written in JAVA and utilizes the ASCAPE framework (see Inchiosa and Parker 2002).

[4]Although our agents' nutritional requirements are denominated in terms of corn production and set to reflect the average human requirements for calories (Allen 1994), we do not infer that the Anasazi met all their caloric requirements with corn. We know that they had a diverse diet, including cultivated corn, squash, and beans, and a host of wild plants and animals, and that an exclusive corn diet could lead to several nutritional problems. For modeling purposes, however, we can subsume these resources and their distribution under a simplified resource space and single proxy (corn) for the agents' nutritional requirements.

TABLE 5.2
Household (agent) Rules

1. A household fissions when a daughter reaches the age of 16.
2. A household moves when the amount of grain in storage in April plus
   the current year's expected yield (based on last year's harvest total)
   falls below the amount necessary to sustain the household through the
   coming year.

   A. Identification of agricultural location:

      The location must be currently unfarmed and uninhabited.
      The location must have potential maize production sufficient for
        a minimum harvest of 160 kg per person per year.[a] Future
        maize production is estimated from that of neighboring sites.
        If multiple sites satisfy these criteria the location closest to the
        current residence is selected. If no site meets the criteria the
        household leaves the valley.

   B. Identification of a residential location:

      i. The residence must be within 1 km of the agricultural plot.
      ii. The residential location must be unfarmed (although it may be
          inhabited, i.e., multihousehold sites permitted).
      iii. The residence must be in a less productive zone than the agricul-
           tural land identified in A.

      If multiple sites satisfy the above criteria the location closest to the
      water resources is selected. If no site meets these criteria they are
      relaxed in order of iii then i.

[a] Allen 1994.

TABLE 5.3
Model Timing—Household "Clocks"

Each household has two internal clocks.
1. One clock tracks the number of years a household is in existence
   and determines when it fissions and dies. A household fissions when
   a daughter marries at age 16 to form a new household. Birth spacing
   is at least 2 years. A household dies once it reaches its death age, a
   parameter drawn randomly from a uniform distribution according to
   model parameters.
2. A second clock runs from April to April and reduces the amount of
   maize in storage by 13.33 kg of maize per month per individual in the
   household.

TABLE 5.4
Base Case Parameterization of the Model

Parameter	Value
Random seed	Varies
Year at model start	A.D. 800
Year at model termination	A.D. 1350
Nutritional need per individual	800 kg
Maximum length of grain storage	2 years
Harvest adjustment	1.00
Annual variance in harvest	0.10
Spatial variance in harvest	0.10
Household fission age	16 years
Household death age	30 years
Fertility (annual probability of fission)	0.125
Grain store given to new household	0.33
Maximum farm to residence distance	1,600 m
Initial corn stocks, minimum	2,000 kg
Initial corn stocks, maximum	2,400 kg
Initial household age, minimum	0 years
Initial household age, maximum	29 years

actual archaeological and environmental data so that the real and simulated population dynamics and residence locations can be visually compared. Time series plots and histograms illustrate annual simulated and actual population numbers, aggregation of population, location and size of residences by environmental zone, the simulated amounts of maize stored and harvested, and the number of households that fission, die out, or leave the valley.

In previous work (Dean *et al.* 2000) we characterized the performance of this model with respect to a "base case" parameterization (table 5.4). Although closely reproducing the qualitative features of the history of demographic changes and settlement patterns in Long House Valley, that model yielded populations that were substantially too large. All attempts to reduce the population in that model by changing agent parameters resulted in premature population collapse.

We modified this earlier model (Dean *et al.* 2000) to incorporate greater levels of both agent and landscape heterogeneity. In the previous model all agents had the same ages for the onset of fertility and death. Here, each agent gets a specific value for these ages when it is born, based on sampling from a uniform distribution. A similar procedure was applied to the household fission rate. These changes introduce six adjustable parameters, namely the endpoints of these

Table 5.5
Optimized Parameter Settings Based on Single "Runs" of the Model

Parameter/Norm	L^1	L^2	L^∞
Minimum death age	26	30	25
Maximum death age	32	39	34
Minimum age, end of fertility	30	28	30
Maximum age, end of fertility	32	30	30
Minimum fission probability	0.125	0.120	0.125
Maximum fission probability	0.129	0.125	0.125
Average harvest	0.60	0.62	0.60
Harvest variance	0.41	0.40	0.40

uniform distributions. For the production landscape, we treated two parameters as variable, the average harvest per hectare and the variance in this harvest.[5]

A systematic search of this eight-dimensional space of parameters yields values that generate model realizations having total populations closest to the historical data, according to several criteria. At each period of the model we compare the number of simulated households at time t, X_t^s, to the historical record, X_t^b. The differences between these two values are cumulated according to an L^p norm, with $p \in \{1, 2, \infty\}$ (Kolmogorov and Fomin 1977). Optimizing the model with respect to the eight adjustable parameters yields distinct "best" configurations, based on which norm was used in the simulation. The search was conducted for the best realizations as well as the best average set of runs.[6] The optimal parameter settings are summarized in tables 5.5 and 5.6 with typical output shown in figs. 5.2 and 5.3.

Simulated population levels closely follow the historical trajectory (fig. 5.2). In the first 200 years the model understates the historical population, whereas the peak population just after A.D. 1100 is somewhat too high in the model. The historical clustering of settlements along the valley zonal boundaries is nicely reproduced in the model (fig. 5.3). Although

---

[5] In the earlier version of the model, all agent heterogeneity was a consequence of local environmental variations.

[6] The model incorporates significant stochasticity, as is typical of agent models generally. Both agent initialization and aspects of agent behavior have stochastic components; therefore distinct runs of the model with different seeds to the random number generator yield distinct histories. For multiple runs of a fixed model, varying only the seeds, a "typical" run is constructed by averaging the realized populations in each period. The resulting typical run is likely never to be encountered in practice, and in some circumstances may not even be feasible, but is useful nonetheless as an idealization.

TABLE 5.6
Optimized Parameter Settings Based on the Average over 15 Runs of the Model

Parameter/Norm	L^1, L^∞	L^2
Minimum death age	30	25
Maximum death age	36	38
Minimum age, end of fertility	30	30
Maximum age, end of fertility	32	38
Minimum fission probability	0.125	0.125
Maximum fission probability	0.125	0.125
Average harvest	0.6	0.6
Harvest variance	0.4	0.4

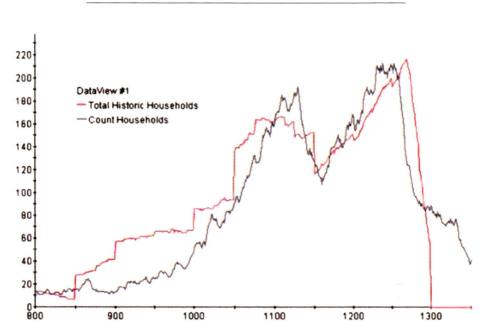

Figure 5.2. Best single run of the model according to the L^1 norm. Other best runs based on other norms yield very similar results. The average run, produced by averaging over 15 distinct runs, looks quite similar to this one as well.

the ability of the model to predict the actual location of settlements varies from year to year, with fig. 5.3 being typical, the progressive movement of the population northward over time, clear in the historical data, is also reproduced in the model.

Long House Valley was abandoned after A.D. 1300, as shown in fig. 5.2. The agent model suggests that even the degraded environment

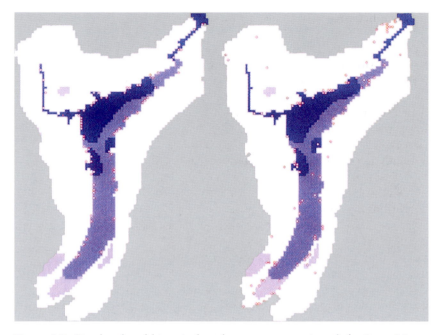

Figure 5.3. Simulated and historical settlement patterns, in red, for Long House Valley in A.D. 1125. North is to the top of the page.

of the 1270–1450 period could have supported a reduced but substantial population in small settlements dispersed across suitable farming habitats located primarily in areas of high potential crop production in the northern part of the valley. The fact that in the real world of Long House Valley, the supportable population chose not to stay behind but to participate in the exodus from the valley indicates the magnitude of sociocultural "push" or "pull" factors that induced them to move (Dean 1969). Thus, comparing the model results with the actual history helps differentiate external (environmental) from internal (social) determinants of cultural dynamics. It also provides a clue—in the form of the population that could have stayed but elected to go—to the relative magnitude of those determinants.

DISCUSSION

As noted, in these initial inquiries our models include only the most basic environmental and demographic specification, permitting calibration with a minimum number of parameters. Introducing more agent and physical heterogeneity, quite accurate results have been obtained. Richer treatments of household characteristics are possible. For example, in calculating mean household values for size, fissioning, and "death," we have envisioned disaggregating the households into individuals of varying ages in the life course.[7] Similarly, the average caloric values used can be adjusted for age of individuals within the household. Nonuniform distributions can be explored. It is, however, interesting that even without implementing these refinements, the output from the current model closely reproduces the record of the archaeological survey.

Issues remain regarding the interpretation of our findings that some inhabitants of Long House Valley could have remained after the archaeologically determined date of abandonment. The fact that environmental conditions may not have been sufficient to drive out the entire population suggests that additional push or pull factors impelled the complete abandonment of the valley after 1300. Another possibility that can be modeled in future simulations might be a combination of environmental, demographic, and epidemiological factors. That is, synergistic interactions between nutritional stress and precolonial epidemic disease might have decimated the population beyond what our model indicates. In addition, the depressed population may simply have been insufficient to maintain cultural institutions, precipitating a collective decision to leave the valley (Woods 1994). These are ripe topics for future research.

Conclusions

Our model closely reproduces important spatial and demographic features of the Anasazi in Long House Valley from about A.D. 800 to 1300. To "explain" an observed spatiotemporal history is to specify agents that generate—or grow—this history. By this criterion, our strictly environmental account of the evolution of this society during this period goes a long way toward *explaining* this history.

References

Allen, Lindsay H. 1994. Nutritional Influences on Linear Growth: A General Review. *European Journal of Clinical Nutrition* 48, suppl. 1: S75–S89.
Bak, Per, and Kim Sneppen. 1993. Punctuation Equilibrium and Criticality in a Simple Model of Evolution. *Physical Review Letters* 24:4083–86.

---

[7]Using reasonable estimations based on model life tables (Swedlund 1994) and fertility schedules (Weiss 1973) for horticultural subsistence populations would create a reasonable set of propensities, or probabilities, that can be used in future simulations.

Canup, Robin M., and Erik Asphaug. 2001. Origin of the Moon in a Giant Impact Near the End of the Earth's Formation. *Nature* (London) 412: 708–12.

Dean, Jeffrey S. 1969. *Chronological Analysis of Tsegi Phase Sites in Northeastern Arizona*. Tucson: University of Arizona Press.

Dean, Jeffrey S., Robert C. Euler, George J. Gumerman, Fred Plog, Richard H. Hevly, and Thor N. V. Karlstrom. 1985. Human Behavior, Demography, and Paleoenvironment on the Colorado Plateaus. *American Antiquity* 50: 537–54.

Dean, Jeffrey S., George J. Gumerman, Joshua M. Epstein, Robert L. Axtell, Alan C. Swedlund, Miles T. Parker, and Stephen McCarroll. 2000. Understanding Anasazi Culture Change through Agent-Based Modeling. In *Dynamics in Human and Primate Societies: Agent-Based Modeling of Social and Spatial Processes*, ed. Timothy A. Kohler and George J. Gumerman. New York: Oxford University Press.

Dean, Jeffrey S., Alexander J. Lindsay, Jr., and William J. Robinson. 1978. Prehistoric Settlement in Long House Valley, Northeastern Arizona. In *Investigations of the Southwestern Anthropological Research Group: An Experiment in Archaeological Cooperation*, ed. Robert C. Euler and George J. Gumerman. Flagstaff: Museum of Northern Arizona.

Epstein, Joshua M., and Robert L. Axtell. 1996. *Growing Artificial Societies: Social Science from the Bottom Up*. Washington, DC: Brookings Institution Press; Cambridge: MIT Press.

Euler, Robert C., and George J. Gumerman, eds. 1978. *Investigations of the Southwestern Anthropological Research Group: An Experiment in Archaeological Cooperation*. Flagstaff: Museum of Northern Arizona.

Forde, C. D. 1931. Hopi Agriculture and Land Ownership. *Journal of the Royal Anthropological Institute of Great Britain and Ireland* 61:357–407.

Gilbert, Nigel, and Rosaria Conte, eds. 1995. *Artificial Societies: The Computer Simulation of Social Life*. London: UCL Press.

Gilbert, Nigel, and Jim Doran, eds. 1994. *Simulating Societies: The Computer Simulation of Social Phenomena*. London: UCL Press.

Glatzmaier, Gary A., Robert S. Coe, Lionel Hongre, and Paul H. Roberts. 1999. The Role of the Earth's Mantle in Controlling the Frequency of Geomagnetic Reversals. *Nature* (London) 401:885–90.

Gumerman, George J. 1984. *A View from Black Mesa: The Changing Face of Archaeology*. Tucson: University of Arizona Press.

———, ed. 1988. *The Anasazi in a Changing Environment*. Cambridge: Cambridge University Press.

Gumerman, George J., and Jeffrey S. Dean. 1989. In *Dynamics of Southwest Prehistory*, ed. Linda S. Cordell and George J. Gumerman. Washington, DC: Smithsonian Institution Press.

Hack, John T. 1942. *Prehistoric Coal Mining in the Jeddito Valley, Arizona*. Cambridge, MA: Peabody Museum.

Inchiosa, Mario E., and Miles T. Parker. 2002. Overcoming Design and Development Challenges in Agent-Based Modeling Using ASCAPE. *PNAS* 99:7304–8.

Kohler, Timothy A., and George J. Gumerman, eds. 2000. *Dynamics in Human and Primate Societies: Agent-Based Modeling of Social and Spatial Processes.* New York: Oxford University Press.

Kolmogorov, A. N., and S. V. Fomin. 1977. *Introductory Real Analysis.* Ed. and trans. Richard A. Silverman. New York: Dover.

Levy, Jerrold E. 1992. *Orayvi Revisited: Social Stratification in an Egalitarian Society.* Santa Fe, NM: School of America Research Press.

Palmer, W. C. 1965. Meteorological Drought. United States Weather Bureau Research Paper 25, U.S. Department of Commerce.

Rohn, Arthur H. 1965. In *Contributions of the Wetherill Mesa Archeological Project*, ed. Douglas Osborne. Salt Lake City: Society for American Archaeology.

Swedlund, Alan C. 1994. Issues in Demography and Health. In *Understanding Complexity in the Prehistoric Southwest*, ed. George J. Gumerman and Murray Gell-Mann. Reading, MA: Addison-Wesley.

Van West, Carla R. 1994. *Modeling Prehistoric Agricultural Productivity in Southwestern Colorado: A GIS Approach.* Pullman: Washington State University, Department of Anthropology.

Weiss, Kenneth M. 1973. *Demographic Models for Anthropology.* Washington, DC: Society for American Archaeology.

Woods, James W. 1994. *Dynamics of Human Reproduction: Biometry, Biology, Demography.* New York: Hawthorne.

Chapter 6

THE EVOLUTION OF SOCIAL BEHAVIOR IN THE
PREHISTORIC AMERICAN SOUTHWEST

GEORGE J. GUMERMAN, ALAN C. SWEDLUND,
JEFFREY S. DEAN, AND JOSHUA M. EPSTEIN*

LONG House Valley, located in the Black Mesa area of northeastern Arizona (USA), was inhabited by the Kayenta Anasazi from circa 1800 B.C. to circa A.D. 1300. These people were prehistoric precursors of the modern Pueblo cultures of the Colorado Plateau. A rich paleoenvironmental record, based on alluvial geomorphology, palynology, and dendroclimatology, permits the accurate quantitative reconstruction of annual fluctuations in potential agricultural production (kg maize/hectare). The archaeological record of Anasazi farming groups from A.D. 200 to 1300 provides information on a millennium of sociocultural stasis, variability, change, and adaptation. We report on a multi-agent computational model of this society that closely reproduces the main features of its actual history, including population ebb and flow, changing spatial settlement patterns, and eventual rapid decline. The agents in the model are monoagriculturalists, who decide both where to situate their fields and where to locate their settlements.

INTRODUCTION

A central question that anthropologists have asked for generations concerns how cultures evolve or transform themselves from simple to more complex forms. Traditional study of human social change and cultural evolution has resulted in many useful generalizations concerning the trajectory of change through prehistory and classifications of types of organization. It is increasingly clear, however, that four fundamental

*The authors' affiliations are as follows: George J. Gumerman: Santa Fe Institute; Alan C. Swedlund: Department of Anthropology, University of Massachusetts; Jeffrey S. Dean: Laboratory of Tree-Ring Research, University of Arizona; and Joshua M. Epstein: Center on Social and Economic Dynamics, The Brookings Institution, and Santa Fe Institute.

This essay was previously published in *Artificial Life* 9(4): 435–444.

problems have hindered the development of a powerful, unified theory for understanding change in human social norms and behaviors over long periods of time.

The first of these problems is the use of whole societies as the unit of analysis. Group level effects, however, must themselves be explained. Sustained cooperative behavior with people beyond close kin is achieved in most human societies, and increasingly hierarchical political structures do emerge through time in many cases. Successful explanation and the possibility of developing fundamental theory for understanding these processes depend on understanding behavior at the level of the individual or the family (DeVore 1988). Among the advantages of such base-level approaches is that they allow specific modeling of peoples' behavioral ranges and norms and their adaptive strategies as community size and structure change.

Second, in addition to subsuming the behavior of individuals within that of larger social units, traditional analyses integrate environmental variability over space. Current research indicates that stable strategies for interpersonal interactions in a heterogeneous, spatially extended population may be very different from those in a homogeneous population in which space is ignored (Lindgren and Nordahl 1994). Most social interactions and relationships in human societies before the recent advent of rapid transportation and communication were local in nature.

Third, cultures have been considered to be homogeneous, tending toward maximization of fitness for their members. Little consideration was given to historical processes in shaping evolutionary trajectories or to nonadaptive aspects of cultural practice.

Finally, most discussions of cultural evolution have failed to take into account the mechanisms of cultural inheritance and the effects of changes in modes of transmission through time (Boyd and Richerson 1985; Cavalli-Sforza and Feldman 1981). Understanding culture as an inheritance system is fundamental to understanding culture change through time.

The Artificial Anasazi project is at the juncture of theory building and experimentation. We use agent-based modeling to test the fit between actual archaeological and environmental data collected over many years and simulations using various rules about how households interact with one another and with their natural environment. By systematically altering demographic, social, and environmental conditions, as well as the rules of interaction, we expect that a clearer picture will emerge as to why the Anasazi followed the evolutionary trajectory we recognize from archaeological investigation. Our long range goal is to develop agent-based simulations to understand the interaction of environment and human behavior and their role in the evolution of culture.

The Study Area

The test area for exploring the use of agent-based modeling for understanding social evolution is the prehistoric American Southwest from about A.D. 200 to 1450 using a culture archaeologists refer to as the Anasazi and a locality called Long House Valley. The Anasazi are the ancestors of the present day Pueblo peoples, such as the Hopi, the Zuni, the Acoma, and the groups along the Rio Grande in New Mexico. A commonly held view is that technological and social complexity coevolve. Anasazi cultural development underscores the interdependence of these aspects of culture. The Anasazi were a technologically simple agricultural society whose major food source was maize supplemented by beans, squash, wild plants, and game. In the A.D. 200 to 1450 period the only major technological changes that are archaeologically verifiable are agricultural intensification (terracing and ditch irrigation) and the introduction of a more efficient system for grinding maize. During this time, however, there is evidence of greatly increased social complexity. Contemporary Pueblo people have complicated social systems made up of sodalities (distinct social associations) including clans, moieties (division of the village into two units), feast groups, religious societies and cults (68 different ceremonial groups have been recorded), war societies, healing groups, winter and summer governments, and village governments. Details of the groups come from historical documents and contemporary ethnographies. The economic, religious, and social realms of Pueblo society are so tightly integrated it is difficult to understand them as separate elements of the society.

Long House Valley, a $180 \, \text{km}^2$ landform in northeastern Arizona, provides a realistic archaeological test of the agent-based modeling of settlement and economic behavior among subsistence-level agricultural societies in marginal habitats. This area is well suited for such a test for a number of reasons. First, it is a topographically bounded, self-contained landscape that can be realistically reproduced on a computer. Second, a rich paleoenvironmental record, based on alluvial geomorphology, palynology, and dendroclimatology, permits the accurate quantitative reconstruction of annual fluctuations in potential agricultural production in kilograms of maize per hectare (Dean *et al.* 2000). Combined, these factors permit the computerized creation of a dynamic resource landscape that accurately replicates actual conditions in the valley from A.D. 200 to the present. The agents of the simulation interact with one another and with their environment on this landscape. Third, tree-ring chronology provides annual calendric dating. Fourth, intensive archaeological research, involving a 100% survey of the area supplemented by limited excavations, creates a database on human behavior during the last 2,000 years that constitutes the real-world target for the modeling (Dean, Lindsay, and Robinson 1978). Finally, historical and ethnographic reports of contemporary Pueblo groups provide anthropological analogs for prehistoric human behavior.

Between roughly 7000 and 1800 B.C., the valley was sparsely occupied by people who depended on hunting and gathering. The introduction of maize around 1800 B.C. began the transition to a food-producing economy and the beginning of the Anasazi cultural tradition, which persisted until the abandonment of the region around A.D. 1300. Long House Valley provides archaeological data on economic, settlement, social, and religious conditions among a localized Anasazi population. These archaeological data provide evidence of stasis, variability, and change against which the agent-based simulation of human behavior on the dynamic, artificial Long House Valley landscape can be judged.

We have tested a large number of hypotheses about the Long House Valley Anasazi (Dean *et al.* 2000; Axtell *et al.* 2000), but we focus on only two issues here: (1) the role of environment in explaining the population dynamics of settlement placement, the large population increase after A.D. 1000, and the complete abandonment of the region around A.D. 1300; and (2) the size of simulated and actual settlements that were selected and abandoned under various environmental, demographic, and social conditions in different years.

METHODS

The Artificial Anasazi Project is an agent-based modeling study based on the Sugarscape model created by Joshua M. Epstein and Robert Axtell (1996). The project was created to provide an empirical, real-world evaluation of the principles and procedures embodied in the Sugarscape model and to explore the ways in which bottom-up, agent-based computer simulations can illuminate human behavior in a real world setting. The landscape (analogous to Sugarscape) is created from reconstructed environmental variables and is populated by artificial agents—in this case households, the basic social unit of local Anasazi society. Agent demographic and marriage characteristics and nutritional requirements are derived from ethnographic studies of historical Pueblo groups and other subsistence agriculturists.

The simulations take place on this landscape of annual variations in potential maize production values based on empirical reconstructions of low- and high-frequency paleoenvironmental variability in the study area. The production values represent as closely as possible the actual production potential of various segments of the Long House Valley environment over the period of study. In general, the reconstructed environment for maize agriculture can be characterized as dramatically improving about A.D. 1000, suffering a deterioration in the mid 1100s, and improving until the late 1200s, when there is a major environmental disruption involving the Great Drought (1276–1299), falling alluvial water table levels, severe floodplain erosion, and changes in the seasonal patterning of precipitation (Dean 1996). On this landscape, the agents of the Artificial Anasazi model play out their lives, adapting to changes in their physical and social environments.

The first step was to enter relevant environmental data, and data on real site location and size. Simulations using these landscapes vary in a number of ways. The initial population of agents (households) can be scattered randomly or placed where they actually existed at some initial year. The simulations reported here were begun with the number of agents (households) actually present in the valley during the initial year with the households distributed randomly across the artificial landscape. The environmental parameters may be left as they were originally reconstructed or adjusted to enhance or reduce maize production. Finally, and most importantly, the rules by which the agents operate may be changed. The simulation has 22 user-controlled variables that govern both agent interactions and interaction with the annually changing environment.

Agent (household) behavior on the production landscape is governed by agent attributes and a set of simple rules entrained sequentially. Standard demographic tables for subsistence agriculturalists are used to determine nutritional requirements, marriage ages and reproduction rates, and household fissioning and longevity. A household (agent) consists of five individuals, two parents and three children, each with nutritional requirements that are represented in the model by 160 kg of maize per person per year for a total requirement of 800 kg of maize per household per year. Because ethnographic data indicate that modern Puebloans try to keep at least two years' worth of corn on hand, our agents attempt to have at least two years' supply (1600 kg) in storage after the harvest in September. An internal clock tracks the amount of maize each household has in storage. This quantity is diminished each month by the amount consumed by the household and is replenished once a year by the amount harvested at the end of the growing season. The amount harvested equals the reconstructed potential production of the household's farmland minus a variable percentage that reflects fallowing, insect damage, and reservation of seed corn. Every April, each household assesses the status of its food supply, adding what it expects to have in storage by harvest time to the predicted yield of its farmplot for the coming growing season based on the previous year's production. If the expected stored amount plus the predicted yield exceeds 1600 kg, the household decides to maintain its current fields and stay where it is. If the sum is less than 1600 kg, the household decides to move to a more productive location where sufficient yield can be expected.

Movement rules for agents are triggered when a new household is created by the marriage of a resident female or when a household determines in April that the amount of stored maize plus the predicted maize production of its current farmplot cannot sustain it for the coming year. Once a household decides to move to a more productive location, it employs three sufficiency criteria for selecting new farmland: (1) the plot must be currently unfarmed; (2) the plot must be currently uninhabited; and (3) the plot must have a minimum estimated potential maize production of 160 kg of maize per household member. There are also three sufficiency criteria for selecting residential sites: (1) the site must be within 2 km of the farmplot; (2) the site must be unfarmed; and (3) the site must be less productive than the selected farmplot. If more than one site meets the sufficiency criteria, the site selected is the one with closest access to domestic water. The fact that potential residential locations need not be unoccupied allows the development of multihousehold settlements.

How closely the simulations mimic the historical data provides the most obvious test of model adequacy, or "generative sufficiency" in the terminology of Epstein (1999). We must ask: Do these exceedingly simple rules for household behavior, when subjected to the parallel computation of other agents and reacting to a dynamic environment, produce the complex behavior that actually did evolve, or are more complex rules necessary? When it is free to vary, does the population trajectory follow the reconstructed curve, and does the population aggregate into villages when we know the population actually did? Does the simulated population crash at A.D. 1300, as we know it did? Do the simulated settlement sizes and population densities closely associated with hierarchy known for the area emerge through time?

Results and Discussion

While potentially enormously informative, agent-based simulations remain theoretical constructs unless their outcomes are independently evaluated against actual cases that involve similar entities, landscapes, and behavior. The degree of fit between the results of a simulation and comparable real-world situations allows the explanatory power of the sociocultural model encoded in the simulation's structure to be objectively assessed. Lack of fit implies that the model is in some way inadequate. Such "failures" are likely to be as informative as successes, because they illuminate deficiencies of explanation and indicate potentially fruitful new research approaches. Departures of real human behavior from the expectations of a model identify potential causal variables not included in the model or specify new evidence to be sought in the archaeological record of human activities.

The most appropriate comparisons between the model and the real world begin at A.D. 400 with the same number of randomly located simulated households as in that year's actual historical situation, as well as the environmental situation as it has been reconstructed for each year. The simulation of household and field locations, as well as the size of each community (the number of households at each site), runs on an annual basis, operating under the movement rules on the changing resource landscape. A map of annual simulated field locations and household residence locations and sizes runs simultaneously with a map of the actual archaeological and environmental data so that the real and simulated population dynamics and residence locations can be compared (figures 6.1, 6.2, 6.3). In addition, time series plots and histograms illustrate annual variation in simulated and actual population numbers, aggregation of population, location and size of residences by environmental zone, simulated amounts of maize stored and harvested, and the number of households that fission, die out, or leave the valley.

Real Long House Valley.   Around 1150, largely in response to changes in productive potential, the inhabitants began to aggregate in localities particularly suitable for farming under the changing hydrologic and climatic conditions. This change in population distribution initiated a trend toward increasing sociocultural complexity, a development driven by problems resulting from increasing settlement size and population density. Among these problems are coordinating the activities of larger groups of people, task allocation, conflict resolution, and the accumulation, storage, control, and redistribution of critical resources such as food and domestic water. An important outcome of this trend was the development of a settlement hierarchy that, by A.D. 1250, involved four levels of organization: the individual habitation site, the central pueblo, the site cluster of 5 to 20 habitation sites focused on a central pueblo, and the valley as a whole. This settlement system is evident in the concentration of sites in favorable localities with empty areas in between, the structured spatial and configurational relationships among sites within clusters, and line-of-sight relationships between the clusters' central pueblos.

Artificial Long House Valley.   The simulation exhibits the demographic markers of the real situation. The greatest similarity is the development of

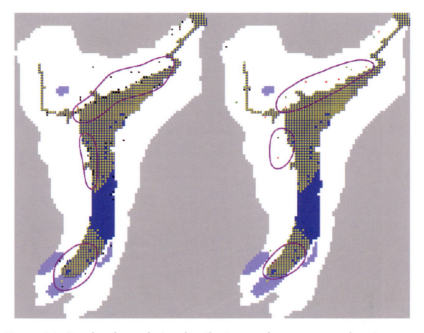

Figure 6.1. Simulated population distribution on the reconstructed environment (right) and the actual situation (left) in A.D. 1170. Yellow hatching on both sides is the simulated land under cultivation. Blue represents the depth of the water table. Darker blue represents higher water table. Colored dots represent settlements. Black = settlements of 5 or fewer households. Green = 6 to 20. Red = 21 or more. Settlements tend to be clustered in the same places, but simulated settlements are more aggregated. The position of the largest simulated settlement is within 100 m of the largest actual settlement—the red dot on the upper arm of the narrow canyon on the left. This is the actual site of Long House, after which the valley was named.

site clusters in the same localities as the actual ones (figures 6.1, 6.2) and the replication of the location and size of the site of Long House itself, indicated by arrows in figure 6.2. In the Artificial Anasazi source code, hierarchy of any kind is not explicitly modeled. However, in the historical record there is an extremely high correlation between organizational hierarchy and settlement clustering. Clustering does emerge from the model, and on this basis we guardedly infer the presence of hierarchy. Rather than producing a site organizational hierarchy in which the population is distributed across several kinds of settlement unit, the simulation tends to pack people into a few large sites that correspond to each real site cluster (figure 6.2). Given the agent rules, this seems a

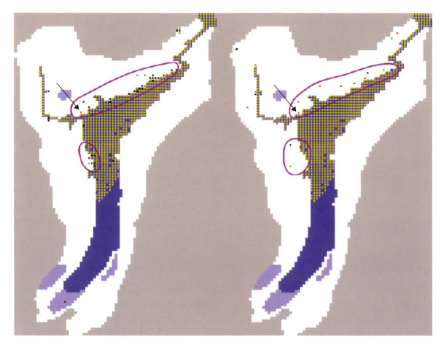

Figure 6.2. Simulated population distribution on the reconstructed environment (right) and the actual situation (left) in A.D. 1270. In both cases the population has begun to move out of the southern part of the valley because of deep erosion and a drop in the water table. Arrows indicate Long House.

reasonable fit, and population size and distribution similarities indicate that the artificial version of the complexity trajectory is in many ways equivalent to the actual situation. As shown by the smaller sites and more scattered settlements in the real valley at A.D. 1100 (figure 6.1), settlement clustering and size growth begin somewhat earlier in the model than in the actual valley. This difference likely is due to lags in the response of the real Anasazi to significant environmental changes.

By A.D. 1170 (figure 6.1), population concentrations have developed in the same localities in both the real and simulated valleys. In both cases, a large unoccupied area has appeared in the middle of the valley, and site density is much reduced along the eastern margin of the valley floor. Also in both cases, the settlement distributions result from combinations of three environmental factors: (1) the valley floor, which is subject to alluvial deposition and erosion and is therefore a poor place to establish residences; (2) arable land near which settlements can be located; and (3) domestic water resources that were concentrated along the north-western margin of the valley floor between A.D. 1130 and 1180 and

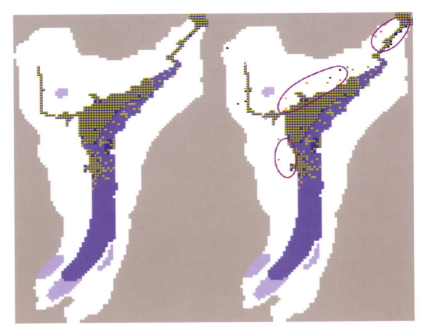

Figure 6.3. Simulated population distribution on the reconstructed environment (right) and the actual situation (left) in A.D. 1305. The actual population has abandoned the valley, but there are still settlements in the simulated version.

after A.D. 1250. Large sites in the simulation are equivalent to groups of small sites in the real world. Early in the process, neither system exhibits a hierarchical settlement structure. By A.D. 1270 (figure 6.2), the actual Long House Valley was the locus of the fully developed settlement organizational hierarchy. This development is evident in the spatial association of sites of different size (see legend) on the left image. The simulation (right image) shows less site size differentiation than the real valley, with most of the population packed into large sites. Nevertheless, some differentiation is evident along the northwestern margin of the valley. In addition, the simulation accurately captures the concentration of sites in the northern part of the valley, the clustering of sites, and the location and size of the largest actual site in the valley, Long House.

Comparing the simulated (figure 6.4) and real time trajectories of site sizes generates some provocative inferences. The number of simulated sites with more than 39 households peaks around A.D. 1100, remains high for nearly two centuries, and drops precipitously at the end of the 13th century, with the largest sites disappearing shortly after A.D. 1300. In contrast, simulated sites with fewer than 40 households maintain a

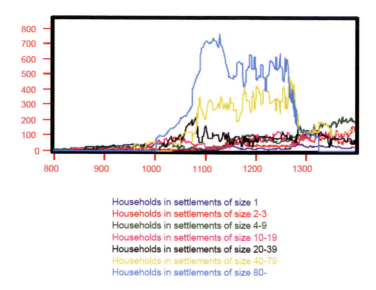

Households in settlements of size 1
Households in settlements of size 2-3
Households in settlements of size 4-9
Households in settlements of size 10-19
Households in settlements of size 20-39
Households in settlements of size 40-79
Households in settlements of size 80-

Figure 6.4. Changes in simulated settlement size. Large settlements (≥80 house-holds) develop rapidly after A.D. 1050, fluctuate in size for 200 years, and disappear abruptly after A.D. 1300. In sharp contrast, the number of smaller sites (4 to 9 households) tends to increase gradually until after A.D. 1300, when it increases more rapidly.

fairly stable profile and increase in number after the late 13th-century population crash and demise of the large settlements. While the rapid decline of the large sites mirrors the Anasazi abandonment of the real valley around A.D. 1300, the persistence of small to medium sites in the simulation contrasts sharply with the abandonment of all real sites at that time.

The different responses by the simulated and real Anasazi to the environmental crisis of the late 13th century have important explanatory implications. It has long been clear (Dean 1969) that even the seriously degraded post-A.D. 1275 environment of the valley could have supported a certain number of people and that the deleterious environmental conditions would not have forced all the Anasazi to depart. A smaller population could have sustained itself by abandoning large settlements and dispersing into smaller communities situated near the few loca-tions that remained agriculturally productive. The Artificial Anasazi do precisely that, the reduced population shifting from large, aggregated communities into smaller settlements (figure 6.4) scattered across the northern part of the valley where isolated pockets of farmable land still exist (figure 6.3). That the real Anasazi employed a different option indicates that environmental degradation was not responsible for the complete abandonment of the valley and that other, undoubtedly social, factors were involved in the final emigration. That these social factors included the unwillingness or inability to forsake the relatively high level of social complexity embedded in the hierarchical settlement system of the late 13th century for a simpler, disaggregated social system is supported by the ready dispersion of the Artificial Anasazi, who, driven primarily by environmental constraints, lacked such cultural inhibitions.

All the evidence indicates that by A.D. 1305, the real Anasazi (figure 6.3, left) had abandoned the valley. The Artificial Anasazi (figure 6.3, right), however, survived by spreading out across the part of the valley that remained productive even under the worsened environmental circumstances of the post-1300 period. This difference accurately reflects the fact that the real Anasazi could have stayed on by farming the northern valley floor and dispersing into medium-size communities (Dean 1969). The environmentally unnecessary total abandonment of the real valley undoubtedly reflects the pull of social factors drawing people to the distant communities established by previous emigrants from Long House Valley. Elements of this social attraction would have included maintaining a large enough pool of potential marriage partners, fulfilling ceremonial and social obligations to their former neighbors, and retaining achieved levels of sociocultural complexity.

CONCLUSION

In summary, agent-based models are laboratories where competing hypotheses and explanations about Anasazi behavior can be tested and judged in a disciplined, empirical way. The simple agents posited here explain important aspects of Anasazi history while leaving other important aspects unaccounted for. Site distribution and density are well approximated by the agent-based simulations. Countless simulations have been run, and the results we report here are quite robust. The hierarchical structure identified in the archaeological context can be more closely approximated with some logical modifications to the settlement rules in the simulations. The explicit modeling of hierarchical social structures is a planned topic of future model development. The departure between real Anasazi and Artificial Anasazi in the final period of settlement is a fascinating challenge. The pattern of abandonment is observed in many regions of the prehistoric Anasazi at approximately this same time.

With agent-based modeling, we can systematically alter the quantitative parameters or make qualitative changes that introduce completely new, and even unlikely, elements into the artificial world of the simulation. In terms of the Artificial Anasazi model, we can experiment with agent attributes, such as fecundity or food consumption, and we can introduce new elements, such as mobile raiders, environmental catastrophes, or epidemics. Actual environmental constraints might have been the trigger to induce many of the Anasazi to abandon the region; however, social or ideological factors were responsible for the complete abandonment of the valley. Demographic and epidemiological models may be utilized to derive additional parameters for the agent-based modeling. We have also considered synergies among variables in the real context that we have not yet experimented with in the modeling efforts. In this analysis, using this bottom-up approach to modeling prehistoric settlement behaviors, we have greatly improved our understanding of the underlying processes involved in the population dynamics.

References

Axtell, Robert L., Joshua M. Epstein, Jeffrey S. Dean, George J. Gumerman, Alan C. Swedlund, Jason Harburger, Shubha Chakravarty, Ross Hammond, Jon Parker, and Miles Parker. 2002. Population Growth and Collapse in a Multiagent Model of the Kayenta Anasazi in Long House Valley. *Proceedings of the National Academy of Sciences* 99, suppl. 3: 7275–79.

Boyd, Robert, and Richerson, Peter J. 1985. *Culture and the Evolutionary Process*. Chicago: University of Chicago Press.

Cavalli-Sforza, L. L., and M. W. Feldman. 1981. *Cultural Transmission and Evolution: A Quantitative Approach*. Princeton: Princeton University Press.

Dean, Jeffrey S. 1969. *Chronological Analysis of Tsegi Phase Sites in Northeastern Arizona*. Tucson: University of Arizona Press.

———. 1996. Demography, Environment, and Subsistence Stress. In *Evolving Complexity and Environmental Risk in the Prehistoric Southwest*, ed. Joseph A. Tainter and Bonnie B. Tainter. Reading, MA: Addison-Wesley.

Dean, Jeffrey S., George J. Gumerman, Joshua M. Epstein, Robert L. Axtell, Alan C. Swedlund, Miles Parker, and Stephen McCarroll. 2000. Understanding Anasazi Culture Change through Agent-Based Modeling. In *Dynamics in Human and Primate Societies: Agent-Based Modeling of Social and Spatial Processes*, ed. Timothy A. Kohler and George J. Gumerman. New York: Oxford University Press.

Dean, Jeffrey S., Alexander J. Lindsay, Jr., and William J. Robinson. 1978. Prehistoric Settlement in Long House Valley, Northeastern Arizona. In *Investigations of the Southwestern Anthropological Research Group: An Experiment in Archaeological Cooperation*, ed. Robert C. Euler and George J. Gumerman. Flagstaff: Museum of Northern Arizona.

DeVore, Irven. 1988. Prospects for a Synthesis in the Human Behavioral Sciences. In *Emerging Syntheses in Science*, ed. David Pines. Reading, MA: Addison-Wesley.

Epstein, Joshua M. 1999. Agent-Based Computational Models and Generative Social Science. *Complexity* 4(5): 41–60.

Epstein, Joshua M., and Robert L. Axtell. 1996. *Growing Artificial Societies: Social Science from the Bottom Up*. Washington, DC: Brookings Institution Press; Cambridge: MIT Press.

Lindgren, Kristian, and Mats G. Nordahl. 1994. Evolutionary Dynamics of Spatial Games. *Artificial Life* 1: 73–104.

GENERATING PATTERNS IN

THE TIMING OF RETIREMENT

WHAT IS THE connection between individual rationality and aggregate efficiency? And what is the role of local interactions and social networks in determining that connection? Regarding the first question, the opening Generative chapter argues that individual rationality is neither necessary nor sufficient for the attainment of macroscopic efficiency. The two are logically independent; neither implies the other. The retirement model furnishes the necessity half of the independence proof: a society of autonomous agents arrives at the economically optimal retirement behavior even though the overwhelming majority do not optimize individually. More prosaically, *the invisible hand does not require rational fingers.* In my own mind, the other half of the independence proof—individual rationality is not *sufficient* for macro efficiency—is given in the trade chapter of *Growing Artificial Societies*, where agents do maximize utility in the orthodox fashion, under evolving preferences. Equilibrium is not attained despite orthodox optimization of utility functions that are themselves orthodox at all times (Cobb-Douglas algebraically).

Turning to its specifics, the retirement model exhibits many of the core themes of the Generative chapter. Here, agents are heterogeneous by age, by social network, and by retirement status. Social interactions are local with most agents playing a coordination game (retire vs. work) with others in their network. In answer to the second question posed initially, local interactions in networks is the mechanism whereby overall optimality is attained in our population of predominantly nonoptimizing individuals. One novel feature of this model, however, is that these networks change over time—they are transient. Bounded rationality is evident in that most agents simply imitate within their dynamic network, or play a random strategy, rather than optimizing in any economic sense. That few agents optimize is, of course, consistent with a wealth of data from psychology and experimental economics. The model thus aims to provide a more plausible microfoundation for an important macroeconomic phenomenon than the optimizing representative agent picture.

The research was motivated by an empirical puzzle brought to our attention by Brookings colleagues Henry Aaron and Gary Burtless.

In truth, neither Rob Axtell nor I was thinking about retirement economics at all. But Henry and Gary quickly convinced us that there was an empirical challenge here and, more intriguing, a promising area for agent-based modeling. Although the model concerns stylized facts, and is "quasi-empirical," I dare say that, with reference to the U.S. data, it is stronger on the observed *dynamics of* retirement norms than the neoclassical efforts to date.

Chapter 7

COORDINATION IN TRANSIENT SOCIAL

NETWORKS: AN AGENT-BASED COMPUTATIONAL

MODEL OF THE TIMING OF RETIREMENT

ROBERT L. AXTELL AND JOSHUA M. EPSTEIN*

THOUGH MOTIVATED by a policy question, this work has theoretical dimensions. There are two related theoretical issues. One is the connection between individual rationality and aggregate efficiency—between optimization by individuals and optimality in the aggregate. The second is the role of social interactions and social networks in individual decision making and in determining macroscopic outcomes and dynamics. Regarding the first, much of mathematical social science assumes that aggregate efficiency requires individual optimization. Perhaps this is why bounded rationality is disturbing to some economists: they implicitly believe that if the individual is not sufficiently rational, it must follow that decentralized behavior is doomed to produce inefficiency. The invisible hand requires rational fingers, if you will.

Experimental economics and psychology have produced strong empirical support for the view that framing effects, as well as contextual and other psychological factors, create a large gap between *homo economicus* and *homo sapiens*.[1] Individual rationality is bounded. The questions we pose here are: Does that matter? How does it matter?

To answer these questions, we have developed a model in which imitation in social networks can ultimately yield high aggregate levels of optimal behavior despite extremely low levels of individual rationality. The *fraction* of agents who are rational in such an imitative system will

---

*Thanks are due George Akerlof, Chris Carroll, Bob Hall, Peyton Young, and participants in the Brookings Work-in-Progress seminar. Research assistance from Trisha Brandon and David Hines is gratefully acknowledged. This research was partially supported by the National Science Foundation, under grant IRI-9725302.

This essay was published previously in *Behavioral Dimensions of Retirement Economics*, edited by Henry J. Aaron. 1999. Washington, D.C.: Brookings Institution Press; New York: Russell Sage Foundation.
[1] See the recent review in Rabin 1998.

definitely affect the *rate* at which a steady state sets in. But the eventual (asymptotic) attainment per se of such a state need not depend on the extent to which rationality is bounded. Perhaps the main issue, then, is not how much rationality there is at the micro level, but how little is enough to generate macro-level patterns in which most agents are behaving "as if" they were rational, and how various social networks affect the dynamics of such patterns. Of particular concern are the puzzling dynamics of retirement.

In 1961 Congress reduced the minimum age at which workers could claim social security benefits from sixty-five to sixty-two. By any measure, this was a major policy shift. Yet it took nearly three decades for the modal retirement age to fall correspondingly. While various explanations are possible, we suggest that imitative behavior and social interactions—factors absent from traditional economic models—may be fundamental in explaining the sluggish response to policy.

For modeling purposes, one can represent retirement decision making (and perhaps a range of other problems) in the following stylized terms. First, there is an initial state of the world in which the individually optimal age at which to take some action is Y. Suddenly, a policy is instituted exogenously. Given this policy, the individually optimal age at which to take the action becomes $Y^* \neq Y$. What one observes, however, is not the instantaneous shift from Y to Y^* that would be predicted assuming universal, fully informed, rational behavior. Rather, there is a long process of patchy social adjustments, in which different clusters of individuals migrate to Y^* at different rates, with some groups perhaps not getting there at all. In our model, the action in question is individual retirement, the exogenously instituted policy is the 1961 congressional reduction in the age of eligibility for social security, and Y and Y^* are sixty-five and sixty-two, respectively. The actual data are plotted in figure 7.1.[2] As noted above, it took nearly three decades for the response—a downward shift in the modal retirement age from sixty-five to sixty-two—to manifest itself. We develop a relatively general model, involving imitation in social networks, that generates such patchy and sluggish dynamics. It is not the only approach possible.[3]

One body of research has sought to explain the data with aggregate models in which a representative agent solves some life-cycle optimization problem.[4] If the goal is simply to fit the data, it is not unreasonable to attribute to agents the capacity to explicitly formulate and solve

---

[2] We thank Gary Burtless for supplying these data.
[3] See, for instance, Burtless 1986.
[4] See, for example, Rust and Phelan 1997; Laibson, Repetto, and Tobacman 1998.

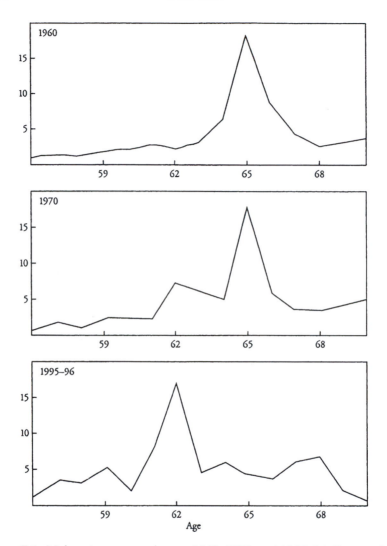

Figure 7.1. Male retirement rate by age, 1960, 1970, and 1995–96. (Source: Gary Burtless, personal communication.)

such dynamic programming problems. However, there is strong empirical evidence that humans do not perform well on problems whose solution involves backward induction (see Camerer 1997). For this reason, these models fail to provide a realistic microeconomic—that is, individualist— account of the phenomenon. We would like to provide such an account.

Our model does not invoke a representative agent but posits a heterogeneous population of individuals. Some of these behave "as if" they were fully informed optimizers, while others, indeed most, do not. Social networks and social interactions, clearly absent from the prevailing literature, play an explicit and central role.

A Model of Retirement Age Norms

The agents in our model fall into three categories. One minority group of agents adopts the (presumably) optimal policy by a process we do not model. Another minority group is composed of randomly behaving agents who retire with a fixed probability once they reach retirement age. The majority of agents are imitators who mimic members of their social networks. For lack of better terminology, we designate these groups "rationals," "randoms," and "imitators," respectively.

Agents, Cohorts, and Social Networks

The agent population is divided into age cohorts ranging from age twenty to age one hundred. Thus, there are eighty-one cohorts. Each contains C agents for a total of $81C \equiv A$ agents. Each agent is assigned a random age of death drawn from $U[60, 100]$.[5] The average death age is thus eighty. When an agent dies, it is replaced by a twenty-year-old agent.[6] In each time period each agent is activated exactly once, and if it is eligible to retire but has not yet done so, decides whether or not to retire.[7]

Agents are heterogeneous by social network; each has its own. A social network is simply a list of other agents, specified randomly and fixed over the agent's lifetime. The number of other agents is set by drawing a random network size, S, from $U[a, b]$. Some of these agents may be younger or older than the agent in question. The extent, E, represents how far the agent's social network extends above and below its own age cohort; E is drawn from $U[0, c]$. Thus one agent might have a social network of seventeen others, ranging in age from five years younger to five years older than itself, while another agent might have a social network consisting of thirteen others who are all within a year of its

---

[5]Certain variables in our model are assigned random values. In all cases below, the random variables are assumed to be uniformly distributed. The uniform (that is, rectangular) distribution on the interval $[a, b]$ is denoted $U[a, b]$.

[6]The number of cohorts, number of agents per cohort, and distribution of death ages are all easily modified in the software that we have created for this model.

[7]In the computational implementation of the model, the order of agent activation is randomized within cohorts in each period. It is commonly held that such randomization is necessary in order to suppress so-called simulation artifacts, that is, spurious correlation in the agent population.

Age

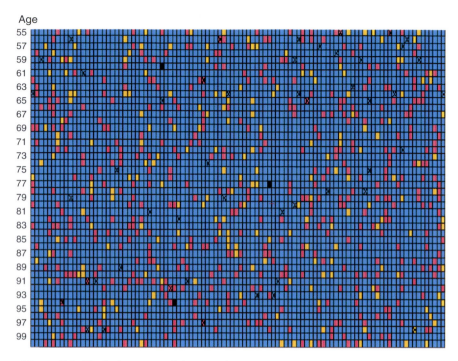

Figure 7.2. Typical agent social networks.

own age. Any two networks may or may not overlap, that is, have agents in common.

At any given time, the set of all social networks constitutes a single random graph, with the agents as nodes and the network relations as directed edges.[8] Figure 7.2 shows a variety of social networks. Each rectangle represents an agent and each row represents an age cohort, with progressively older cohorts arrayed from top to bottom in the figure. Social networks are shown with X's for the three agents who are colored black, in the sixty, seventy-seven, and ninety-four age cohorts. The twenty-four members of the sixty-year-old agent's social network include thirteen younger agents and eleven older agents.

Agent Types

As noted earlier, there are three broad types of agents. Rational agents retire at the earliest possible age allowed by government policy. Random agents retire with probability p each period, once they reach the age of eligibility for retirement.

---

[8] For more on social networks, see Kochen 1989 and Scott 1991.

Imitator agents are the most heterogeneous and interesting. Each imitator has a unique social network. Within this individual network, there is some fraction f of eligible agents who have actually retired. At each instant this is heterogeneous across agents, since the size and composition of networks are agent-specific. Agents are assigned an *imitation threshold*, τ, representing the minimum proportion of members of the agent's social network who must be retired for that agent to retire. Each agent's behavioral rule then amounts to comparing τ with f.[9] If $f \geq \tau$, the agent retires; otherwise, the agent continues working until the following period, when it reevaluates its decision.

Notionally, the imitator agents play a simple coordination game within their social networks.[10] That is, agents derive utility from coordinating their behavior with the members of their social network. At every instant each agent in the population is either working or retired. Since A is the number of agents, call $x \in \{working, retired\}^A$ the state of the population, and x_i agent i's state. Agent i's social network is denoted by N_i. Then the utility that i derives from interacting with the members of its social network in state x, $U_i(x)$, can be written

$$U_i(x) = \sum_{j \in N_i} u(x_i, x_j),$$

where $u(x_i, x_j)$ is the utility of i's interaction with j.

The function u can be thought of as the payoff function of a two-by-two symmetric game:

	work	*retire*
work	w, w	$0, 0$
retire	$0, 0$	r, r

U_i is then the payoff function of the social network game. Note that τ can be expressed in payoff terms. When an agent is young and none of its social peers are retired, $f = 0$, and the agent derives maximum utility from working. However, as its friends begin to retire ($f > 0$), the utility from retiring rises to rf, and the utility from working falls from w to $w(1 - f)$. The agent will decide to retire if f rises to a level such that $rf \geq w(1 - f)$, or equivalently, $f \geq w/(r + w)$; thus, the agent's imitation threshold τ in terms of payoffs is $w/(r + w)$.

---

[9]It makes a difference to the numerical results whether an agent considers all agents in its social network or only those who are eligible to retire. The qualitative character of the results described below, however, does not depend on this distinction.

[10]This development closely follows Young (1998, 3–4).

In this social network game, then, how does the shift to earlier retirement diffuse through coupled heterogeneous networks? And how do the dynamics vary with key parameters, such as the number of rational agents, the distribution of imitation thresholds, and the probability that a random player will retire when eligible? We resolve these questions quantitatively below by appeal to an agent-based computational model.[11] Before delving into detailed analysis of the model, however, a brief introduction to the general approach is in order.

AGENT-BASED COMPUTATIONAL MODELS

Compactly, in agent-based computational models, a population of data structures representing individual agents is instantiated and permitted to interact.[12] One then looks for systematic regularities—often at the macro level—to emerge from the local interactions of the agents. The shorthand for this is that macroscopic regularities "grow" from the bottom up. No equations governing the overall social structure are stipulated in multiagent computational models, thus avoiding any aggregation or misspecification bias. Typically, the only equations present are those used by individual agents for decision making. Different agents may have different decision rules and different information; usually, no agents have global information, and the behavioral rules involve bounded computational capacities—the agents are "simple." This relatively new methodology facilitates the modeling of agent heterogeneity, boundedly rational behavior, nonequilibrium dynamics, and spatial processes.[13] A particularly natural way to implement agent-based models is through "object-oriented" programming. Our object-oriented implementation of the present model is described in the appendix.

ESTABLISHMENT OF AN AGE SIXTY-FIVE NORM:
TWO REALIZATIONS OF THE MODEL

We begin our analysis by describing in detail two particular realizations of our model, one with a relatively large fraction of rational agents and the other with relatively few. Because the model involves stochastic elements, each realization is essentially unique, even for fixed numerical

---

[11] Coordination games on *fixed* social networks have been studied by Blume (1995) and Young (1998). But because our networks are transient, their analytical results do not apply.

[12] For extended discussions of the agent-based computational approach, see Epstein and Axtell 1996; Axelrod 1997.

[13] For more on the comparative advantages of this modeling technique, see Epstein and Axtell 1996.

values of all parameters. While we do characterize large numbers of realizations statistically below, we first focus on individual realizations, in order to build up some intuition about how the model works.

In all runs of the model described below, each cohort consists of $C = 100$ agents. Therefore the population size, A, is 8100. The size of each individual's social network is set by drawing a random number from $U[10, 25]$. Each agent's network extends up to five age cohorts above and below its own. Imitating agents have a homogeneous imitation threshold, τ, of 0.5, meaning that 50 percent of the members of an agent's social network must be retired before that agent will retire. Random agents retire with probability $p = 0.5$ each period, once they are eligible. The government age of eligibility for retirement is sixty-five, and there is no age of forced retirement.

In the first realization, 15 percent of the agents are rationals, 80 percent are imitators, and 5 percent are randoms. Animation 7.1 portrays the evolution of retirement in this society and conveys a sense of how imitation propagates the retirement decision through social networks.[14] As in figure 7.2, each agent is a rectangle. Agents are arrayed across the page by cohort and down the page by increasing age. Retired agents are shown in red and dead agents are colored white. Among the unretired agents, the pink agents are rationals, the blue agents are imitators, and the few yellow agents are randoms.

It is worthwhile to explain exactly how to "read" an animation. At the start, there are one hundred agents in each of eighty-one age cohorts, of which the eldest forty-six are displayed. So, the top row of the animation represents one hundred agents of age fifty-five. Call the upper left-hand agent Tom. In matrix notation, at time $t = 1$, Tom is cell $(1, 1)$. At $t = 2$, Tom is the cell immediately below: $(2, 1)$. In general, at time t, Tom is cell $(t, 1)$. A change in color indicates that an agent has either retired or died.

In animation 7.1, notice that a uniform retirement age of sixty-five quickly sets in, despite the fact that only a fairly small minority (15 percent) of the population arrives at this decision rationally. Figure 7.3 gives a time-series plot of the fraction of agents eligible for retirement who actually are retired. Note that this trajectory is essentially monotone. Within the first six periods, essentially all of the eligible population retires. In the second realization, the mix of agent types is changed: now only 5 percent of the agents are rationals and 90 percent are imitators. Animation 7.2 is a typical result. Note that the older cohorts show extensive fluctuation in retirement levels before the system converges to full retirement at age sixty-five. It is as if retirement

---

[14]QuickTime™ movies are available on the CD accompanying this book.

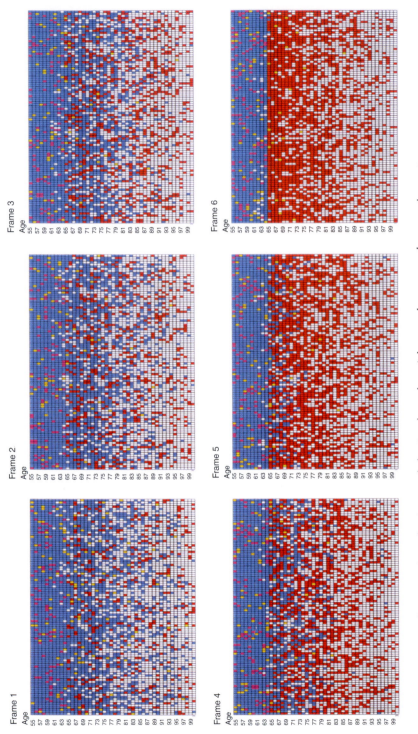

Animation 7.1. Rapid propagation of retirement behavior through social networks, no mandatory retirement age.

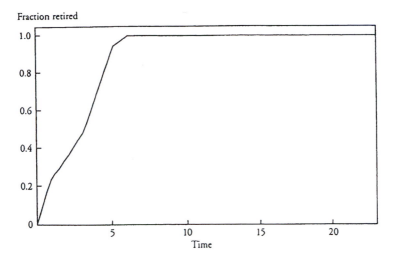

Figure 7.3. Fraction of eligible agents retired over time, typical realization, 15 percent rational agents.

"percolates up" from older to younger agents. Figure 7.4 gives the time series of the fraction of agents eligible for retirement who are retired. It takes a long time for the absorbing state to be achieved in this case. Notice that now the trajectory is not monotone.

Some Sensitivity Analysis

Each of the realizations described above yields interesting qualitative information about the model. However, in order to characterize the model's overall behavior quantitatively, it is necessary to make many realizations for a particular set of parameters and progressively build up a statistical portrait of the solution space computationally. That is, the intrinsic stochasticity of the model can be approximately characterized through a sufficiently large number of realizations. Once this is done for a particular configuration, one can study the effect of varying parameters. We first define a base case configuration of the model:

Parameter	Value
Agents/cohort (C)	100
Rational agents	10 percent
Imitative agents	85 percent
Imitation threshold (τ)	0.50
Social network size (S)	U[10, 25]
Network age extent (E)	U[0, 5]
Random agents	5 percent
p	0.50

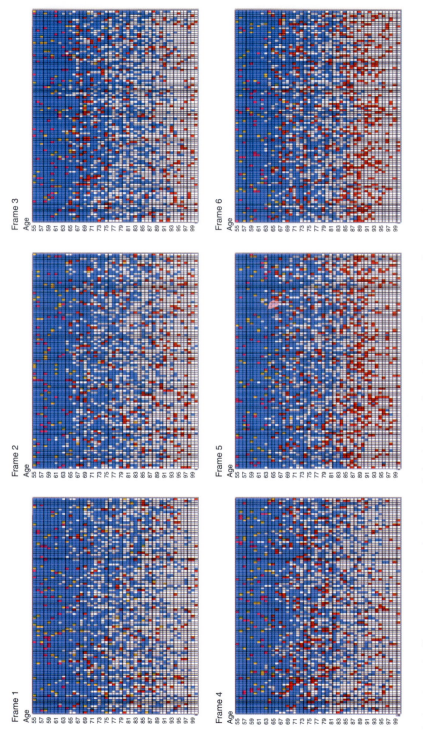

Animation 7.2. Slow propagation of retirement behavior through social networks, no mandatory retirement age.

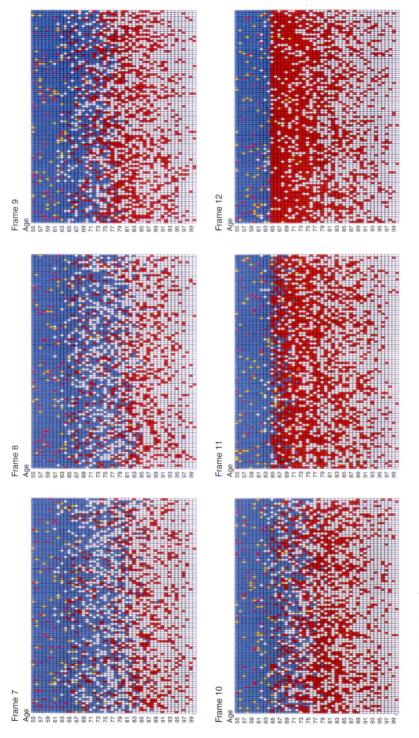

Animation 7.2. Continued.

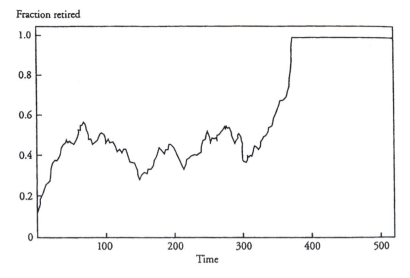

Figure 7.4. Fraction of eligible agents retired over time, typical realization, 5 percent rational agents.

We study the effect of each of these parameters on the time required for the age sixty-five retirement norm to emerge, the *transition time*. The first parameter, the number of agents per cohort (C), was found to have no effect on the average transition time for C > 100. So we begin our exploration of the model by varying the relative proportions of the three agent types—rationals, imitators, and randoms—keeping all other parameters as in the base case. We performed fifty realizations for each configuration of the model and estimated mean transition times along with standard deviations. Figure 7.5 shows the average transition times for three levels of randomly behaving agents as a function of the fraction of rationals (and hence imitators). Note that the ordinate is in logarithmic coordinates; error bars are ±1 standard deviation and are asymmetrical, due to the logarithmic scale.

Reducing the proportion of rationals, while holding constant the proportion of randoms, increases transition time. When randoms comprise 0 or 5 percent of the population, certain minimum proportions of the population must be rational for a retirement age norm to arise. For a given fraction of rationals, the transition time decreases as the proportion of randoms increases. Notice that the variances increase rapidly with transition times.

Our next sensitivity analysis concerns the effect of the imitation threshold (τ) on transition time. Since social networks are composed of individuals, the fraction of agents in a given network who are engaged

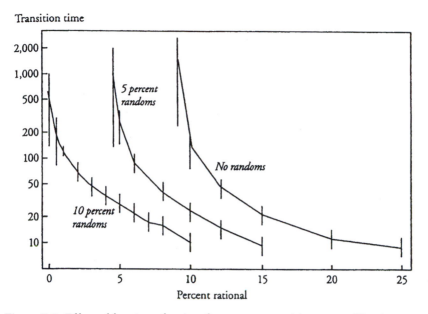

Figure 7.5. Effect of fraction of rational agents on transition to age 65 retirement norm, by fraction of randomly behaving agents.

in some behavior can only take on certain discrete values. That is, small changes in τ may have no effect on agent decision making, and thus no effect on transition times. For example, imagine that all agents have social networks of size ten. Clearly, increasing τ from 0.55 to 0.58 has no effect; agents either have five or fewer retired agents in their network or they have six or more. Only when τ is moved across a discrete boundary, such as from 0.58 to 0.62, is there an effect.

Therefore instead of studying the dependence of transition times on the average imitation threshold—surely a very "lumpy" dependence— we investigate the effect of making the threshold progressively more heterogeneous in the agent population while holding the average value of τ constant. Figure 7.6 shows how transition times depend on the standard deviation in the imitation threshold, with the average threshold fixed at 0.50. Once again, the ordinate is in logarithmic coordinates. Increasing the variance in the threshold decreases the average transition time. The reason is that in high-variance populations there are relatively more agents with low thresholds, and these agents quickly retire, leading the rest of the population to retire quickly as well. Note that when there is low variance in the imitation threshold, there is significantly more variance in transition time.

Transition time

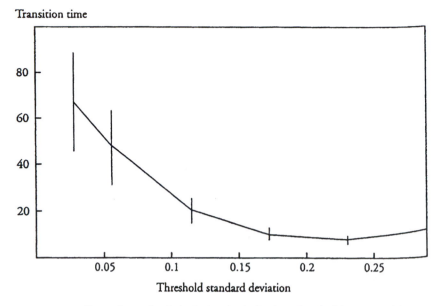

Figure 7.6. Effect of standard deviation in imitation threshold on transition to age 65 retirement norm.

The three panels of figure 7.7 explore the dependence of transition time on the size of agent networks (S). The separate (and opposite) effects of changing the average size and the size variance are presented in the first two panels. They are combined to produce the overall effect shown in the third panel. In particular, the first panel describes the effect of increasing network size, holding variance constant. Note that the time required to transit to a uniform retirement age increases very rapidly with increasing social network size; in large networks, it is difficult for a new norm to become established. The second panel gives the dependence of the transition time on the dispersion (the population standard deviation) in social network size, holding the average size constant. In this case, as the variance increases the transition time decreases, although this is a relatively weak effect. The reason for this is that the small networks catalyze the transition to a new norm, and as the variance increases, there are more small networks.

These two effects are combined in the third panel, where the abscissa, call it $\bar{S}$, represents the maximum size of any social network; that is, the size (S) of an agent's network is set by drawing a random number from U[10,$\bar{S}$]. As $\bar{S}$ increases, both the average social network size and the variance rise, and the two competing effects on transition time given in the first two panels of the figure play out, yielding the third panel.

Transition time

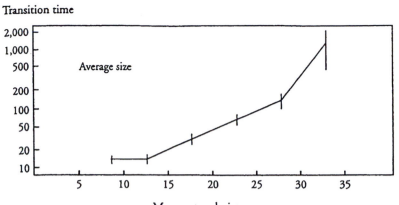

Mean network size

Transition time

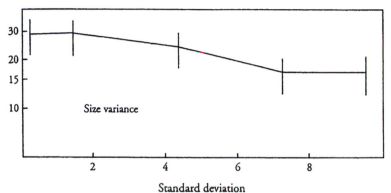

Standard deviation

Transition time

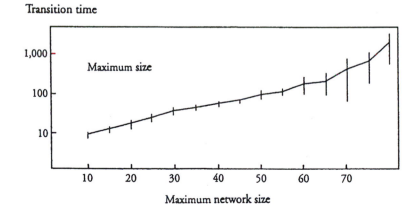

Maximum network size

Figure 7.7. Effect of size of social network on transition to age 65 retirement norm.

Transition time

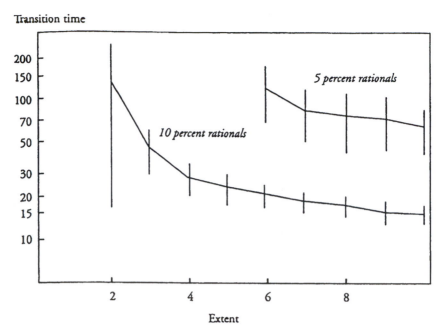

Figure 7.8. Effect of extent of social network on transition to age 65 retirement norm, by fraction of rational agents.

Overall, the general effect is that the transition time increases very rapidly with $\bar{S}$.

Next, we consider the effect of increasing the extent of agent social networks in the age cohort dimension (E). Throughout the discussion above, the maximum extent of a social network has been five cohorts above and below an agent's own cohort. Figure 7.8 shows the effect of varying this parameter. Note that increasing the extent of agent social networks in the age dimension decreases transition times. The reason for this is that networks having greater extent include older agents, who are more likely to be retired.

Dynamics and "As If"

Notice that in figure 7.6 the only variable affected by the fraction of rationals is the transition time. The attainment, per se, of the age sixty-five retirement norm is compatible with any rationality fraction above a critical level. So while in establishing the social norm, the system does behave "as if" all agents are rational, it also behaves "as if" none are! However, in taking a long time to achieve the norm it does not behave "as if" all agents are rational; indeed, it behaves as if most are not.

RESPONSE TO POLICY CHANGE: SHIFTING THE RETIREMENT AGE FROM SIXTY-FIVE TO SIXTY-TWO

Above, our model permitted agents to retire when they wished; no mandatory retirement age was in effect. We now require that all agents retire at age seventy. This increases the speed at which the age sixty-five retirement norm is established. We wish to investigate the effect of policy interventions on retirement age norms. Therefore once the norm of sixty-five is established, we throw a "policy switch" and lower the retirement age to sixty-two, mimicking the 1961 policy change by Congress. In our model, this switch means only that rationals claim benefits at age sixty-two and that randoms and imitators *may* receive benefits at age sixty-two. We measure how long it takes for a new retirement age norm to become established. Keep in mind that when the age of eligibility for social security benefits was lowered from sixty-five to sixty-two, it took nearly thirty-five years for a new norm to emerge (see figure 7.1). In animation 7.3, with rationals and randoms each constituting 5 percent of the population, τ distributed on U[0.5, 1.0], and the social networks as in the base case, a new norm emerges after some twenty periods. In short, the model replicates the sluggish adjustment that in fact occurred, at least qualitatively.

We have made many realizations of this model, varying the number of rationals in the population. The results are shown in figure 7.9. Note that the transition time to the age sixty-two norm falls as the fraction of rational agents increases. Based on this parameterization of the model, a new norm is instituted in about thirty-five periods if between 1 and 4 percent of the population responds rationally—that is to say, immediately—to the new policy. The sensitivity analyses described in figures 7.6 through 7.8 indicate how the speed of adjustment depends on other parameters in the model. In particular, we expect the time required to adjust to a policy shock to rise for less variance in the imitation threshold (τ) and for increases in the average size and extent of social networks.

This use of the agent-based computational model as a kind of laboratory in which alternative policies can be studied seems to us a fertile application of the technology, and one that has not been systematically exploited.

TWO SUBPOPULATIONS, LOOSELY COUPLED BY SOCIAL NETWORKS

Some subgroups in society may be better informed and educated than others. Such differences can affect the relative rates at which these

Frame 1

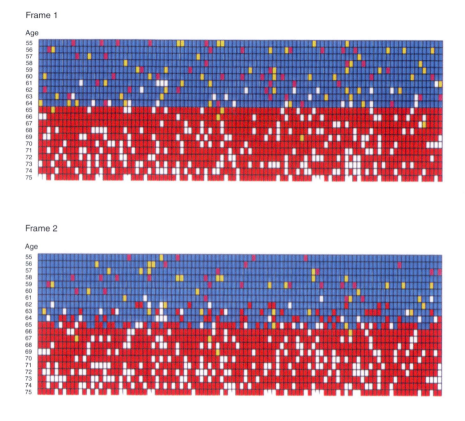

Frame 2

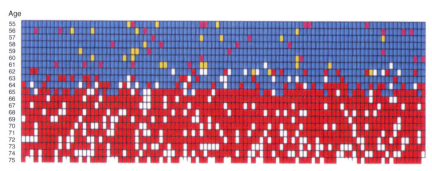

Frame 3

Animation 7.3. Propagation of retirement behavior through social networks, mandatory retirement age of 70 and policy change from earliest retirement age of 65 to 62.

Frame 4

Age

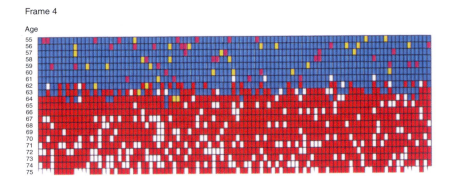

Frame 5

Age

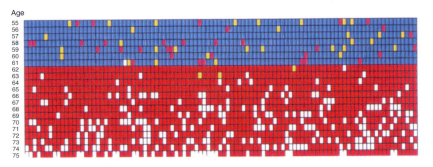

Frame 6

Age

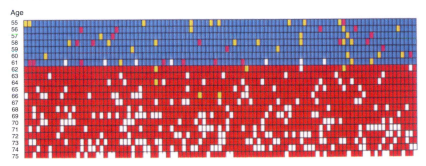

Animation 7.3. Continued.

Transition time

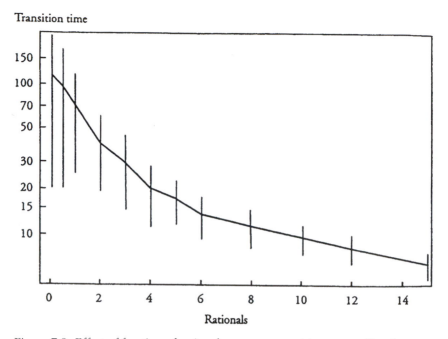

Figure 7.9. Effect of fraction of rational agents on transition to age 62 retirement norm.

communities adopt various norms. In animation 7.4 the agents have been divided into two distinct subpopulations of the same size: the agents on the left do not include any rational agents, while of the agents on the right, 10 percent are rationals. Other parameters are as in the base case configuration. The two subpopulations are coupled through their social networks as follows: 10 percent of each agent's network belongs to the other subpopulation, with the remainder belonging to its own group. We term this quantity—10 percent—the *coupling* between subpopulations. Even this rather loose coupling is sufficient for the group containing some rationals to pull the other into conformity with its retirement norm, as shown in animation 7.4.

We have studied this general effect by systematically varying the extent of coupling between subpopulations and measuring the times required for each group to reach a retirement norm of age sixty-five from an initially unretired state. The results are shown in figure 7.10; each point is an average of over fifty realizations. Note that very little coupling is needed for the nonrational subpopulation to be pulled into conformity with the more rational subpopulation.

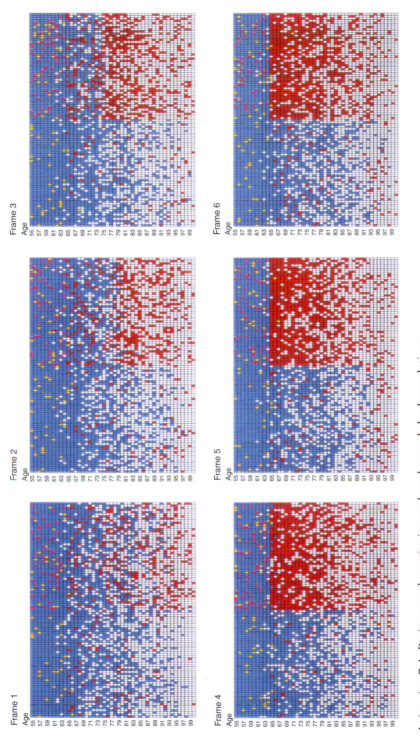

Animation 7.4. Retirement dynamics in two loosely coupled subpopulations.

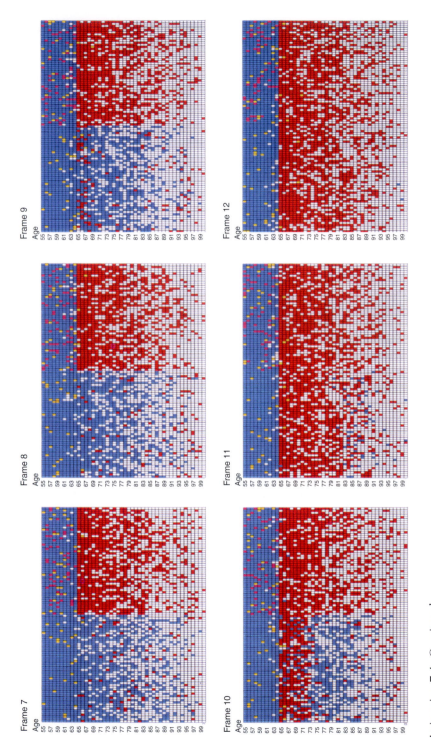

Animation 7.4. Continued.

Transition time

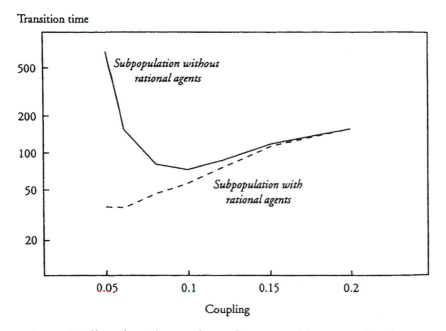

Figure 7.10. Effect of social network coupling on transition to age 65 retirement norm.

CONCLUSIONS

With social network interactions and imitative dynamics, very little individual rationality may be needed for society as a whole ultimately to exhibit optimal behavior. More pointedly, there is a large literature, experimental and theoretical, devoted to the question: how rational are individual humans? From the perspective of network imitation, it may not matter. Moreover, the nonequilibrium dynamics and the social patchiness of a response to policy will depend on both the size and the structure of networks. It is not clear how one would adapt the representative agent approach to study either of these dependencies. However, they are naturally explored within the agent-based computational framework.

This chapter has barely scratched the surface of a rich and promising area of study. Many fruitful avenues for future research suggest themselves, both analytical and computational. On the analytical side, it would be extremely useful to have—for the transient networks we describe—theorems analogous to those of Lawrence Blume and H. Peyton Young, which give conditions under which social norms will be established eventually for static networks (Blume 1995; Young 1998). Furthermore, it would be desirable to have formal expressions for the way in which transition time distributions depend on model parameters such as the fraction of imitators and size of social networks.

Computationally, it would be useful to extend our model of retirement age norms to include income shocks and imitative consumption behavior over the life cycle, as in the agent-based model of Christopher Carroll and Todd Allen (1997). We suspect that doing so would add more heterogeneity to the outcomes observed in our model. Furthermore, such a model would provide a useful laboratory in which to explore new theoretical ideas, like the effect of hyperbolic discounting, as well as to experiment with policy alternatives, like increasing the retirement age or privatizing social security.

While we have interpreted this model as applying to retirement, it could be applied to a wide range of settings in which social interactions mediate purely rational behavior. Obvious candidates include contagion behavior in markets, migration to different health plans, or the diffusion of technological innovations. In reality, these phenomena occur in social networks, while most existing models treat them either as occurring in "perfectly mixed" environments or via local interactions on regular lattices or other highly specialized topologies. The agent-based computational approach is well suited to studying such processes with any topology of interactions.

APPENDIX A. IMPLEMENTATION OF THE MODEL: AGENTS AS OBJECTS

There are many ways to computationally implement agent-based models. This can be done in any modern programming language, or with any of several mathematical or simulation software packages. However, since the model is stated in terms of individual agents, there is one idea from modern computer science that renders the implementation both transparent and efficacious: object-oriented programming.

Objects are contiguous blocks of memory that contain both data (so-called *instance variables*) and functions for modifying these data (the object's so-called *methods*). This ability of objects to hold both data and functions operating on data is called *encapsulation*.[15] Agent-based models are very naturally implemented using objects, by interpreting an object's data as an agent's state information and the object's functions as the agent's rules of behavior. A population of agents who have the same behavioral repertoire but local state information is then conveniently

---

[15] Other features of the object model, such as inheritance and polymorphism, seem to be less relevant to agent-based computational models than encapsulation.

implemented as multiple instantiations of a single agent object type or class.[16]

We have implemented the model described in the text using object-oriented programming. Not only are individual agents objects, but so too are cohorts, albeit objects of a different class than agents. In fact, it has proven convenient for the population of cohorts as a whole to be an object as well.

The agent object has a variety of state variables and behavioral methods. An agent's state information includes type (rational, imitator, or random), age, current employment status (working or retired), and the age at which it will die ("death age").[17] All of this information is stored locally, in the agent object. Each agent also keeps track of some number of other agents who are identified as its social network. These data are maintained in a social network object, described below. The agent's main decision in the present model is whether or not to retire: this is the agent object's basic *method*. This agent object specification is summarized as follows:

Pseudo-Code Block 1: Agent Object

```
OBJECT agent;
    type;
    age;
    death_age;
    alive_or_dead;
    social_network;
    working_or_retired;
    next_agent_in_agent_list;
    FUNCTION initialize;
    FUNCTION retirement_decision;
    FUNCTION draw.
```

In practice, it makes sense to implement as private some of these data and methods—that is, accessible only to the agent to whom they belong—while others are public, although this is not essential.[18]

Each social network is also conveniently implemented as an object. The size of each social network is data local to that object, as is an array

---

[16] For a discussion of the distinction between object and agent, see Jennings, Sycara, and Wooldridge 1998.

[17] The agent is assumed not to know its death age.

[18] Private data and methods are accessible only to the agent to whom they belong, unless other objects are given special access privileges.

of pointers to (that is, memory addresses of) the agents who constitute the network. Methods associated with this object include routines for determining how many agents in the network are eligible to retire, and how many are actually retired. This social network object specification is summarized as follows:

Pseudo-Code Block 2: Social Network Object

OBJECT social_network;
   size;
   array_of_agents;
   FUNCTION initialize;
   FUNCTION number_eligible_to_retire;
   FUNCTION number_retired;
   FUNCTION fraction_retired_of_eligible;
   FUNCTION draw.

Cohorts are also implemented as objects. The size of the cohort is kept as local data, as is an array of agents who constitute the cohort. The methods of this object are primarily data-gathering and statistical routines, useful in characterizing the behavior of the cohort overall. The cohort object is summarized as follows:

Pseudo-Code Block 3: Cohort Object

OBJECT cohort;
   size;
   array_of_agents;
   FUNCTION initialize;
   FUNCTION average_social_network_size_among_agents_in_cohort;
   FUNCTION number_retired;
   FUNCTION fraction_retired_of_eligible;
   FUNCTION draw.

The population of cohorts is also an object. Similar to the cohort object, it is merely an array of entities (here, cohorts), together with data gathering and statistical methods for discerning the state of the population overall.

Putting all of this together, the agent-based computational model amounts to

1. initializing all agents, social networks, and cohorts;
2. choosing an agent at random and incrementing its age;
3. checking to see if the agent has achieved its death age; if yes, then go to 2; else

4. having the agent decide whether to retire;
5. repeat 2 through 4 for all agents;
6. periodically gather and report statistics on the population.

This algorithm is summarized as follows:

Pseudo-Code Block 4: Overall Model

```
PROGRAM retirement;
    initialize agents;
    initialize social networks;
    initialize cohorts;
    repeat:
        select an agent at random;
        increment its age;
        if age < death_age then do
            retirement_decision;
        get statistics on the agents and cohorts;
    until user terminates.
```

The object model is largely responsible for the relatively compact description of this code.[19]

REFERENCES

Axelrod, Robert. 1997. *The Complexity of Cooperation.* Princeton: Princeton University Press.

Blume, Lawrence. 1995. The Statistical Mechanics of Strategic Interaction. *Games and Economic Behavior* 5:387–424.

Burtless, Gary. 1986. Social Security, Unanticipated Benefit Increases, and the Timing of Retirement. *Review of Economic Studies* 53:781–805.

Camerer, Colin. 1997. Progress in Behavioral Game Theory. *Journal of Economic Perspectives* 11(4): 167–88.

Carroll, Christopher, and Todd Allen. 1997. Learning about Intertemporal Choice. Paper presented at the Santa Fe Institute Workshop on Local Interactions Models in Economics, Santa Fe, NM.

Epstein, Joshua M., and Robert L. Axtell. 1996. *Growing Artificial Societies: Social Science from the Bottom Up.* Washington, DC: Brookings Institution Press; Cambridge: MIT Press.

---

[19]The actual source code is less than 2,000 lines of C++ and compiles in the CodeWarrior environment for the Macintosh. A Java implementation is available on the CD accompanying this book.

Jennings, Nicholas R., Katia Sycara, and Michael Wooldridge. 1998. A Roadmap
    of Agent Research and Development. *Autonomous Agents and Multi-Agent
    Systems* 1(1): 7–38.

Kochen, Manfred, ed. 1989. *The Small World*. Norwood, NJ: Ablex.

Laibson, David I., Andrea Repetto, and Jeremy Tobacman. 1998. Self-
    Control and Retirement Savings: Do 401(k)'s Help? Working paper, Harvard
    University.

Rabin, Matthew. 1998. Psychology and Economics. *Journal of Economic Litera-
    ture* 36(1): 11–46.

Rust, John, and Christopher Phelan. 1997. How Social Security and Medicare
    Affect Retirement Behavior in a World of Incomplete Markets. *Econometrica*
    65(4): 781–831.

Scott, John. 1991. *Social Network Analysis*. Newbury Park, CA: Sage.

Young, H. Peyton. 1998. Diffusion in Social Networks. Working paper, Brook-
    ings Institution.

Prelude to Chapter 8

GENERATING CLASSES WITHOUT CONQUEST

THE preceding chapter concerned norms of retirement. We turn now to norms of a different, overtly discriminatory, type. In the Classes model, initially meaningless "tags" acquire socially organizing salience: tag-based discriminatory norms arise, and persist for long periods. The Classes model illustrates once more a core theme of the Generative chapter: that simple rules of individual behavior can generate—or map up to—macroscopic regularities, in this case, class structures. Of course, social classes can arise through outright conquest and subjugation. But, it is notable that they can "self-organize" as well. Indeed, a nice title for chapter 8 would have been "Classes without Conquest."

In some respects, this self-organizing (no conquest) model seems particularly disturbing in that a discriminatory—and, in welfare terms, suboptimal—social order emerges in a population of individuals, *all of whom are behaving rationally*. In this respect, the result is reminiscent of mutual defection in the one-shot Prisoner's Dilemma Game (discussed in the subsequent chapter).

The generativist motto, we recall from chapter 1, is "If you didn't grow it, you didn't explain it." The distinction between existence and attainment—the focus of chapter 3—arises again here. While it has been proved analytically that the state with highest long-run (asymptotic) probability is the equity norm, we demonstrate computationally that the waiting time to transition from inequitable states to the equitable one can be astronomically long. In particular, the waiting time scales exponentially in a number of variables. In such cases, the transient, out-of-equilibrium dynamics are of fundamental interest, and can be studied rigorously in the multi-agent system. The model thus illustrates a useful hybrid of analytical asymptotic equilibrium analysis and computational nonequilibrium analysis.

Finally, it is worth stating outright that we do not engage in any concerted defense of the term *classes* in this research, and will readily grant that scholars steeped in the vast sociological literature on that topic may justifiably bridle at our cavalier usage. Indeed, Tom Schelling, upon seeing the model, suggested that—in the spirit of understatement—we rename it the Sneetches model, after Dr. Seuss's wonderful tale of baseless discrimination, which begins as follows:

> Now, the Star-Belly Sneetches
> Had Bellies with stars.
> The Plain-Belly Sneetches
> Had none upon thars.[1]

> Those stars weren't so big. They were really so small
> You might think such a thing wouldn't matter at all.

But, of course, just as in our model, a baseless tag matters a great deal, as Seuss recounts. For example,

> When Star-Belly children went out to play ball,
> Could a Plain-Belly get in the game...? Not at all.
> You could only play if your bellies had stars
> And the Plain Belly children had none upon thars.

Moreover, while the Sneetches ultimately—"asymptotically"—arrive in the equity norm, just as in our model the inequitable regime is long-lived:

> When the Star-Belly Sneetches had frankfurter roasts
> Or picnics or parties or marshmallow toasts,
> They never invited the Plain-Belly Sneetches,
> They left them out cold, in the dark of the beaches,
> They kept them away. Never let them come near.
> And that's how they treated them year after year.

Well, you get the idea; Tom's point is well taken. Whether what follows is a better model of Sneetches or classes, I therefore leave for you, the reader, to judge.

---

[1] Dr. Seuss, *The Sneetches and Other Stories* (New York: Random House Children's Books, 1961).

Chapter 8

THE EMERGENCE OF CLASSES IN A MULTI-AGENT

BARGAINING MODEL

Robert L. Axtell, Joshua M. Epstein, and H. Peyton Young*

Introduction

Norms are self-enforcing patterns of behavior: it is in everyone's interest to conform given the expectation that others are going to conform. Many spheres of social interaction are governed by norms: dress codes, table manners, rules of deference, forms of communication, reciprocity in exchange, and so forth. In this chapter we are interested in norms that govern the distribution of property. In particular, we are concerned with the contrast between *discriminatory norms*, which allocate different shares of the pie according to gender, race, ethnicity, age, and so forth, and *equity norms*, which do not so discriminate. An example of a discriminatory norm is the practice of passing on inherited property to the eldest son (primogeniture). Another is the custom, once common in the southern United States, that blacks should sit in the back of the bus. A third is the notion that certain categories of people (e.g., women, blacks) should receive lower compensation than others doing the same job, and in other cases that they not be given the job at all. These kinds of discriminatory norms can lead to significant differences in economic class, that is, long-lived differences in property rights based on characteristics that are viewed as socially salient.[1]

---

*For valuable comments the authors thank Sam Bowles, Jeff Carpenter, Steve Durlauf, Nienke Oomes, John Roemer, John Rust, Thomas Schelling, John Steinbruner, Leigh Tesfatsion, Frank Thompson, Erik Olin Wright, and seminar participants at Brookings, Davis, Michigan, the Santa Fe Institute, Stanford, and the University of Massachusetts (Amherst). Steven McCarroll's research assistance was invaluable. Support from the National Science Foundation under grant IRI-9725302 is gratefully acknowledged. Additional support was provided to the Center on Social and Economic Dynamics by the MacArthur Foundation.

This essay was published previously in *Social Dynamics*, edited by Steven N. Durlauf and H. Peyton Young. 2001. Cambridge, Mass.: MIT Press.

[1] For other models of classes see Roemer 1982 and Cole, Mailath, and Postlewaite 1998. This chapter differs from these by focusing on the dynamic process by which classes emerge, rather than on the equilibrium conditions that sustain them.

In this chapter we study the question of how such classes can emerge and persist, given a norm-free, classless world initially. The framework combines concepts from evolutionary game theory on the one hand and agent-based computational modeling on the other. The essential idea is to show how norms can emerge spontaneously at the social level from the decentralized interactions of many individuals that cumulate over time into a set of social expectations. Due to the self-reinforcing nature of the process, these expectations tend to perpetuate themselves for long periods of time, even though they may have arisen from purely random events and have no a priori justification. We show that social expectations gravitate to one of three conditions: (i) an equity norm in which property is shared equally among claimants, and there are no "class" distinctions; (ii) a discriminatory norm in which the claimants get different amounts based on observable characteristics that have become socially salient (but are fundamentally irrelevant); and (iii) fractious states in which norms of distribution have failed to coalesce, resulting in constant disputes and missed opportunities. In both the first and second cases, society functions efficiently in the sense that no property is wasted. There is no equity-efficiency tradeoff, just a difference in the way property rights are distributed. The third case, by contrast, is highly inefficient and may involve substantial inequality as well.

The long-run probability of being in these three different regimes can be computed using techniques from stochastic dynamical systems theory (Freidlin and Wentzell 1984; Foster and Young 1990; Young 1993a, 1998; Kandori, Mailath, and Rob 1993). But these methods are less helpful in characterizing the short- and intermediate-run behavior of these processes. Here agent-based computational techniques can play a central role, by identifying regimes that are long-lived on intermediate time scales, though not necessarily stable over very long time scales (Epstein and Axtell 1996; Axtell and Epstein 1999).

Overview of the Model

Our model of class formation is based on Young's evolutionary model of bargaining (Young 1993b). The model is bottom-up in the sense that norms emerge spontaneously from the decentralized interactions of self-interested agents.[2] In each time period two randomly chosen agents interact, bargaining over shares of available property. Their behavior, and their expectations about others' behavior, evolve endogenously based

---

[2] We use the term "emergent" as defined in Epstein and Axtell 1996 to mean simply "arising from the local interactions of agents." The term and its history are discussed at length in Epstein 1999.

on prior experiences. These expectations may be conditioned on certain visible characteristics or "tags" that serve to differentiate people. These tags have no *inherent* social or economic significance—they are merely distinguishing features, such as dark or light skin, or brown or blue eyes. Over time, however, they can acquire social significance due to path dependency effects. It might happen, for example, that blue-eyed people get a larger share of the pie than brown-eyed people due to a series of chance coincidences. The existence of these precedents causes the expectation to develop that blue-eyed people generally get more than brown-eyed people, and *a discriminatory norm* or class system emerges. Alternatively, an *equity norm* can develop in which the tags have no significance, and both sides get equal shares.

It can be shown that, asymptotically, the equity norm is more stable than any discriminatory norm. In other words, starting from arbitrary initial conditions, society is more likely to be at or near an equal sharing regime than an unequal or discriminatory one if we wait long enough. Nevertheless, metastable regimes can emerge that are discriminatory and inequitable, yet persist for substantial periods of time. These inequitable regimes correspond, roughly speaking, to situations where a discriminatory intergroup norm divides society into upper and lower classes, while a different, intragroup norm causes dissension within one (or both) of the classes. Based on many realizations of the agent-based computational model, we estimate the time it takes to exit from these discriminatory regimes as a function of the number of agents, the length of agents' memory, and the level of background noise. In this case, the waiting time increases *exponentially in memory length and the number of agents*, and can be immense even for relatively modest values of the parameters. The contrast between asymptotic and nonequilibrium results illustrates how analytical and computational methods complement one another in studying a given social dynamic.

BARGAINING

We begin by modeling a bargaining process between individual agents. Consider two players, A and B, each of whom demands some portion of a "pie," which we take as a metaphor for a piece of available property. The exact nature of the property need not concern us here. For simplicity, however, we shall suppose that the property is divisible, and that both parties have an equal claim to it a priori.[3] A posteriori differences in claim will emerge endogenously from the process itself.

---

[3] Indivisible forms of property, such as a bus seat, can be made divisible by giving the claimants equal a priori chances at being the occupant.

TABLE 8.1
The Nash Demand Game

	H	M	L
H	0,0	0,0	**70,30**
M	0,0	**50,50**	50,30
L	**30,70**	30,50	30,30

To specify the process, we must first delineate how agents solve the one-shot bargaining problem. A standard way of modeling this situation is the *Nash demand game*: each party gets his demand if the *sum* of the two demands is not more than 100 percent of the pie; otherwise each gets nothing. For instance, if employers and employees demand more than 100 percent of total revenues, negotiations break down.

To simplify the analysis, we shall suppose that each agent can make just three possible demands: low (30 percent of the pie), medium (50 percent), and high (70 percent).[4] For example, if row demands H and column plays M, their demands sum to 120 and each gets nothing. The payoffs (in percentage share) from all combinations of demands are shown in table 8.1.

This yields a coordination game in which there are exactly three pure-strategy Nash equilibria, shown in bold: (L, H), (M, M), and (H, L). While various theories have been advanced that identify a particular equilibrium as being most plausible a priori (e.g., Harsanyi and Selten 1988), we do not find these equilibrium selection theories to be especially compelling. Instead of assuming equilibrium, we wish to explore the process by which equilibrium emerges (if indeed it does) at the aggregate level, from the repeated, decentralized interactions of individuals.

The Model with One Agent Type

We begin by studying this question for a population of agents who are indistinguishable from one another, but who have different experiences (life histories) that condition their beliefs. Then we consider a population consisting of two distinct types of agents, who are differentiated by a visible "tag" (dark or light skin, brown or blue eyes) that has no intrinsic economic significance, but on which agents may condition their behavior. In the latter case, long-lived discriminatory norms can develop purely by historical chance, while this does not happen in the case

---

[4]The more general case is considered in Young 1993b.

of homogeneous agents. But in both situations, fractious regimes can emerge in which society fails to develop any coherent norm for long periods of time.

Let the population consist of N agents. Each time period consists of $[N/2]$ "matches." In each match, one pair of agents is drawn at random from the population, and they play the game in table 8.1.[5] Each agent's data about its world—its beliefs—are based on experience from previous plays. In particular, every agent remembers the demands—H, M, or L—played by each of her last m opponents, where m is *memory length*.[6] The concatenation of all agent memories defines the current *state* of the society. Behaviorally, each agent forms an expectation about her opponent's demands. She assumes that the probability of the current opponent demanding L, M, or H is equal to the relative frequency with which her previous opponents made these demands in the last m interactions. But with some relatively small probability, ε, she selects her demand randomly. Her behavior is thus a kind of "noisy best reply" to her past experience:

- With probability $1 - \varepsilon$ an agent makes a demand that maximizes her expected payoff given her expectations about the opponent's behavior. If several demands maximize expected payoff, they are chosen with equal probability.
- With probability ε the agent does not optimize but chooses one of the three demands, H, M, or L, at random.

These rules for matching, belief formation, and behavior define a particular *social dynamic* as a function of the population size N, memory length m, and error rate ε. Notice that it is a Markov process, because there is a well-defined probability of moving from any given state s to any other state s' in the next period.

In this model, agents' beliefs evolve according to their particular experiences. Thus, at any given time, the beliefs can be highly heterogeneous because agents will have had different histories of interactions with others. Importantly, moreover, these beliefs may be inconsistent with the actual state of the world. A given agent's experiences may not be representative of behavior in the whole population. For example, one agent, say A, might by chance have been matched against opponents who demanded H in each of the last m periods. Thus A will believe

---

[5]Some agents may be active more than once in a particular period, while others are inactive. On average, agents are active once per period.

[6]Some agents may have larger memories than others; that is, m may be a random variable in the agent population.

that the next opponent is likely to demand H, so she is very likely to demand L (which is a best reply to H). But another agent, say B, may have been matched against opponents who always demanded M; for this agent it makes sense to demand M. The reality, however, could be that most people in the population actually plan to demand L, in which case the beliefs of both A and B are at variance with the facts. Moreover, if A is matched against B in the next round, they will make the demands (L, M) with high probability, which is not an equilibrium of the one-shot bargaining game.

A *social norm* is a self-perpetuating state in which players' memories, and hence their best replies, are unchanging. In other words, it is a rest point or equilibrium of the dynamical system when the error term $\varepsilon = 0$. Consider, for example, the state in which everyone's experience is that opponents always demand M. Then everyone believes that her next opponent will play M. Given these beliefs, M is a best response. Assuming there are no errors ($\varepsilon = 0$), both sides demand M in the next period. Thus, agents' beliefs about opponents turn out to be correct, and this situation perpetuates itself from one period to the next. This is the *equity norm* in which everyone *expects* the other to demand one-half, and as a result everyone *does in fact* demand one-half. Note that this social norm involves no tradeoff between equity and efficiency: the solution is equitable because both sides get equal shares of each pie, and it is efficient because there is no rearrangement of shares that makes all agents better off. It can be verified that, when there are no observable differences among agents, the equity norm is the unique equilibrium of the Nash demand game and is the unique rest point of the unperturbed social dynamic.

Simplex Representation of Agent States

We represent the state of the agent population on a simplex with three differently shaded regions, as shown in figure 8.1. At each time, every agent occupies a position on the simplex that is determined by the content of her memory. For example, an agent who has encountered only agents playing L is located at the lower-right vertex of the simplex (labeled "low"). The shading within the simplex represents the best reply strategy *given* the agent's memory. That is, since each agent best replies to her memories, an agent's location on the simplex can be though of as representing her *expectation* about her opponents' play. In the white region, L is the best reply since memory configurations here are dominated by H. In the dark gray zone, the opposite occurs—memories are dominated by Hs—so L is the best reply. Agents in the light gray zone have memories for opponents playing M, so it is best for them to play M as well.

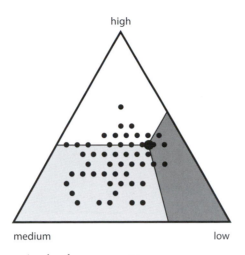

Figure 8.1. Memory simplex for one agent type.

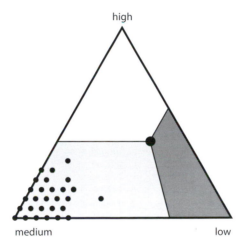

Figure 8.2. Convergence to the equity norm.

Starting from different initial states, we can examine various realiza-tions of the process.[7] Suppose, for example, that $N = 100$, $m = 10$, and $\varepsilon = 0.1$, and the initial state is random about the point of indifference between the three strategies. After eighty periods the process can evolve to the situation shown in figure 8.2.[8] In this new state, all agents

---

[7] A working version of this model is available on this book's CD.

[8] There are less than one hundred dots shown in the figures because some agents have the same memory state.

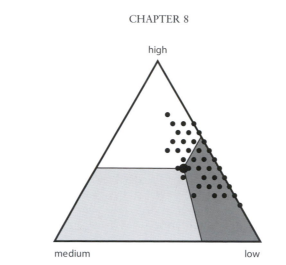

Figure 8.3. Emergence of a fractious state.

have encountered frequent demands of M in the past, and thus they expect their opponents to play M in the next period. Given this expectation, M is the best response. Hence most agents play M next period, which reinforces the expectation of M. However, by a process we do not model, agents occasionally deviate from best reply and play either H or L. This may occur due to random errors, conscious experimentation, simple imitation, or for any number of other reasons. This is analogous to mutation in biological models and serves to create variety in the population.[9]

If the process is allowed to continue from the state shown in figure 8.2, the probability is high that most agents will remain in the light gray region for quite a long period of time. This is because the equity norm has a large basin of attraction, and even substantial deviations caused by random "mutations" in individual behavior may not be enough to tip society into a fundamentally different regime. Nevertheless such tipping events will eventually occur, and they can lead to regimes that have a fundamentally different character.

Such inequitable regimes may also emerge right away when we start from a different initial state. Figure 8.3 illustrates this for one realization of the process, showing the state after 150 periods.

In this *fractious* state, people at each instant are either aggressive or passive; they have not learned to compromise. If, in one's experience,

---

[9]Each matched agent chooses randomly with probability $\varepsilon = 0.20$. However, there is a one-third chance that the random choice will in fact be the best reply, hence the probability that an "error" is realized is $0.1333\ldots$

a sufficient proportion of one's opponents are aggressive (demand H), then it is better to submit (play L) than to offer to share equally, and conversely. (It can be checked, in fact, that M is *never* a best response for someone who has never experienced an opponent who played M.) This fractious state persists in excess of 10^9 time periods, though it is neither equitable nor efficient. There is frequent miscoordination in which the players either demand too much (both play H) or demand too little (both play L) and end up leaving part of the pie on the table. In the state shown in figure 8.3, the average share of pie per person in each period is only about one-quarter, or about half the expected share under the equity norm. But while this is an inefficient state, it does not exhibit classes, because agents frequently migrate between zones, sometimes demanding H and sometimes demanding L.

Transitions between Regimes

Using asymptotic methods, it can be shown that when m and N are sufficiently large, the probability of being in the equity region is substantially higher than being in the fractious region if one waits long enough and the error rate ε is small. In the terminology of evolutionary game theory, the equity norm is *stochastically stable* (Foster and Young 1990). The intuitive reason is that it takes much longer to undo the equity norm once it is established than to undo the fractious regime once it is in place. However, the *inertia* of the system—the waiting time to reach the stochastically stable regime—can be very large indeed. Suppose that we start the agent society off in the fractious regime with $N = 10$, $\varepsilon = 0.10$, and compute the expected number of periods to transit to a neighborhood of the equity norm (i.e., to a state where all agents have at least $(1 - \varepsilon)m$ instances of M in their memories). As figure 8.4 shows, the waiting time increases exponentially in memory length. For example, when $m = 13$ it takes in excess of 10^5 periods on average for the fractious regime to be displaced in favor of the equity norm.

Similarly, the transit time increases exponentially with population size, as shown in figure 8.5.

Hence, although the equity norm is stochastically stable, the agent-based computational model reveals that—depending on the number of agents and the memory length—the waiting time to transit from the fractious regime to the equity norm may be astronomically long.[10]

---

[10] It is important to note that the expected waiting time depends crucially on the geometry of the interaction structure. In this model we have assumed that agents are paired at random from the whole society. In reality, agents interact in social networks in which

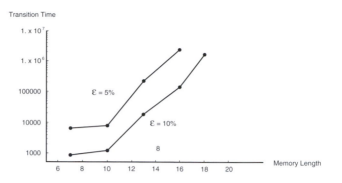

Figure 8.4. Transition time between regimes as a function of memory length, $N = 10$, various ε.

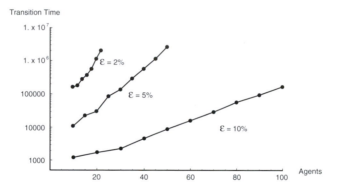

Figure 8.5. Transition time between regimes as a function of population size, $m = 10$, various ε.

Broken Ergodicity

In figure 8.4, for $m = 18$, the expected number of time periods the society must wait in order to move from the split regime to the equity norm is $O(10^6)$. In human societies, a million interactions per agent

---

there are both local (neighborhood) and global (long-range) interactions. The existence of such neighborhood structures can greatly reduce the dependence of the social learning process on population size (Ellison 1993; Young 1998). Intuitively, the reason is that a local switch in regime—say from fractious to equitable—may be relatively easy because it involves only a small number of agents (the local population size is small). Agent behavior in local interaction models is, however, quite different than in the model described here. Agents repeatedly interact with the same agents in such models, and memory plays no essential role, namely, interactions are not anonymous.

is not realizable. So how are we to interpret such large interaction requirements?

Dynamical systems that are formally ergodic but that possess sub-regions of the state space that confine the system with high probability over a long time scale are said to display *broken ergodicity*[11] with respect to that time scale.[12] Call $R_{trans}(m, N, \varepsilon)$, the rate of transition from the split state to the equity norm. For example, from figure 8.4, for $m = 18$, this rate would be approximately 10^{-6}. Now, say that the lifetime of the society is $T \ll 1/R_{trans}(m, N, \varepsilon)$. Then, to a first approximation the probability of regime transition $Pr_{trans}(T, m, N, \varepsilon) = T R_{trans}(m, N, \varepsilon)$. A system has *effective* broken ergodicity if $Pr_{trans}(T, m, N, \varepsilon) < p_0$, where p_0 is some small level of significance, say 0.001. Clearly, the exponential dependence of transition times on memory length and population size implies that our model society displays broken ergodicity.

We can summarize these results as follows. *Occasional random choices create noise in the system, which implies that no state is perfectly absorbing. However, there are two regions of the state space—one equitable, the other fractious—that are very persistent: once the process enters such a region, it tends to stay there for a long period of time. A particular implication is that, while there is only one pure equilibrium of the game (corresponding to the equity norm), it may be difficult for decentralized decision makers to discover this equilibrium from certain initial conditions. Put differently, the computation of the equity norm by a decentralized society of agents is "hard" in the sense that it takes exponential time to achieve it from some states.*[13]

Two Agent Types: The "Tag" Model

Thus far agents have been indistinguishable from one another. Even though they have different experiences that lead them to act differently, they look the same to others. Let us now suppose that agents carry a distinguishing tag (e.g., light or dark).[14] The tag is completely meaningless in that agents are identical in competence; for example, they

---

[11]The authors thank Kai Nagel and Maya Paczuski for suggesting the relevance of this concept to our results.

[12]For a review article on broken ergodicity see Palmer 1989.

[13]The view of social systems as distributed computational devices and the associated characterization of various social problems as computationally hard are developed more fully in Epstein 1999 and Epstein and Axtell 1996; see also Shoham and Tennenholtz 1996 and DeCanio and Watkins 1998.

[14]For different uses of tags and taglike devices in agent-based models, see Epstein and Axtell 1996, Holland 1996, and Axelrod 1997.

have the same amount of memory and follow the same behavioral rule
(conditional on experience). However, the presence of the tag allows
agents to condition their behavior on the tag of their opponents. To
be specific, assume that each agent records in his memory the tag of
his opponent and the demand that he made. Faced with a new dark
opponent, the agent demands an amount that maximizes the expected
payoff against his remembered distribution of dark opponents.[15] Faced
with a light opponent, the agent plays a best reply against his remembered
distribution of light opponents. All of this happens with high probability,
but with some small probability $\varepsilon > 0$ agents make random demands. In
this model, the social possibilities are richer than before, since equity or
fractiousness can prevail both between and within types.

To fix ideas, assume for the moment that there is no noise in the
agents' strategy choice ($\varepsilon = 0$). Define an *intergroup equilibrium* as a state
in which each agent in the light group demands x against members in
the dark group, each agent in the dark group demands $1 - x$ against
each opponent from the light group, and this is true for every previous
encounter that each agent remembers. An *intragroup equilibrium* is a
state in which everyone demands one-half against members of his own
group, and this is true for every previous encounter that each agent
remembers.

Using methods from perturbed Markov process theory (Young 1993a),
it can be shown that when m and N/m are sufficiently large, then
the unique stochastically stable state corresponds to the particular case
where $x = 1/2$; that is, equity prevails both between and within groups.
When ε is sufficiently small, this state or something close to it will be
observed with very high probability in the long run. But, as before,
there exist fractious states and inequitable norms that have considerable
staying power. Furthermore, the dynamics governing the emergence (and
dissolution) of intergroup norms differs from that governing intragroup
norms.

To study these dynamics computationally, we shall represent events
on two simplexes: the one on the right corresponds to agent memory
states when playing agents of the *opposite* type—it depicts the intergroup
dynamics—while the one on the left displays agent memories for playing
agents of the same type—the intragroup dynamics. Black dots refer to
dark agents, gray dots to light ones. In each run, there are a total of one
hundred agents, fifty of each type. All agents have memory length twenty
and the noise level $\varepsilon = 0.1$. The initial state differs between the runs in
order to illustrate the effects of path dependency.

---

[15] In the event that an agent has no memory of blue opponents, it picks a random strategy.

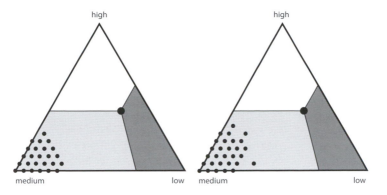

Figure 8.6. Equity between and within types.

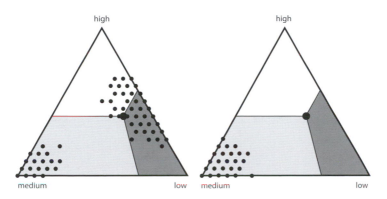

Figure 8.7. Equity between, but not within, types.

Figure 8.6 illustrates our first case. Starting from random initial conditions, it depicts the state of the system at time $t = 150$.

At this point the process has reached a state where something close to the equity norm prevails both between and within groups. In particular, the process is in the basin of attraction of the equity norm for dark against dark, light against light, and light against dark. Average payoffs in this regime are high, because most agents succeed in dividing the pie rather than fighting over it.

Figure 8.7 tells a different story. Starting from different random initial conditions, it shows the system at $t = 150$. Internally, the darks (black dots) have come close to the equity norm while the lights (gray dots) are still in a fractious state. However, something close to the equity norm prevails *between* the lights and the darks.

Classes

Yet another history unfolds in figure 8.8. In this case the process evolves fairly rapidly (after 225 periods) to a state in which the equity norm holds within each group, whereas a discriminatory norm governs relations between the two groups. When agents meet others of their own type, most of them expect to divide the pie in half. But when a dark agent meets a light agent, the darks act aggressively and the lights act passively. The result is that, on average, the payoff to dark agents (70) is over twice as high as it is to light agents (30). In other words, *class distinctions have emerged endogenously*. Once established, such class structures can persist for very long periods of time. The reason is that lights have come to expect that darks will be very demanding, so it is rational to submit to their demands. Similarly, darks have come to expect that lights will submit, so it is rational to take advantage of them.

The final case is to us the most interesting and disturbing. Starting from a different random initial state, society evolves after 260 periods to the state shown in figure 8.9. As evident in the right (inter-type) simplex, the darks dominate the lights. However, from the left simplex, it is clear that the equity norm prevails within the dominant darks while the lights are a fractious society. This, then, is the picture of a *divided underclass oppressed by a unified elite*. This result seems particularly disturbing in that every individual is behaving rationally—playing the best reply strategy—and yet the social outcome is far from optimal. Even though this regime does not correspond to a coordination equilibrium of the bargaining game (unlike fig. 8.8), it may nevertheless persist for long periods of time.

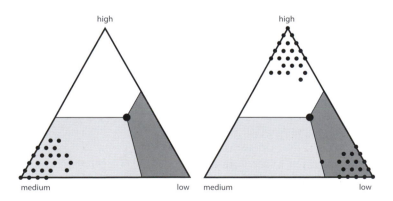

Figure 8.8. Equity within, but not between, types.

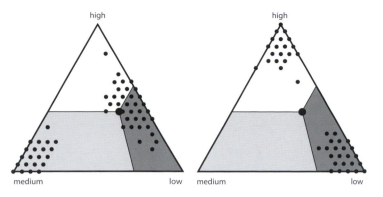

Figure 8.9. Equity above, division below.

Transition Dynamics

Figures 8.4 and 8.5 above show how the transition time from the fractious configuration to the equity state depends on the population size and memory length, for various values of ε. A similar analysis is possible for the classlike configurations displayed in figure 8.9. That is, we can start the system off in a configuration with classes and measure how long it takes to transit to the equity norm, as a function of the model parameters. We have not executed such analyses for a simple reason: even for model configurations that should be hospitable to such transitions (e.g., 10 agents of each type, $m = 10$, and $\varepsilon = 0.1$), these events are very rare, and thus difficult to systematically investigate. This is in sharp contrast to earlier results where $O(10^3)$ periods were sufficient on average for such equity transitions to occur. The "basin of attraction" of the classlike configuration is much deeper than the fractious outcome, and the transition times are correspondingly longer. It is an open problem to estimate analytically the expected duration of these transient regimes as a function of the parameters of the process.

SUMMARY

Although class systems can certainly arise through outright coercion (Wright 1985), we have argued that various kinds of social orders—including segregated, discriminatory, and class systems—can also arise through the decentralized interactions of many agents in which accidents of history become reinforced over time. In these path-dependent dynamics, society may self-organize around distinctions that are quite arbitrary from an a priori standpoint. Above, initially meaningless "tags" acquire socially organizing salience: tag-based classes emerge. Asymptotically, equity norms have an advantage over discriminatory norms. Computational analysis indicates, however, that long-lived regimes may emerge that are very far from equitable and may be highly inefficient as well.

APPENDIX: STATE SPACE OF THE MULTIAGENT BARGAINING GAME

This model admits a Markovian formulation. Briefly, call ξ the set of all possible individual memory configurations—each one a string of length m (the memory length) recording the demands (H, M, or L) made by an agent's opponents in the most recent m periods played. In a population of N agents, the state space Z of this process is the set of all possible N-tuples of ξ. The random matching and strategy choice rules then determine a Markov chain with fixed transition probabilities—that is, a $|Z| \times |Z|$ transition matrix, dependent on N, m, and the noise level ε.

The origin of the broken ergodicity displayed by this model for seemingly modest configurations—10 to 100 agents, each of whom has memory length $O(10)$—arises from the enormous dimension of the state space, Z. For memory length m and three strategies, the number of distinct memory configurations is 3^m. Generally, for S strategies there are S^m memory configurations. For N agents, since individual memories are independent, $|Z| = 3^{Nm}$; S^{Nm} generally. Therefore, the $|Z| \times |Z|$ transition matrix will have 3^{2Nm} entries, S^{2Nm} generally. However, because any individual's memory configuration can only be converted into nine others in a single interaction (S^2 others generally), the transition matrix is sparse—there are only 3^{2N} transitions possible for each state, thus only $3^{2N} \times 3^{Nm} = 3^{N(m+2)}$ entries in the transition matrix are nonzero; generally, $S^{N(m+2)}$. Table 8.A.1 gives numerical values for these various quantities as a function of m, for a population of ten agents ($N = 10$).

Even for this relatively small population size, most of these quantities are enormous. As a practical matter, a state-of-the-art workstation is not even capable of holding the $m = 2$ state vector in memory, since this would require some 6 gigabytes of RAM at two bytes per entry,

TABLE 8.A.1

	3^m	3^{Nm}	$3^{N(m+2)}$	$S^{Nm}; S = 5$
$m = 2$	9	$\approx 3 \times 10^9$	$\approx 1 \times 10^{19}$	$\approx 1 \times 10^{14}$
$m = 7$	2187	$\approx 3 \times 10^{33}$	$\approx 9 \times 10^{42}$	$\approx 8 \times 10^{48}$
$m = 10$	59,049	$\approx 5 \times 10^{47}$	$\approx 2 \times 10^{57}$	$\approx 8 \times 10^{69}$
$m = 20$	3,486,784,401	$\approx 3 \times 10^{95}$	$\approx 9 \times 10^{104}$	$\approx 6 \times 10^{139}$

TABLE 8.A.2

	3^m	3^{Nm}	$3^{N(m+2)}$	$S^{Nm}; S = 5$
$m = 2$	9	$\approx 3 \times 10^{95}$	$\approx 1 \times 10^{190}$	$\approx 1 \times 10^{139}$
$m = 7$	2187	$\approx 3 \times 10^{334}$	$\approx 9 \times 10^{429}$	$\approx 8 \times 10^{489}$
$m = 10$	59,049	$\approx 5 \times 10^{477}$	$\approx 2 \times 10^{572}$	$\approx 8 \times 10^{698}$
$m = 20$	3,486,784,401	$\approx 3 \times 10^{954}$	$\approx 9 \times 10^{1049}$	$\approx 6 \times 10^{1397}$

a conceivable although untypically large quantity of memory (ca. 1999). Furthermore, the corresponding (sparse) transition matrix is so large that it could not be stored by conventional means—its entries would therefore have to be computed as needed.

The situation is vastly worse for a population size of one hundred. Table 8.A.2 gives the number of memory configurations, dimension of the state space, and the size of the sparse transition matrix, this time for $N = 100$. These quantities are unimaginably large. However, it turns out that it is possible to shrink these sizes significantly. This is because the best reply (BR) rule of the type employed here does not use any information on the order in which past opponents' strategies were encountered. That is, for $m = 6$, memory string (H, H, H, L, L, L) is equivalent to (L, L, L, H, H, H) for purposes of BR; in each the frequency of L and H is 0.5. Because the order of an agent's memories is unimportant—at least to this variant of BR—the number of BR-distinct memory configurations is much smaller than S^m. This permits significant reduction in sizes of the state space and transition matrix of the overall Markov process. Let us call Z, where $|Z| = 3^{Nm}$, the naive state space. We explore this smaller (aggregated) state space, Z', presently.[16]

Call n_L, n_M, n_H, the number of low, medium, and high memories, respectively, that some particular agent possesses. Because these must sum to m, n_H can be written as $m - n_L - n_M$. Thus, the pair (n_L, n_M) gives all information needed by the agent in order to execute BR. Now, since each $n_{(.)} \in [0, 1, \ldots, m]$, the number of distinct memory states is simply $(m+1)+m+(m-1)+\cdots+1 = (m+1)(m+2)/2 = (m^2 + 3m + 2)/2$; for $m = 10$, the total is 66. So, $|Z'| = [(m+1)(m+2)]^N/2^N$; for $N = m = 10$ the state space has $66^{10} \approx 1.6 \times 10^{18}$ dimensions, which is smaller than the naive state space from table 8.A.1 of 5×10^{47} by approximately 3×10^{29}. A dense transition matrix for a state space of this size is $[(m+1)(m+2)]^{2N}/4^N$ in size. But for the problem at hand this is yet

---

[16] For more on aggregating Markov processes, see Howard 1971.

sparse—each state can be converted into only nine others—and thus only $3^{2N}[(m+1)(m+2)]^N/2^N$ entries need to be stored; for $N=m=10$, there are some 5×10^{27} nonzero entries, which is a massive reduction from the 3×10^{52} entries of the transition matrix associated with the naive state space.

Unfortunately, the vast reduction in the size of the state space in going from Z to Z' does not make the problem tractable computationally. In particular, consider the case of $S=3$, $N=m=10$. In this instance, there are only two recurrent communication classes (see Young 1993a, 68 for definition), one in which all agents are in state (M, M, M, M, M, M, M, M, M, M), the unperturbed equity norm—call it H_1, $|H_1|=I$—and one in which each agent has some combination of (only) Ls and Hs in memory—call this H_2, and note that there are at most $2^{100} \approx 1.3 \times 10^{30}$ of these states in the naive state space, while in Z', $|H_2|=11^{10} \approx 2.6 \times 10^{10}$. Since $|Z'|=66^{10}$, the number of states outside of both H_1 and H_2 is $66^{10}-11^{10}-1 \approx 66^{10} \approx 1.6 \times 10^{18}$. Finding the path with least total resistance between H_1 and H_2 is indeed a shortest path problem (Young 1993a, 69), but in 1.6×10^{18} vertices with approximately $9^{10} \approx 3.5 \times 10^9$ times that many edges, namely, 5.5×10^{27} total. Now, shortest path problems can be solved in an amount of time linear in the number of vertices + edges (cf. Bertsekas and Tsitsiklis 1989), but a problem of this magnitude is far beyond the scope of conventional computation.

REFERENCES

Axelrod, Robert. 1997. The Dissemination of Culture: A Model with Local Convergence and Global Polarization. *Journal of Conflict Resolution* 41:203–26.
Axtell, Robert L., and Joshua M. Epstein. 1999. Coordination in Transient Social Networks: An Agent-Based Computational Model of the Timing of Retirement. In *Behavioral Dimensions of Retirement Economics*, ed. Henry J. Aaron. Washington, DC: Brookings Institution Press; New York: Russell Sage Foundation.
Bertsekas, Dimitri P., and John N. Tsitsiklis. 1989. *Parallel and Distributed Computation*. Englewood Cliffs, NJ: Prentice-Hall.
Cole, Harold L., George J. Mailath, and Andrew Postlewaite. 1998. Class Systems and the Enforcement of Social Norms. *Journal of Public Economics* 70:5–35.
DeCanio, Stephen J., and William E. Watkins. 1998. Information Processing and Organizational Structure. *Journal of Economic Behavior and Organization* 36:275–94.
Ellison, Glenn. 1993. Learning, Local Interaction, and Coordination. *Econometrica* 61:1047–71.
Epstein, Joshua M. 1999. Agent-Based Computational Models and Generative Social Science. *Complexity* 4(5): 41–60.

Epstein, Joshua M., and Robert L. Axtell. 1996. *Growing Artificial Societies: Social Science from the Bottom Up*. Washington, DC: Brookings Institution Press; Cambridge: MIT Press.

Foster, Dean P., and H. Peyton Young. 1990. Stochastic Evolutionary Game Dynamics. *Theoretical Population Biology* 38:219–32.

Freidlin, M., and A. Wentzell. 1984. *Random Perturbations of Dynamical Systems*. New York: Springer Verlag.

Harsanyi, John C., and Reinhard Selten. 1988. *A General Theory of Equilibrium Selection in Games*. Cambridge: MIT Press.

Holland, John H. 1996. *Hidden Order: How Adaptation Builds Complexity*. Cambridge, MA: Perseus.

Howard, Ronald A. 1971. *Dynamic Probabilistic Systems*. New York: Wiley.

Kandori, Michihiro, George J. Mailath, and Rafael Rob. 1993. Learning, Mutation, and Long-Run Equilibria in Games. *Econometrica* 61:29–56.

Nash, John. 1950. The Bargaining Problem. *Econometrica* 21:128–40.

Palmer, R. 1989. Broken Ergodicity. In *Lectures in the Sciences of Complexity*, ed. Daniel L. Stein. Santa Fe Institute Studies in the Sciences of Complexity, vol. 1. Reading, MA: Addison-Wesley.

Roemer, John E. 1982. *A General Theory of Exploitation and Class*. Cambridge: Harvard University Press.

Shoham, Y., and M. Tennenholtz. 1996. On Social Laws for Artificial Agent Societies: Off-Line Design. In *Computational Theories of Interaction and Agency*, ed. Philip E. Agre and Stanley J. Rosenschein. Cambridge: MIT Press.

Wright, Erik Olin. 1985. *Classes*. Verso: London.

Young, H. Peyton. 1993a. The Evolution of Conventions. *Econometrics* 61(1): 57–84.

———. 1993b. An Evolutionary Model of Bargaining. *Journal of Economic Theory* 59(1): 145–68.

———. 1998. *Individual Strategy and Social Structure*. Princeton: Princeton University Press.

GENERATING ZONES OF COOPERATION IN

THE PRISONER'S DILEMMA GAME

THE famous two-person Prisoner's Dilemma (PD) is widely assumed to raise the problem of cooperation in an arresting way (no pun intended). Two strategies are available to each player: cooperate or defect. In the one-shot game, the dominant strategy is to defect, even though higher payoffs would accrue to each individual were they to cooperate. As when Oedipus solves the riddle of the Sphinx, individual rationality leads to a suboptimal outcome. It is a very elegant puzzle and leads to a core question in social science: How can cooperation evolve in populations whose pairwise interactions have the PD payoff structure?

One gambit has been to introduce repeated play with memory. A "strategy" then specifies a contestant's behavior given some history of interaction. The well-known Tit for Tat (TFT) strategy, for example, is a memory-one strategy: open with cooperation, but then play whatever strategy your opponent played in your last interaction. TFT (not at all uniquely) will sustain cooperation (though it is obviously defeated by Always Defect).[1] There is a large literature on cooperation-supporting strategies in the iterated PD with memory. Of course, the higher the memory, the greater the computational load. It strains credulity that humans could ever arrive at "solutions" involving high memory, so as explanations of observed human cooperation, they are not very appealing. Truth be known, the cognitive load of even one-shot play strains my credulity.

My main reason for saying so is this: I have taught the one-shot PD at Princeton. And it requires a lecture to convey, to attentive and analytical students, exactly what the dominant strategy is. Nonetheless, behavior violating it (cooperation), in populations of inattentive and highly nonanalytical folks, is seen as a central anomaly for social science. Why? We're not puzzled if the man on the street can't work out "mate in two moves" chess endings. Why are we puzzled that people don't work out the dominant strategy in the one-shot prisoner's dilemma? Obviously, if people *know* that the optimal strategy is defect, and they play cooperate

---

[1] Defect wins on play number one, and ties TFT thereafter (since both then defect eternally).

anyway, that is anomalous; but they don't know. To me, therefore, the real challenge is to build a model in which cooperation does emerge, but in which people aren't challenged to think much—indeed, at all.

Now, classical evolutionary game theory relieves the cognitive load completely. In the so-called replicator dynamics, a society of thoughtless hardwired agents can evolve precisely the strategy that a perfectly rational player would adopt in the one-shot PD. That is very elegant. But, that strategy is to *defect*, and we are still left with no explanation for cooperation, which, again, is not very appealing.

So the problem, as I saw it in developing this model, was as follows: posit the most extreme form of bounded rationality, as in the replicator dynamics, but somehow (unlike the replicator dynamics) generate cooperation. The solution is a variation I dubbed the Demographic Game, so-called because it involves spatial, evolutionary, and population dynamics.

Events transpire on a lattice (a topological torus). Agents continually move to random unoccupied sites within their finite vision, playing a fixed inherited strategy of cooperate (C) or defect (D) against neighbors. Unlike the Classes model, agents have no "tags" and are indistinguishable to one another. While payoff orderings are PD, they are *negative* for mutual defectors and those playing C against D, and positive for mutual cooperators and those playing D against C. Importantly, payoffs *accumulate* in this model. If these cumulative payoffs exceed some threshold, agents clone offspring of the same strategy onto neighboring sites and continue play. If accumulated payoffs go negative, agents are assumed to die and are removed from circulation. On these assumptions, *spatial zones of cooperation emerge*. Their stability under various assumptions (e.g., mutation rates) is explored. But here is one way to evolve zones of cooperation in populations of locally interacting boundedly rational agents. Of the points emphasized in the Generative chapter, it is perhaps the role of space that looms largest here.

Given that one evolutionary model (the replicator dynamics) returns the classic result (defect) while another equally evolutionary model (the demographic game) returns cooperate, I am left wondering whether evolutionary theory will end up saving economics or burying it. Time will tell.

Since its original publication, there has been some nice mathematical work formalizing the Demographic PD and proving theorems in agreement with the computational results I report.[2]

---

[2] Victor Dorofeenko and Jamsheed Shorish, "Dynamical Modeling of the Demographic Prisoner's Dilemma," Economic Series 124, Institute for Advanced Studies, Vienna. November, 2002.

A natural line of future work would be to develop a classification theory for demographic games generally. In a new appendix to the chapter, I present the demographic version of the two-person symmetric coordination game. If we imagine a "rules of the road" interpretation with left-hand and right-hand driving being the only strategic options, traditional game theory offers no way to adjudicate between the equally attractive pure strategy Nash equilibria: drive on the left and drive on the right. The replicator dynamics for the symmetric coordination game predicts convergence to one or the other (i.e., the mixed strategy interior equilibrium is unstable). By contrast to both the traditional and evolutionary game theory pictures, the demographic coordination game eventuates in *spatial maps divided into connected regions within each of which a specific norm prevails, with "accidents" on the region boundaries*. The agent-based demographic games approach again yields a novel result.

Chapter 9

ZONES OF COOPERATION IN DEMOGRAPHIC

PRISONER'S DILEMMA

Joshua M. Epstein*

THE EMERGENCE of cooperation in prisoner's dilemma (PD) games is generally assumed to require repeated play (and strategies such as Tit for Tat, involving memory of previous interactions) or features ("tags") permitting cooperators and defectors to distinguish one another. In the demographic PD, neither assumption is made: Agents with finite vision move to random sites on a lattice and play a fixed culturally-inherited zero-memory strategy of cooperate (C) or defect (D) against neighbors. Agents are indistinguishable to one another—they are "tagless." Positive payoffs accrue to agents playing C against C, or D against C. Negative payoffs accrue to agents playing C against D, or D against D. Payoffs accumulate. If accumulated payoffs exceed some threshold, agents clone offspring of the same strategy onto neighboring sites and continue play. If accumulated payoffs are negative, agents die and are removed. Spatial zones of cooperation emerge.

THE PRISONER'S DILEMMA

The prisoner's dilemma (PD) game raises the problem of cooperation in a stark form. Two strategies are available to each player: cooperate or defect. The payoff to mutual cooperation (R) exceeds the payoff to mutual defection (P). But the highest payoff (T) goes to one who defects against a cooperator, while the lowest payoff (S) goes to one who cooperates against a defector. The letters T, R, P, and S are used to

The author is a Senior Fellow in Economic Studies, The Brookings Institution, Washington, D.C., 20036, USA, and a member of the External Faculty, Santa Fe Institute, Santa Fe, NM, 87501, USA.

The author thanks Steven McCarroll and Ross Hammond for research assistance and Henry Aaron, Robert Axelrod, Robert Axtell, Jack Hirshleifer, Kristian Lindgren, Mats Nordahl, Miles Parker, Daniel Shapero, John Steinbruner, and Peyton Young for valuable discussions. For production assistance, he thanks Trisha Brandon and David Hines.

This essay was published previously in *Complexity* 4(2): 36–48.

denote the Temptation to defect, the Reward for mutual cooperation, the Punishment for mutual defection, and the Sucker's payoff accruing to a sole cooperator. With $T > R > P > S$, the PD payoff matrix is

$$
\begin{array}{cc}
 & \begin{array}{cc} C & \quad D \end{array} \\
\begin{array}{c} C \\ D \end{array} & \left(\begin{array}{cc} (R,\,R) & (S,\,T) \\ (T,\,S) & (P,\,P) \end{array} \right)
\end{array}
\tag{1}
$$

In a one-shot game, the dominant strategy is D for both players. Rationality yields an outcome that, from the individual's perspective, is suboptimal. Indeed, the PD is seen as posing a fundamental problem in social science (and biology): How can cooperation evolve in populations whose bilateral interactions are governed by the PD?

REPEATED GAMES

One important line of attack has involved repeated play. This is a bit of a misnomer—it really means repeated play *with memory*. The distinction is crucial for this research. If A plays B repeatedly with no recollection of previous engagements (as in the demographic game below), that is *not* a repeated game. A "strategy" in a repeated game is a rule that specifies a player's behavior given some history of interactions against the opponent in question. For example, the well-known Tit for Tat (TFT) strategy is: cooperate on the first move; thereafter, adopt whatever strategy your opponent played in your last interaction. TFT is a *memory one* strategy and will sustain cooperation (Axelrod 1984). The iterated PD must be of indeterminate length or there is regress to pure defection. An extensive literature has developed cooperation-supporting strategies for the repeated PD.[1] An important issue is the computational complexity of various strategies.[2] These can involve large memory and what, for humans, might be considered high computational loads. (Memory can be defined as the number of states in a finite automaton that implements the strategy. See Papadimitriou and Yannakakis 1994.)

---

[1]Axelrod 1984; Albin and Foley 1997; Binmore 1992; Guttman 1996; Hirshleifer and Martínez Coll 1988; Lindgren and Nordahl 1994b; Lomberg 1996; Martínez Coll and Hirshleifer 1991; Miller 1996; Rubinstein 1986.

[2]For a survey, see Kalai 1990. See also Rubinstein 1986; Abreu and Rubinstein 1988; Ben-Porath 1990; Binmore 1987, 1988; Binmore and Samuelson 1992; Deng and Papadimitriou 1994; Gilboa 1988; Neyman 1985; Papadimitriou 1992.

EVOLUTIONARY GAMES

At the other end of the cognitive requirements spectrum is classical evolutionary game theory (Smith 1982; Weibull 1995). Here, agents do not consciously optimize over strategic alternatives. Rather, they inherit a fixed strategy (a phenotype) and then replicate depending on that strategy's payoff (fitness). One elegant feature of these *replicator dynamics* is that their evolutionarily stable strategies (ESSs) can correspond to the strategies that would be adopted by fully informed rational players of the game. (For a thorough development of this equilibrium concept, see Binmore 1992.) For the PD specifically, the classical replicator dynamics (see below) lead straight to a world of pure defection— precisely the dominant strategy in one-shot rational play. In a social science context, the basic evolutionary reasoning is nicely set forth by Frank (1988).

His exposition begins with the assumption "that everyone in the population is one of two types—cooperator or defector. A cooperator is someone who, possibly through intensive cultural conditioning, has developed a heritable capacity to experience a moral sentiment that predisposes him to cooperate. A defector is someone who either lacks this capacity or has failed to develop it" (Frank 1988). From a modeling standpoint, then, individuals are "hard wired" to execute a fixed strategy. Now, Frank continues, "suppose, for argument's sake, that cooperators and defectors look exactly alike, thus making it impossible to distinguish the two types. In this hypothetical ecology ... individuals will pair at random .... The expected payoffs to both defectors and cooperators therefore depend on the likelihood of pairing with a cooperator, which in turn depends on the proportion of cooperators in the population" (Frank 1988). (In the model below, agents look indistinguishable to one another, but do not pair at random.) If cooperators comprise a fraction c of the population and individuals are paired randomly, then the probability of a given cooperator being paired with another cooperator is c; the probability of her being randomly paired with a defector is $(1 - c)$. Expected payoffs in the two cases are $E[C] = cR + (1 - c)S$ and $E[D] = cT + (1 - c)P$. For expository purposes, we will assume the payoffs below. (Frank uses different numerical values. These do not affect Frank's argument and will be used for illustrative purposes throughout. These values satisfy the common additional condition that $2R > T + S$.)

$$
\begin{array}{cc}
 & \begin{array}{cc} C & D \end{array} \\
\begin{array}{c} C \\ D \end{array} & \left(\begin{array}{cc} (5, 5) & (-6, 6) \\ (6, -6) & (-5, -5) \end{array} \right)
\end{array}
\qquad (2)
$$

Then, we have average returns of $E[D] = -5 + 11c$ for defectors and $E[C] = -6 + 11c$ for cooperators. For all c in $[0, 1]$, $E[D] > E[C]$ and Frank's summary claims apply: "Since defectors always receive a higher payoff here, their share of the population will grow over time. Cooperators, even if they make up almost all of the population to begin with, are thus destined for extinction. When cooperators and defectors look alike, genuine cooperation cannot emerge" (Frank 1988). (This is one motivation for introducing features, "tags," allowing cooperators and defectors to distinguish one another. On tags, see Holland 1995. See also Tesfatsion 1995.)

The same conclusion follows in the closely related classical replicator dynamics (Smith 1982; Weibull 1995; Taylor and Jonker 1978; Hofbauer and Sigmund 1988; Samuelson 1997; Young 1993; Young and Foster 1991). Here, the relative frequency z_i of a strategy i grows in accordance with

$$\frac{dz_i}{dt} = z_i \left[(\text{Return to pure } i) - (\text{Average Return}) \right] \qquad (3)$$

For a symmetric payoff matrix A, this becomes

$$\frac{dz_i}{dt} = z_i [(Az)_i - z^T Az], \qquad (4)$$

where, for n strategies,

$$z = (z_1, \ldots, z_n).$$

For our two-strategy PD, this z vector is simply $(x, 1-x)$, where $x \in [0, 1]$ is the relative frequency of defection. For the symmetric payoff matrix (1) above, A takes the form

$$A = \begin{pmatrix} R & S \\ T & P \end{pmatrix}$$

and, expanding (4), the dynamics are given by

$$\frac{dx}{dt} = (T - R)x + (P + 2R - S - 2T)x^2 + (T + S - P - R)x^3,$$

a cubic whose equilibria are $x = 0$, $x = 1$, and $x = (R-T)/(R-T+P-S)$. For PD payoff orderings, $x = 0$ (defection frequency zero) has eigenvalue $T - R > 0$, making it unstable, while $x = 1$ (universal defection) has eigenvalue $S - P < 0$ making it asymptotically stable. With the numerical

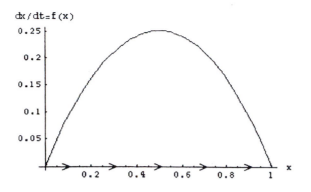

Figure 9.1. Phase diagram for PD replicator dynamics.

assumptions from (2), A becomes

$$A = \begin{pmatrix} 5 & -6 \\ 6 & -5 \end{pmatrix} \tag{5}$$

With this A matrix, the phase diagram for the replicator dynamics (4) on [0,1] is given in figure 9.1.

Just as in Frank's example, even the slightest perturbation from pure cooperation ($x = 0$) results ultimately in pure defection ($x = 1$). (In a more deliberate development, one would distinguish the evolutionarily stable strategies from the asymptotically stable equilibria. However, for $n = 2$, they are the same. See Binmore 1992.)

DEMOGRAPHIC GAMES

I wish to extend this evolutionary literature with a class of agent-based models that I call *demographic games*. In the demographic PD, (1) agents inherit a fixed strategy of cooperate or defect and (2) agents are indistinguishable to one another, just as in the models above. Yet, *cooperation can emerge and endure.*

Demographic games seems an appropriate name for this class of models because they involve spatial, evolutionary, and population dynamics. (I am using the term "demography" in its broad sense, as including migration.) The space on which agents interact is a 30-by-30 lattice of sites. Periodic boundary conditions obtain. Visually, agents who exit the lattice on the right/bottom reenter from the left/top. (Topologically, the space is a torus.) Agents move around this space, interact with von Neumann neighbors, and have offspring. (They are not fixed cellular automaton sites, as in Nowak and May 1992; Nowak, May, and Sigmund 1995; Feldman and Nagel 1993; Lindgren and Nordahl 1994a.) Each agent is an object whose main attributes are vision, wealth, age, and strategy. Vision is the distance an agent can see, looking north, south, east, or west. In the evolutions presented below, vision is one. Agents are born with a strategy of cooperate (C) or defect (D). The agents' sole rule of behavior is as follows: *Choose a random unoccupied site within your vision; go there and play your strategy against each neighbor.* (If there is no unoccupied site within the agent's vision, it remains in place and plays all current neighbors.) The agent and the neighbor receive payoffs from some game matrix, here the PD bimatrix (1), where $T > R > P > S$. For the demographic PD, we introduce negative payoffs. (A discussion of this assumption and an equivalent formulation involving non-negative payoffs are presented below.) Specifically, we require that $T > R > 0 > P > S$, as in (2), which we shall use here.

Payoffs (wealths) accumulate. Since our payoff matrix has negative entries (Rapoport 1996), an agent's accumulated wealth may go negative. In that event the agent "dies"—it is removed from play and its wealth disappears. By the same token, if an agent's accumulated wealth exceeds some positive threshold and there is an unoccupied site within the agent's vision, the agent has an offspring, who begins life on one of these sites with a nominal initial endowment subtracted from the parent's wealth. (By contrast with this use of a simple threshold, the standard replicator dynamic assumes that a phenotype's [i.e., a strategy's] frequency grows as the difference between the strategy's fitness and the population's average payoff. For the runs discussed here, the threshold is 10 units of accumulated payoff and offspring inherit an initial endowment of 6 units. An agent's initial age is a random integer between one and the maximum age.) Here (with vision one) these are neighboring sites. Progeny are born with a fixed strategy; in the zero mutation case, it is the parent's strategy. (Oliphant [1994] also studies the evolution of cooperation in a spatial model of the noniterated PD. That interesting model differs from the demographic game in fundamental respects, including the use of a genetic algorithm, a fitness-biased probability of individual play, fixed population size [no population dynamics], and gaussian interaction probabilities on a different [one-dimensional] space.)

Perhaps it is worth emphasizing that, in adopting this assumption of a fixed agent strategy, we are not claiming that human strategies are literally hard-wired genetically. Rather, for modeling purposes, we are assuming that they are culturally transmitted from parents to children— vertically transmitted—with high fidelity, like certain religious or ethnic affiliations, tastes, and native tongues. (On cultural transmission, see Cavalli-Sforza and Feldman 1981; Boyd and Richerson 1985.)

Below we consider the effect of degradation (mutation) in this vertical transmission fidelity. In a more elaborate model, horizontal (intragenerational) transmission would also be included. (For models including horizontal transmission, see Epstein and Axtell 1996; Axelrod 1997b.) However, here we follow Shubik's injunction: "Start with radical simplification . . . do not reject the simplest models because they are *a priori* too simple. Reject them when a quick investigation shows that the phenomenon of interest to you cannot appear at this level of simplicity" (Shubik 1996). The phenomenon of interest here is persistent cooperation, and we demonstrate that an extremely simple spatial model with vertical transmission suffices to generate it.

Run 1. No Maximum Age, Zero Mutation

For our first run of the model, we impose no upper bound on agent lifespans. Initially, 100 agents are assigned random fixed strategies (C or D) and random initial positions. Numerical and other assumptions are collected in table 9.1. The five panels of figure 9.2 show the spatial situation at selected times illustrative of the main points. Cooperators are colored blue, defectors red. A time step represents one cycle through

TABLE 9.1
Run 1 Assumptions[a]

T	6
R	5
P	−5
S	−6
Space	30 × 30 torus
Neighborhoods	von Neumann
Vision	1
Mutation Rate	0
Metabolism	0
Maximum Lifetime	None
Accumulation Needed to Clone	10
Initial Endowment to Offspring	6
Initial Number of Agents	100
Initial Locations	Random
Initial Strategies	Random
Updating	Asynchronous
Call Order	Random

[a]Departures from these assumptions are noted in the text.

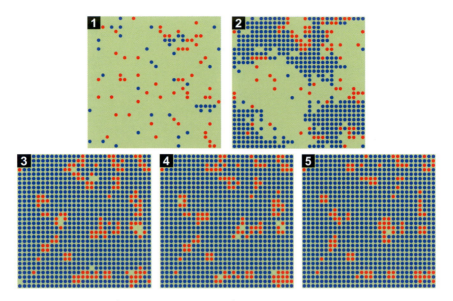

Figure 9.2. An evolution to cooperative dominance.

the agent list, which is randomized after every cycle. (Randomization of the agent call order works as follows: Agent objects are held in a doubly linked list and are processed serially. If there are N agents, a pair of agents is selected at random and the agents swap positions in the list. This random swapping is done $N/2$ times after each cycle.) The agents are updated asynchronously. On the importance of asynchronous updating, see Huberman and Glance 1993. (Regarding the Huberman and Glance [1993] critique of Nowak and May [1992], my model shares with Nowak and May the assumption that agents play zero-memory strategies of always cooperate or always defect. It differs in many other respects, including agent movement, population dynamics, mutation, accumulations and negative payoffs, and, notably, asynchronous updating. As Huberman and Glance point out, Nowak and May's main results on the persistence of cooperation depend crucially on synchronous updating. When, *ceteris paribus*, Huberman and Glance introduce asynchronous updating into the Nowak and May model, the result is convergence to pure defection. Like Huberman and Glance, my model employs asynchronous updating, but—by mechanisms different from Nowak and May's—it generates persistent cooperation for many parameter settings. In that sense, it supports the general result obtained by Nowak and May, that cooperation can flourish in populations of simple agents.)

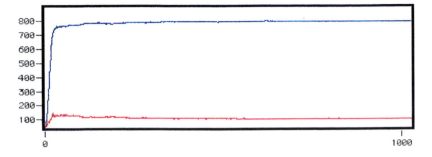

Figure 9.3. Typical aggregate population time series. Note: Blues are cooperators, reds are defectors.

Panel 1 of figure 9.2 gives the early situation, with a few random agents scattered about the lattice. As cooperators randomly encounter one another, the positive payoffs associated with their interactions accumulate and they "clone" cooperator offspring onto neighboring sites. Neighborhoods of cooperation are thus formed. The first of these have clearly taken shape by $t = 15$, as seen in panel 2. By $t = 50$, a stable ratio of cooperators to defectors (approximately 5 to 1), has set in, as shown in panel 3. And this slightly noisy equilibrium persists. Panels 4 and 5 give the spatial configuration at $t = 100$ and $t = 1000$. In stark contrast to the replicator dynamics picture, cooperators are not annihilated. Indeed, they endure and predominate.

The time series showing aggregate cooperator and defector population dynamics is given in figure 9.3.

A phase diagram of these same data is given in figure 9.4. The x and y coordinates of each point correspond respectively to the cooperator and defector populations at that time. We see direct attraction to a (dark) invariant region of phase space corresponding to the noisy equilibrium noted earlier.

The basic point, then, is that *cooperation can emerge and flourish in a population of tagless agents playing zero-memory fixed strategies of cooperate or defect in this demographic setting.*

STATISTICAL ANALYSES

Moreover, for the given payoffs this result is robust to random effects. "The analysis of a single run can be misleading. In order to determine whether the conclusions from a given run are typical it is necessary to do several dozen simulation runs using identical parameters (using different random number seeds).... The ability to do this is one major

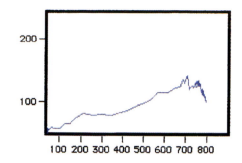

Figure 9.4. Phase diagram.

TABLE 9.2
Statistics

Measure	Cooperators	Defectors
Range	(752, 806)	(93, 148)
Mean	779	121
Standard deviation	15	15
95% CI for the mean	(773, 784)	(115, 126)

advantage of simulation: The researcher can rerun history to see whether the particular patterns observed in a single run are idiosyncratic or typical" (Axelrod 1997a).

Table 9.2 gives the result of such a statistical analysis. (On the statistical analysis of simulation output, see Law and Kelton 1991; Feldman and Valdez-Flores 1996.) The model was rerun 30 times with the same initial conditions and parameter values, but with a different random seed each time. In each run, data were sampled at $t = 500$ (after the essentially steady state had set in). Cooperator levels ranged from 752 to 806, while defector levels ranged from 93 to 148. Sample distributions for both populations are tightly clustered about their means, as reflected in the small standard deviations; 95 percent confidence intervals for the mean, in turn, are narrow. For the *given* payoffs, cooperator persistence is robust.

How sensitive is this core result to variations in the payoffs themselves? To explore this, a more extensive analysis is conducted. For simplicity, we maintain the symmetries assumed in payoff matrix (2), namely $S = -T$ and $P = -R$. We also assume integer-valued payoffs. To "sweep" the payoff space, we then let T range from its minimum possible value of 2 up to some maximum, here set at 10. For each such T value, R ranges from 1 up to $T - 1$. Hence, for a given maximum T, there are

$T(T - 1)/2$ payoff vectors. Table 9.3 gives the results for all 45 payoff vectors entailed by our maximum T value of 10. For each payoff vector $(T, R, -R, -T)$, the same analysis as in table 9.2 was conducted, running 30 times out to $t = 500$ with different random seeds each time. Table 9.3 gives the ranges, means, standard deviations, and 95 percent confidence intervals for the means for cooperators and defectors for each such sample of 30.

The essential pattern is clear. When R (the reward to mutual cooperation) is close to T (the defector's payoff against a cooperator), cooperators dominate. This preponderance wanes as R is systematically reduced, though cooperators persist. However, when R falls far enough (e.g., to 2 in the $T = 9$ cases) cooperators are wiped out. And ultimately, so too are the defectors since, with zero mutation and negative payoffs to mutual defection, societies of pure defection are doomed in the long run. (It is important to distinguish between convergence in the space of *strategies* and convergence of the *population* dynamics. The strategy distribution can and (for R sufficiently low) does converge to pure defection. However, with negative payoffs and zero mutation, a nontrivial *population* of pure defectors is not sustainable, though it may persist for many cycles after the demise of cooperation. As shown below, with non-negative payoffs, populations of pure defection can be sustained indefinitely.)

RUN 2. MAXIMUM AGE, ZERO MUTATION

In Run 1, agents could live forever. A cooperator living for a long time in the interior of a cooperative zone could amass vast *wealth*. An enormous number of encounters with defectors would then be needed to eliminate him. This is not entirely unrealistic: Once a human becomes extremely wealthy, a great many small setbacks are required before he or she is ruined financially. But, just to explore whether our results depend sensitively on this assumption, let us impose a maximum lifetime of 100 on the agents. (This actually ends up being slightly favorable to the defectors since the richest possible cooperator would then have 500 while the richest possible defector would have 600.) All other settings are as in Run 1. The time series for this case is shown in figure 9.5.

The time series for this realization is clearly more oscillatory than Run 1. Not surprisingly, the cross-sectional statistics for 30 runs (sampling again at $t = 500$) also exhibit greater variability, and for the defectors a reduction in the mean, as displayed in table 9.4. Clearly, cooperation persists as before. Now, as noted, we see in figure 9.5 the

TABLE 9.3
Payoff Sensitivity for Run 1

Payoffs (T, R)	Cooperators				Defectors			
	Range	Mean	SD	95% CI	Range	Mean	SD	95% CI
10, 9	(772, 845)	809	19	(802, 816)	(43, 109)	77	17	(71, 83)
10, 8	(707, 802)	748	27	(738, 757)	(82, 169)	132	23	(124, 140)
10, 7	(568, 717)	654	38	(641, 668)	(154, 280)	198	30	(188, 209)
10, 6	(312, 604)	469	60	(447, 490)	(191, 329)	274	30	(264, 285)
10, 5	(150, 383)	258	51	(240, 277)	(202, 319)	270	26	(261, 279)
10, 4	(0, 598)	231	151	(177, 285)	(0, 287)	199	74	(172, 225)
10, 3	(0, 1)	0	0	(0, 0)	(0, 1)	0	0	(0, 0)
10, 2	(0, 0)	0	0	(0, 0)	(0, 0)	0	0	(0, 0)
10, 1	(0, 0)	0	0	(0, 0)	(0, 0)	0	0	(0, 0)
9, 8	(766, 850)	806	21	(799, 814)	(39, 117)	81	19	(74, 88)
9, 7	(669, 793)	728	34	(716, 740)	(86, 190)	146	28	(136, 156)
9, 6	(490, 768)	604	58	(583, 625)	(102, 303)	225	38	(212, 239)
9, 5	(268, 460)	374	44	(358, 390)	(252, 323)	290	20	(283, 297)
9, 4	(98, 442)	203	78	(175, 231)	(159, 296)	235	36	(222, 248)
9, 3	(0, 382)	35	89	(3, 67)	(0, 284)	38	84	(8, 68)
9, 2	(0, 0)	0	0	(0, 0)	(0, 0)	0	0	(0, 0)
9, 1	(0, 0)	0	0	(0, 0)	(0, 0)	0	0	(0, 0)
8, 7	(744, 879)	807	30	(796, 818)	(14, 142)	80	27	(70, 89)
8, 6	(626, 787)	721	36	(708, 734)	(98, 231)	153	30	(142, 163)
8, 5	(460, 604)	530	39	(516, 544)	(209, 312)	263	22	(255, 271)
8, 4	(128, 387)	259	57	(239, 279)	(227, 313)	271	22	(263, 279)
8, 3	(0, 513)	93	113	(53, 134)	(0, 279)	113	99	(77, 148)
8, 2	(0, 0)	0	0	(0, 0)	(0, 0)	0	0	(0, 0)
8, 1	(0, 0)	0	0	(0, 0)	(0, 0)	0	0	(0, 0)
7, 6	(739, 852)	797	28	(787, 807)	(43, 133)	88	25	(79, 97)
7, 5	(542, 782)	668	44	(652, 684)	(101, 271)	187	33	(175, 199)
7, 4	(323, 547)	430	55	(410, 449)	(244, 345)	286	26	(277, 295)
7, 3	(0, 370)	126	101	(90, 162)	(0, 267)	153	76	(126, 180)
7, 2	(0, 0)	0	0	(0, 0)	(0, 0)	0	0	(0, 0)
7, 1	(0, 0)	0	0	(0, 0)	(0, 0)	0	0	(0, 0)
6, 5	(752, 806)	779	15	(773, 784)	(93, 148)	121	15	(115, 126)
6, 4	(524, 658)	587	33	(576, 599)	(193, 278)	241	22	(233, 248)
6, 3	(120, 344)	266	53	(247, 285)	(199, 320)	280	29	(270, 291)
6, 2	(0, 482)	24	92	(0, 57)	(0, 166)	13	37	(0, 27)
6, 1	(0, 0)	0	0	(0, 0)	(0, 0)	0	0	(0, 0)
5, 4	(689, 810)	741	33	(729, 752)	(79, 179)	136	28	(126, 146)
5, 3	(379, 574)	473	46	(456, 489)	(243, 329)	283	24	(274, 291)
5, 2	(0, 589)	125	155	(70, 180)	(0, 260)	114	89	(82, 145)
5, 1	(0, 0)	0	0	(0, 0)	(0, 0)	0	0	(0, 0)
4, 3	(613, 814)	710	47	(693, 727)	(76, 231)	159	36	(146, 172)
4, 2	(171, 380)	282	56	(262, 302)	(176, 322)	265	32	(254, 376)
4, 1	(0, 0)	0	0	(0, 0)	(0, 0)	0	0	(0, 0)
3, 2	(516, 699)	624	38	(610, 637)	(164, 284)	221	30	(210, 232)
3, 1	(0, 123)	4	22	(0, 12)	(0, 197)	7	35	(0, 20)
2, 1	(256, 471)	361	47	(344, 377)	(234, 337)	280	21	(273, 288)

Note: We assume $S = -T$ and $P = -R$. For each parameter setting, 30 runs were conducted, using different random seeds each time. Values were computed at $t = 500$. All figures are rounded to the nearest integer. Numbers less than 0.5 are rounded down to zero.

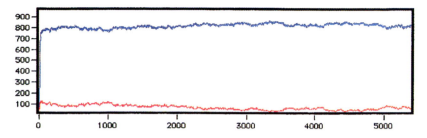

Figure 9.5. Typical population time series with maximum age.

TABLE 9.4
Statistics

Measure	Cooperators	Defectors
Range	(708, 846)	(45, 160)
Mean	784	99
Standard deviation	29	25
95% CI for the mean	(773, 794)	(90, 108)

emergence of some oscillatory dynamics. We can amplify these by reducing one of the entries in our payoff matrix.

RUN 3. MUTUAL COOPERATION PAYOFF REDUCED FROM 5 TO 2

Specifically, with mutation still at zero, and leaving all else fixed as before, let us begin to "dial down" the mutual cooperation payoff (the value of R), first from 5 to 2. A typical realization is shown in figure 9.6. Cooperators do worse than in figure 9.5, defectors do better, and the oscillations are now more evident.

RUN 4. MUTUAL COOPERATION PAYOFF REDUCED TO 1

This oscillatory dynamic is more pronounced if, *ceteris paribus*, we reduce R further to 1. The panels of figure 9.7 summarize the spatial story. Panel 1 of figure 9.7 shows the initial agents in their random starting positions. By $t = 40$, cooperative neighborhoods have begun to take shape, as shown in panel 2. Cooperators dominate by $t = 60$, as shown in panel 3. However, the cooperative zone is bordered by defectors. These surrounding defectors gradually "eat away" at the cooperative region and by $t = 160$ have nearly annihilated it, as shown

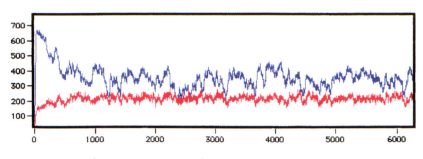

Figure 9.6. Population time series with $R = 2$.

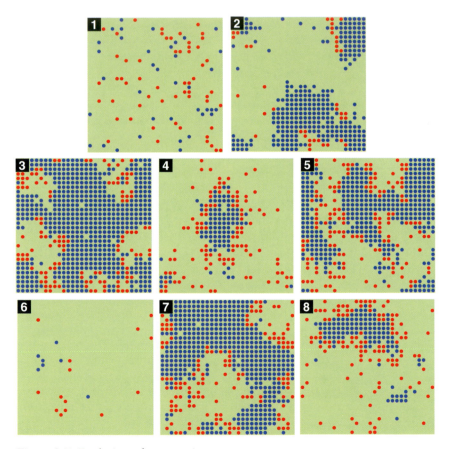

Figure 9.7. Evolution of cooperative zones.

in panel 4. In time, the defectors will have few cooperators with whom to interact; most interactions are then with other defectors. But since these interactions carry negative payoff, the quarrelsome defectors are no sooner the majority than they begin to kill one another off, making way for a resurgence of cooperation, evident by $t = 240$, shown in panel 5. This is a typical cycle. Another such cycle is shown in panels 6, 7, and 8, corresponding to times 560, 625, and 700. A time series of this evolution is shown in figure 9.8. Figure 9.9 offers a phase diagram of the same dynamics. We see, essentially, a perturbed limit cycle. The dynamics are reminiscent of predator-prey cycles. (The amplitude of the cycles depends on the lattice size.) And, indeed, defectors "feed" on cooperators in the sense that they require interactions with cooperators

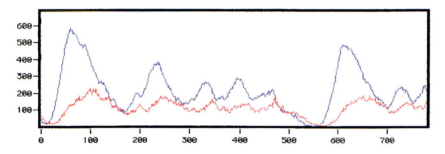

Figure 9.8. Oscillatory dynamics.

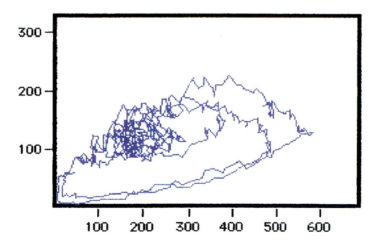

Figure 9.9. Phase diagram.

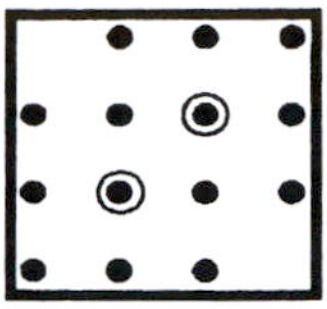

Figure 9.10. Two surrounded cooperators.

to accumulate positive payoff, quite as if they were predators and cooperators were prey. In panel 4 of figure 9.7, we clearly see a cooperative neighborhood ringed by defector parasites. In the simulation proper, it looks very much like a predatory red species "eating through" a blue prey concentration. But defectors need cooperators for a less obvious reason, also illustrated in this run: *Cooperators function to separate defectors from one another spatially.* Because defector-defector interactions carry negative payoff, defector neighborhoods seldom take shape and are highly self-destructive—and hence ephemeral—when they do. Indeed, their self-destruction is what clears the way for the resurgence of cooperative zones.

I have not attempted to count cooperative zones over the course of this run. A rough index of cooperative blotchiness is a count of cooperators all of whose eight Moore neighbors are also cooperators; I call these *surrounded cooperators.* For example, the 14-cooperator configuration in figure 9.10 would count as 2 (circled) surrounded cooperators. The time series for this index is plotted in figure 9.11 and reflects the rise and fall of cooperative zones in Run 4.

In contrast to Runs 1 and 2, different random seeds *do* produce different outcomes for these *same* payoffs. While the oscillatory dynamic is robust to such variation, the oscillations are extreme enough that qualitatively different long-run outcomes are possible. In some cases, cooperators and defectors coexist. In others, cooperators inherit the earth, while in others they are the first to go extinct (followed by the defectors, since the latter's mutual interactions carry strictly negative payoffs). In those realizations where cooperators inherit the earth, they do so for an interesting and counterintuitive reason. With the higher R value (the payoff to mutual cooperation) of 5, the cooperators never "thin out" enough spatially to let the defectors annihilate each other, while with $R = 1$ they may, paving the way for their own monopoly. In such cases, cooperators ultimately do better with a low payoff ($R = 1$) than with a high one ($R = 5$)!

This counterintuitive result is reminiscent of altruistic behavior in which one generation (here the generation of cooperators that thins out) sacrifices itself for the long-run benefit of the species. But it is not robust.

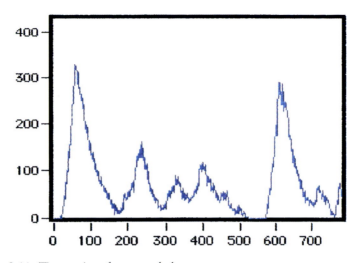

Figure 9.11. Time series of surrounded cooperators.

Run 5. Maximum Age, Mutation

Let us now restore our initial value of $R = 5$, the payoff to mutual cooperation, and explore the sensitivity of our basic result—cooperator persistence—to mutation, defined as the probability that an agent will have a strategy different from its parent's (e.g., that a cooperator's offspring will be a defector).

In this spatial setting, cooperator mutation can introduce defector "invaders" into the very heart of cooperative zones. One might well imagine that these will then spread until cooperators are annihilated. In fact, cooperator persistence will withstand very high mutation rates. In figure 9.12, all else is as in Run 2 (finite lives with maximum age of 100), except that a 50 percent mutation rate is assumed. (The same qualitative result obtains at a mutation rate of 25 percent, with cooperators averaging around 350 and defectors around 400.) While the dynamics are much more oscillatory than in the first case, and the cooperator-defector ratio closer to 1, the long-term persistence of cooperation is intact through 10 thousand cycles.

Simulation and Sufficiency

Simulation is a particularly direct tool when the aim is to establish that some set of micro assumptions is *sufficient to generate* a macro phenomenon of interest. (On generative sufficiency, see Epstein and

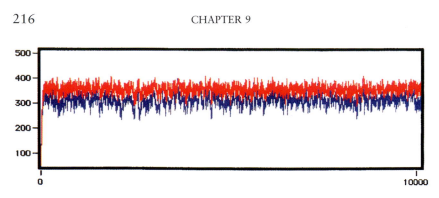

Figure 9.12. Cooperation withstands mutation.

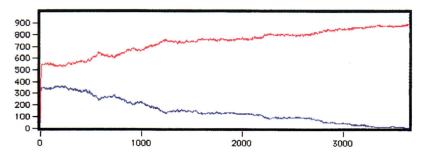

Figure 9.13. Space and non-negative payoffs.

Axtell 1996, 1997.) In this case, the macro phenomenon is the persistence of cooperation. A widely accepted view is that "when cooperators and defectors look alike, genuine cooperation cannot emerge" (Frank 1988). In the simulations above, cooperators and defectors have no "tags"; they do "look alike" in the relevant sense that they are indistinguishable *to one another.* Yet, on certain micro-assumptions, cooperation emerges and endures. This basic sufficiency result is robust and withstands the introduction of a maximum age and high mutation rates. Of course, there are variations it will not withstand. Specifically, if we replace space and local interactions with soup (equiprobable random agent pairings), then the system runs to pure defection despite negative payoffs, quite consistent with the related replicator dynamics reviewed earlier. Moreover, space alone is not sufficient to ensure cooperation; with maximum age of 100, zero mutation, and all payoffs shifted up by 6, so that $T = 12$, $R = 11$, $P = 1$, and $S = 0$, the spatial system again converges to pure defection, as shown in figure 9.13. And on these assumptions, the society of pure defection is sustained.

Negative Payoffs Unnecessary but Appealing

Negative payoffs (for P and S) are not necessary for cooperation to persist, or even monopolize. (For instance, with all payoffs hiked by ten (so that $T = 16$, $R = 11$, $P = 5$, $S = 4$) and the maximum lifetime reduced from 100 to 10 cycles, and using the same random seed, we again see an evolution to cooperative monopoly.) However, with all payoffs non-negative, the only way for agents to die is through an exogenous "old age" (i.e., a random maximum age assigned when the agent is initialized). With negative payoffs, agents can die through an endogenous series of infelicitous social interactions. To "bottom-up" modelers, this endogeneity is an appealing feature of negative payoffs. However, it should be noted that, since defector-defector interactions then carry negative payoffs, a society composed purely of defectors is not viable in the long run. Of course, with mutation, defectors may persist— by spawning cooperators to "save them from themselves," as it were. Negative payoffs admit an alternative interpretation, which will lead to an interesting conjecture.

An Alternative Formulation of Negative Payoffs

At the outset, we assumed a payoff ordering:

$$T > R > 0 > P > S. \tag{6}$$

Negative per-period payoffs can accumulate until an agent's aggregate holding goes negative, at which point it dies. Equivalent mathematically, but quite different conceptually, is a set-up with all payoffs non-negative (as is perhaps more typical in the literature), but with some fixed "global" metabolic rate imposed on all agents after every interaction. (By metabolic rate, I mean a fixed decrement to accumulated payoff per cycle.) For example, if we add $|S|$ to all payoffs in (2) and denote the new values with asterisks, we have

$$T^* > R^* > P^* > S^* = 0. \tag{7}$$

Now, taking as the unstarred values the entries in matrix A $(6, 5, -5, -6)$, we obtain (adding $|S| = 6$ to each), the nonnegative payoffs $12, 11, 1, 0$. With metabolism set at zero, these are the *ex post* payoffs as well and we generate the immediately preceding run (figure 9.12), in which cooperators are annihilated. However, with the global metabolic rate hiked to 6, we recover, in effect, our initial payoffs and generate the run in which cooperators are *not* annihilated.

Necessity the Mother of Cooperation?

Now, one can think of this global metabolic rate as a kind of "environmental" or "selective" pressure. When the environment is hospitable—when selection pressure is weak—there is no particular advantage to cooperation, and defectors rule. However, as we turn up the selection pressure, we find that cooperators are more persistent. This makes some sense. After all, if there's an abundance of food lying around, we don't need to cooperate agriculturally. A conjecture, then, is that, in the demographic setting we have been exploring, *cooperation acquires selective advantage as selection pressures increase.* Or, more prosaically, necessity is the mother of cooperation.

CONCLUDING THOUGHTS

Perhaps, in conclusion, it deserves emphasis that the results reported above involve a particular space (a 30-by-30 torus) and a limited set of numerical assumptions—payoffs, maximum ages, mutation rates, initial populations levels, and so forth. The only claims that can be advanced definitively are that certain ensembles of assumptions are *sufficient to generate cooperative persistence* on the time scales explored in the research. As it happens, that is fairly notable given the existing literature. While some sensitivity analysis was conducted, it would of course be worthwhile to further sweep the parameter space of the model, exploring the robustness of the main results and, if possible, to assess their generality mathematically. Meanwhile, the sufficiency results and basic dynamics were deemed notable enough, and the conjecture interesting enough, to warrant wider dissemination and discussion.

Implementation

The source code was written in C++ in the Metrowerks CodeWarrior development environment for the Macintosh. Readers interested in running the model under their own assumptions—payoffs, mutation rates, maximum lifetimes, and so forth—may do so using this book's CD.

REFERENCES

Abreu, D., and A. Rubinstein. 1988. The Structure of Nash Equilibrium in Repeated Games with Finite Automata. *Econometrica* 56(6): 1259–81.
Albin, P. S., and D. K. Foley. 1997. The Evolution of Cooperation in the Local Interaction Multi-Person Prisoners' Dilemma. Working paper, Department of Economics, Barnard College, Columbia University.
Axelrod, R. 1984. *The Evolution of Cooperation.* New York: Basic Books.

————. 1997a. Advancing the Art of Simulation in the Social Sciences. Santa Fe Institute Working Paper 97-05-048.

————. 1997b. The Dissemination of Culture: A Model with Local Convergence and Global Polarization. *Journal of Conflict Resolution* 41:203–26.

Ben-Porath, E. 1990. The Complexity of Computing Best Response Automata in Repeated Games with Mixed Strategies. *Games and Economic Behavior* 2(1): 1–12.

Binmore, K. G. 1987. Modeling Rational Players, I. *Economics and Philosophy* 3:179–214.

————. 1988. Modeling Rational Players, II. *Economics and Philosophy* 4:9–55.

————. 1992. *Fun and Games: A Text on Game Theory.* Lexington, MA: D.C. Heath.

Binmore, K. G., and L. Samuelson. 1992. Evolutionary Stability in Repeated Games Played by Finite Automata. *Journal of Economic Theory* 57:278–305.

Boyd, R., and P. J. Richerson. 1985. *Culture and the Evolutionary Process.* Chicago: University of Chicago Press.

Cavalli-Sforza, L. L., and M. W. Feldman. 1981. *Cultural Transmission and Evolution: A Quantitative Approach.* Princeton: Princeton University Press.

Deng, X., and C. H. Papadimitriou. 1994. On the Complexity of Cooperative Solution Concepts. *Mathematics of Operations Research* 19:257–66.

Epstein, J. M., and R. L. Axtell. 1996. *Growing Artificial Societies: Social Science from the Bottom Up.* Washington, DC: Brookings Institution Press; Cambridge: MIT Press.

————. 1997. Artificial Societies and Generative Social Science. *Artificial Life and Robotics* 1(1): 33–34.

Frank, R. H. 1988. *Passions within Reason: The Strategic Role of the Emotions.* New York: Norton.

Feldman, B., and K. Nagel. 1993. Lattice Games with Strategic Takeover. In *Lectures in Complex Systems*, ed. L. Nadel and D. Stein. Reading, MA: Addison Wesley.

Feldman, R. M., and C. Valdez-Flores. 1996. *Applied Probability and Stochastic Processes.* Boston: PWS Publishing.

Gilboa, I. 1988. The Complexity of Computing Best-Response Automata in Repeated Games. *Journal of Economic Theory* 45:342–52.

Guttman, J. M. 1996. Rational Actors, Tit-for-Tat Types, and the Evolution of Cooperation. *Journal of Economic Behavior and Organization* 29(1): 27–56.

Hirshleifer, J., and J. C. Martínez Coll. 1988. What Strategies Can Support the Evolutionary Emergence of Cooperation? *Journal of Conflict Resolution* 32(2): 367–98.

Hofbauer, J., and K. Sigmund. 1988. *The Theory of Evolution and Dynamical Systems.* Cambridge: Cambridge University Press.

Holland, J. H. 1995. *Hidden Order: How Adaptation Builds Complexity.* Reading, MA: Helix.

Huberman, B., and N. Glance. 1993. Evolutionary Games and Computer Simulations. *Proceedings of the National Academy of Sciences* 90(16): 7715–18.

Kalai, E. 1990. Bounded Rationality and Strategic Complexity in Repeated Games. In *Game Theory and Applications*, ed. T. Ichiishi, A. Neyman, and Y. Tauman. San Diego: Academic Press.

Law, A. M., and W. D. Kelton. 1991. *Simulation Modeling and Analysis*. 2nd ed. New York: McGraw-Hill.

Lindgren, K., and M. G. Nordahl. 1994a. Cooperation and Community Structure in Artificial Ecosystems. *Artificial Life* 1:15–37.

———. 1994b. Evolutionary Dynamics of Spatial Games. *Physica D* 75:292–309.

Lomberg, B. 1996. Nucleus and Shield: The Evolution of Social Structure in the Iterated Prisoner's Dilemma. *American Sociological Review* 61(2): 278–307.

Martínez Coll, J. C., and J. Hirshleifer. 1991. The Limits of Reciprocity. *Rationality and Society* 3(1): 35–64.

Maynard Smith, J. 1982. *Evolution and the Theory of Games*. Cambridge: Cambridge University Press.

Miller, J. H. 1996. The Coevolution of Automata in the Repeated Prisoner's Dilemma. *Journal of Economic Behavior and Organization* 29(1): 87–112.

Neyman, A. 1985. Bounded Complexity Justifies Cooperation in the Finitely Repeated Prisoners' Dilemma. *Economics Letters* 19:227–29.

Nowak, M. A., and R. M. May. 1992. Evolutionary Games and Spatial Chaos. *Nature* 359:826–29.

Nowak, M. A., R. M. May, and K. Sigmund. 1995. The Arithmetics of Mutual Help. *Scientific American* 272(6): 76–81.

Oliphant, M. 1994. Evolving Cooperation in the Non-iterated Prisoner's Dilemma: The Importance of Spatial Organization. In *Artificial Life IV*, ed. R. A. Brook and P. Maes. Cambridge: MIT Press.

Papadimitriou, C. H. 1992. On Players with a Bounded Number of States. *Games and Economic Behavior* 4:122–131.

Papadimitriou, C. H., and M. Yannakakis. 1994. On Complexity as Bounded Rationality. *Proceedings of the Twenty-Sixth Annual ACM Symposium on the Theory of Computing*, Montreal, Quebec, Canada.

Rapoport, A. 1996. Prisoner's Dilemma: Reflections and Recollections. *Simulation and Gaming* 26(4): 489–503.

Rubinstein, A. 1986. Finite automata Play the Repeated Prisoners' Dilemma. *Journal of Economic Theory* 39:83–96.

Samuelson, L. 1997. *Evolutionary Games and Equilibrium Selection*. Cambridge: MIT Press.

Shubik, M. 1996. Simulations, Models and Simplicity. *Complexity* 2(1): 60.

Taylor, P., and L. Jonker. 1978. Evolutionarily Stable Strategies and Game Dynamics. *Mathematical Biosciences* 40:145–56.

Tesfatsion, L. 1995. A Trade Network Game with Endogenous Partner Selection. Economic Report 36, Iowa State University, Department of Economics.

Weibull, J. W. 1995. *Evolutionary Game Theory*. Cambridge: MIT Press.

Young, H. P. 1993. The Evolution of Conventions. *Econometrica* 61:57–84.

Young, H. P., and D. Foster, 1991. Cooperation in the Short and in the Long Run. *Games and Economic Behavior* 3:145–56.

Additional Resources

Ashlock, D., M. D. Smucker, E. A. Stanley, and L. Tesfatsion. 1996. Preferential Partner Selection in an Evolutionary Study of Prisoner's Dilemma. *BioSystems* 37:99–125.

Axelrod, R., and W. D. Hamilton. 1981. The Evolution of Cooperation. *Science* 211:1390–96.

Axelrod, R. The Evolution of Strategies in the Iterated Prisoner's Dilemma. 1987. In *Genetic Algorithms and Simulated Annealing*, ed. L. Davis. London: Pitman; Los Altos, CA: Morgan Kaufman.

Blume, L. E. Population Games. 1997. In *The Economy as a Complex Evolving System II*, ed. W. B. Arthur, S. Durlauf, and D. Lane. Menlo Park, CA: Addison-Wesley.

Casti, J. L. 1992. *Reality Rules: Picturing the World in Mathematics*. New York: John Wiley.

Rapoport, A. 1966. *Two-Person Game Theory*. Ann Arbor: University of Michigan Press.

Schelling, T. C. 1978. *Micromotives and Macrobehavior*. New York: Norton.

GENERATING NORM MAPS IN

THE DEMOGRAPHIC COORDINATION GAME

ONE NATURAL research question is this: What happens if, keeping everything else (movement, space, cloning) as is, we simply change the payoff structure from PD to the symmetric two-person coordination game? Does the demographic variant produce an outcome different from both one-shot rational play and the replicator dynamics? The answer is, again, yes.

For expository purposes, assume the payoff matrix below:

	Left	*Right*
Left	1	−3
Right	−3	1

We imagine two drivers approaching one another on a road. If each drives on his left, they pass without incident, whereas a failure to coordinate in this sense eventuates in a collision, and negative payoff. The pure strategy pairs—(L, L) and (R, R)—are both Nash equilibria.

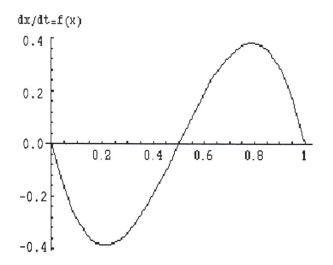

Figure 9.A.1. Replicator dynamics phase portrait for coordination game.

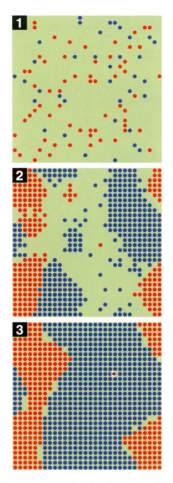

Run 9.A.1. Evolution of norm maps in the demographic coordination game.

Traditional game theory does not offer a compelling account of how either is attained in populations of interacting agents.[1] Hence, as before, we turn to evolutionary game theory.

The phase diagram for the replicator dynamics, assuming the payoff matrix above, is shown in figure 9.A.1.

There are three equilibria: 0, 1/2, and 1. Since $df/dx > 0$ at the interior equilibrium and is strictly negative at 0 and 1, we see that

---

[1] Young (1996) offers an elegant account using Best Reply to Recent Sample Evidence. However, as noted (somewhat self-critically) in the prelude to chapter 10, this learning scheme does assume that agents continue sampling long after a norm is firmly entrenched.

the interior equilibrium is unstable, and the coordination equilibria are attractors. Hence, the replicator dynamics predicts convergence to either zero (everyone driving on the left) or one (everyone driving on the right).

In the demographic variant, however, we find quite a different result: spatial maps divided into connected regions within each of which a specific norm prevails, with "accidents" on the region boundaries. A full animation is provided on the CD. The frames shown in run 1, however, are typical.

Agents are, of course, colored by their norm. Events progress from an initial random distribution (frame 1) to a spatial map of the sort described above (frame 3).

To generate norm maps of this sort using the CD, open the Prisoners Dilemma Ascape applet.[2] When the Ascape Control Bar appears, click on the Globe icon (for global variables) and a dialog box will pop up. In it, change Death Age to 1000, Mutation Rate to 0.0, and the Payoffs to $[1, -3, -3, 1]$. Leave all other settings at their default values. Then click Done, and restart the model. You are unlikely to get the exact run captured here, because Ascape uses a new random seed for every run. But they should agree qualitatively, generating spatial norm maps with accidents on the boundaries.

Reference

Young, H. P. The Economics of Convention. 1996. *Journal of Economic Perspectives* 10(2): 105–22.

---

[2] See the Read Me file on the CD for further details.

GENERATING THOUGHTLESS CONFORMITY

TO NORMS

LET ME TRY to get at something big with something small. Here are two small experiments: (1) Stand four inches too close to someone when conversing. (2) The next time someone at a conference table makes a clever academic quip, try laughing heartily for ten full seconds after the obligatory polite chuckles have subsided. In both cases, you'll feel uncomfortable. Why? Because we have a very finely tuned sense of what's inside a norm and what's outside it. But we don't *think about* that boundary in any conscious way; not, that is, until it is crossed.

We don't think about how to walk normally, but we can detect remarkably slight departures from a normal human gait, or a normal human face. We don't think about how to talk normally, but we'd notice immediately if someone put a one-second silence after every third word. That which is normal is maintained as a kind of homeostatic target, and we only notice when the observed value wanders outside the acceptable range, just like a thermostat. Perhaps we are endowed with something like a *normostat*.

Societies may differ in the normostat's target value (what constitutes normalcy), and a given social group can change its value (as when fashions change) endogenously. But once a norm is entrenched—once the normostat is set to a value—we don't think about it, until a violation is observed. That is at once understandable and, arguably, adaptive. Clearly, if we spent the entire day consciously *deciding* whether or not to conform to every norm of locution, dress, gesture, hygiene, social comportment, and so on, we'd have no time left to accomplish anything. Thoughtless conformity to entrenched norms has an adaptive computation-saving aspect. But it is also very disturbing that we can just as thoughtlessly adopt discriminatory norms: racial biases, xenophobias, and so on.

A great deal of social science is concerned with conscious choices and decisions. *Homo economicus* collects information, arrays alternatives, and makes conscious strategic choices. He's *thinking*. To me, however, one of the more remarkable aspects of *homo sapiens* is how much of our social behavior is entirely thoughtless. Indeed, as argued in the Generative chapter, the major patternings in the fabric of humankind are of this sort: We didn't *decide* to identify with our particular ethnic group; we didn't consider the alternatives and *select* our particular native tongue; we didn't lay out a range of religions and *pick* orthodox Judaism; we didn't *choose*, in any meaningful sense, to prefer ragas to sonatas, or pancakes to monkey brain for breakfast.

Not all of the above examples are equally momentous. And none of them is specifically modeled in the next chapter. Rather, I try to get at the general phenomenon of thoughtless conformity in a simple two-norm setting.

As emphasized in the Generative chapter, space and heterogeneity are typical of agent models, and they are important here. Agents are arranged on a one-dimensional ring. Although they differ by norm, they are most heterogeneous by their search radius—that is, by how much they are thinking about what norm to adopt. In regions of the space where there is norm variation (i.e., on norm boundaries), agents are worrying about how to behave. In zones where one norm or another is monolithic (i.e., norm interiors), there is no thinking. Norm innovators (represented as noise) can jog the thoughtless into worrying, a process that can eventuate in local "tipping" from one norm to another. The dynamics can then exhibit what Young has identified as hallmarks of complex systems: local conformity (norm patches), global diversity, and punctuated equilibria.[1]

Best Reply to Adaptive Sample Evidence

The essential point of the model is to generate thoughtlessness itself. Hence, the number of sites an agent samples in figuring out how to behave is adaptive. It increases in the face of variation (observed norm violations), and, most importantly, it shrinks—in principle, to zero—once a norm is entrenched. Agents play a coordination game with others in this adaptive search radius. Hence, I have dubbed the learning dynamic *Best Reply to Adaptive Sample Evidence*. Now the attentive reader will notice that this departs from precisely the scheme used in the earlier Classes model (chapter 8). There, agents played Best Reply to Sample Evidence, but the sample size was fixed, not adaptive. In a sense, then, this chapter criticizes that one. So be it. As Emerson said, "A foolish consistency is the hobgoblin of little minds."[2] Best Reply misses something important in cases where *thoughtless* conformity (not merely conformity) is the "emergent" phenomenon of interest.

---

[1] H. P. Young, *Individual Strategy and Social Structure: An Evolutionary Theory of Institutions* (Princeton: Princeton University Press, 1998).

[2] Ralph Waldo Emerson, "Self-Reliance," in *Essays by Ralph Waldo Emerson: First Series* (New York: Houghton, Mifflin, 1904), 57.

Unobservables

The chapter introduces a graphical technique also exploited in the subsequent chapter's Civil Violence models. Two screens are used. On the left is the screen of observable action: one's norm (e.g., left-hand driving) in the present model; one's rebellious state (active or not) in the next. On the right screen is an *unobservable*: one's search radius (how much one is thinking) in the present model; one's level of political grievance in the next. Unobservable theoretical entities have a long and venerable history in science, and I do not shrink from using them in this and other chapters. Contrary to popular inductivist myth, moreover, it often occurs in science that theory *precedes* observation. Electromagnetic theory predicted the existence of radio waves, which were only later observed. Further examples abound.

Without theory, it is not obvious what data to collect. In any case, when deep in a norm, the adaptive agents you are about to meet (and, in my view, many humans) collect no data at all!

Chapter 10

LEARNING TO BE THOUGHTLESS: SOCIAL NORMS AND INDIVIDUAL COMPUTATION

JOSHUA M. EPSTEIN*

THIS PAPER EXTENDS the literature on the evolution of norms with an agent-based model capturing a phenomenon that has been essentially ignored, namely that individual thought—or computing—is often inversely related to the strength of a social norm. Once a norm is entrenched, we conform thoughtlessly. In this model, agents learn how to behave (what norm to adopt), but—under a strategy I term Best Reply to Adaptive Sample Evidence—they also learn how much to think about how to behave. How much they are thinking affects how they behave, which—given how others behave—affects how much they think. In short, there is feedback between the social (inter-agent) and internal (intra-agent) dynamics. In addition, we generate the stylized facts regarding the spatio-temporal evolution of norms: local conformity, global diversity, and punctuated equilibria.

TWO FEATURES OF NORMS

When I'd had my coffee this morning and went upstairs to get dressed for work, I never considered being a nudist for the day. When I got in my car to drive to work, it never crossed my mind to drive on the left. And when I joined my colleagues at lunch, I did not consider eating my salad barehanded; without a thought, I used a fork.

The point here is that many social conventions have *two* features of interest. First, they are self-enforcing behavioral regularities (Lewis 1969; Axelrod 1986; Young 1993a, 1995). But second, once entrenched, we conform *without thinking about it*. Indeed, this is one reason why social norms are useful; they obviate the need for a lot of individual

---

*For valuable discussions the author thanks Peyton Young, Miles Parker, Robert Axtell, Carol Graham, and Joseph Harrington. He further thanks Miles Parker for translating the model, initially written in C++, into his Java-based ASCAPE environment. For production assistance, he thanks David Hines.

This essay was published previously in *Computational Economics*, vol. 18, no. 1, August 2001, pp. 9–24.

computing. After all, if we had to go out and sample people on the street to see if nudism or dress were the norm, and then had to sample other drivers to see if left or right were the norm, and so on, we would spent most of the day figuring out how to operate, and we would not get much accomplished. Thoughtless conformity, while useful in such contexts, is frightening in others—as when norms of discrimination become entrenched. It seems to me that the literature on the evolution of norms and conventions has focused almost exclusively on the first feature of norms—that they are self-enforcing behavioral regularities, often represented elegantly as equilibria of n-person coordination games possessing multiple pure-strategy Nash equilibria (Young 1993a, 1995; Kandori, Mailith, and Rob 1991).

GOALS

My aim here is to extend this literature with a simple agent-based model capturing the second feature noted above, that *individual thought—or computing—is inversely related to the strength of a social norm*. In this model, then, agents learn how to behave (what norm to adopt), but they also learn *how much to think about* how to behave. How much they are thinking affects how they behave, which—given how others behave—affects how much they think. In short, there is *feedback* between the social (inter-agent) and internal (intra-agent) dynamics. In addition, we are looking for the stylized facts regarding the spatio-temporal evolution of norms: local conformity, global diversity, and punctuated equilibria (Young 1998).

AN AGENT-BASED COMPUTATIONAL MODEL

This model posits a ring of interacting agents. Each agent occupies a fixed position on the ring and is an object characterized by two attributes. One attribute is the agent's "norm," which in this model is binary. We may think of these as 'drive on the right (R) vs. drive on the left (L).' Initially, agents are assigned norms. Then, of course, agents update their norms based on observation of agents within some sampling radius. This radius is the second attribute and is typically heterogeneous across agents. An agent with a sampling radius of 5 takes data on the five agents to his left and the five agents to his right. Agents do not sample outside their current radius. Agents update, or "adapt," their sampling radii incrementally according to the following simple rule:

RADIUS UPDATE RULE

Imagine being an agent with current sampling radius of r. First, survey all r agents to the left and all r agents to the right. Some have L (drive on the left) as their norm and some have R(drive on the right). Compute the relative frequency of Rs at radius r; call the result $F(r)$. Now, make the same computation for radius $r + 1$. If $F(r + 1)$ *does not* equal $F(r)$, then increase your search radius to $r + 1$.[1] Otherwise, compute $F(r - 1)$. If $F(r - 1)$ *does* equal $F(r)$, then reduce your search radius to $r - 1$. If neither condition obtains (i.e., if $F(r + 1) = F(r) \neq F(r - 1)$), leave your search radius unchanged at r.

Agents are "lazy statisticians," if you will. If they are getting a different result at a higher radius $(F(r+1) \neq F(r))$, they increase the radius—since, as statisticians, they know larger samples to be more reliable than smaller ones. But they are also lazy. Hence, if there's no difference at the higher radius, they check a lower one. If there is no difference between that and their current radius $(F(r - 1) = F(r))$, they reduce. This is the agent's *radius update rule*. Having updated her radius, the agent then executes the Norm Update Rule.

NORM UPDATE RULE

This is extremely simple: *match the majority within your radius*. If, at the updated radius, Ls outnumber Rs, then adopt the L norm. In summary, the rule is: *When in Rome, do as the (majority of) Romans do, with the (adaptive) radius determining the "city limits."* This rule is equivalent to Best Reply to sample evidence with a symmetric payoff matrix such as:

	L	R
L	(1,1)	(0,0)
R	(0,0)	(1,1)

Following Young (1996), we imagine a coachman's decision to drive on the left or the right. "Among the encounters he knows about, suppose that more than half the carriages attempted to take the right side of the road. Our coachman then predicts that, when he next meets a carriage on the road, the probability is better than 50-50 that it will go right. Given this expectation, it is best for him to go right also (assuming that the payoffs are symmetric between left and right)." The coachman "calculates the observed frequency distribution of left and right, and uses this to predict the probability that the next carriage he meets will go left or right. He then chooses a best reply," which Young terms "best reply

---

[1] When we say "not equal" we mean the difference lies outside some tolerance, T. That is, $|F(r + 1) - F(r)| > T$ for inequality, and $|F(r + 1) - F(r)| \leq T$ for equality. For our basic runs, $T = 0.05$.

to recent sample evidence." Best reply maximizes the expected utility (sum of payoffs) in playing the agent's sample population.[2]

The departure introduced here is that each individual's sample size is itself *adaptive*.[3] In particular, as suggested earlier, once a norm of driving on the left is established (firmly entrenched) real coachmen don't calculate anything—they (thoughtlessly and efficiently) drive on the left. So, we want a model in which "thinking"—individual computing—declines as a norm gains force, and effectively stops once the norm is entrenched. Of course, we want our coachmen to start worrying again if suddenly the norm begins to break down. Of the many adaptive individual rules one might posit, we will explore the radius update rule set forth above.

Overall, the individual's combined (norm and search radius) updating procedure might appropriately be dubbed *Best Reply to Adaptive Sample Evidence*.

NOISE

Finally, there is generally some probability that an agent will adopt a random norm, a random *L* or *R*. We think of this as a "noise" level in society.

Graphics

With this set-up, there are two things to keep track of: the evolution of social norm patterns on the agent ring, and the evolution of individual search radii. In the runs shown below, there are 191 agents. They are drawn at random and updated asynchronously. Clearly, each agent's probability of being drawn k times per cycle (191 draws with replacement) has the binomial distribution $b(k; n, 1/n)$, with $n = 191$. Agents who are not drawn keep their previous norm. After 191 draws—one cycle—the new ring is redrawn below the old one (as a horizontal series of small contiguous black and white dots), so time is progressing down the page. There are two Panels. The left Panel shows the evolution of norms, with *L*-agents colored black and *R*-agents colored white. With the exception of Run 4, each entire Panel displays 300 cycles (each cycle,

---

[2] For arbitrary payoff matrices, Best Reply is *not* equivalent to the following rule: Play the strategy that is optimal against the most likely type of opponent (i.e., the strategy type most likely to be drawn in a single random draw from your sample). For our particular set-up, these are both equivalent to our "match the majority" update rule. These three rules part company if payoffs are not symmetric.

[3] In Best Reply models, the sample size is fixed for each agent, and is equal across agents. See Young 1995.

again, being a sequence of 191 random calls.) The right window shows the evolution of search radii, using grayscale. Agents are colored black if $r = 1$, with progressively higher radii depicted as progressively lighter shades of gray.

Runs of the Model

We present seven basic runs of this model, and some statistical and sensitivity analysis. Once more, we are looking for the stylized facts regarding the evolution of norms: Local conformity, global diversity, and punctuated equilibria (Young 1998). But we wish also to reflect the rise and fall of individual computing as social norms dissolve and become locked in.

Run 1. Monolithic Social Norm, Individual Computing Dies Out

For this first run, we set all agents to the L norm (coloring them black) initially and set noise to zero. We give each agent a random initial search radius between 1 and 60 (artificially high to show the strength of the result in the monolithic case). There is no noise in the decision-making. The uppermost line (the initial population state) of the right graph (191 agents across) is multi-shaded, reflecting the random initial radii. Let us now apply the radial update rule to an arbitrary agent with radius r. First look out further. We find that $F(r + 1) = F(r)$, since all agents are in the L norm (black). Hence, try a smaller radius. Since $F(r - 1) = F(r)$, the agent reduces from r to $r - 1$. Now, apply the norm update rule. At this new radius, match the majority. Clearly, this is L (black), so stay L. This is the same logic for all agents. Hence, on the left panel of figure 10.1, the L social norm remains entrenched, and, as shown in the right panel, individual "thinking" dies out—radii all shrink to the minimum of 1 (colored black).

Run 2. Random Initial Norms, Individual Computing at Norm Boundaries

With noise still at zero, we now alter the initial conditions slightly. In this, and all subsequent runs, the initial maximum search radius is 10. Rather than set all agents in the L norm initially, we give them random norms. In figure 10.2, we see a typical result.

In the left panel, there is rapid lock-in to a global pattern of alternating local norms on the ring. In the right panel, we see that deep in each local norm, agents are colored black: there is no individual computing, no

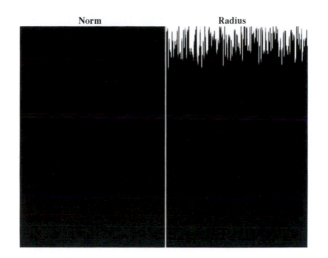

Figure 10.1. Monolithic norms induce radial contraction.

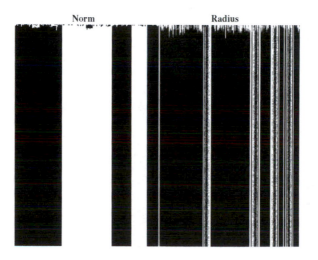

Figure 10.2. Local conformity, global diversity, and thought at boundaries.

"thinking," as it were. By contrast, agents at the boundary of two norms must worry about how to behave, and so are bright-shaded.[4] (For future reference notice that, since there are two edges for each local norm—each stripe on the left panel—the *average* radius will stabilize around different

---

[4]For the particular realization shown in figure 10.2, the average radius settles (after the initial transient phase) to around 3.

values from run to run, depending on the number of different norms that emerge.)

Run 3. Complacency in New Norms

In the 1960's, people smoked in airplanes, restaurants, and workplaces, and no one gave it much thought. Today, it is equally entrenched that smoking is prohibited in these circumstances. The same point applies to other social norms (e.g., revolutions in styles of dress) and to far more momentous political ones (e.g., voting rights, segregation of water fountains, lunch counters, and seats on the bus). After the "revolution" entirely new norms prevail, but once entrenched, people become inured to them; they are observed every bit as thoughtlessly (in our sense) as before. I often feel that the same point applies to popular beliefs about the physical world; these represent a procession of conventions rather than any real advance in the average person's grasp of science. For example, if you had asked the average 14th Century European if the earth were round or flat, he would have said "flat." If, today, you ask the average American the same question, you will certainly get a different response: "round." But I doubt that the typical American could furnish more compelling reasons for his correct belief than our 14th Century counterpart could have provided for his erroneous one. Indeed, on this test, the "modern" person will likely fare worse: at least the 14th Century "norm" accorded with intuition. Maybe we are going backward! In any event, there was no "thinking" in the old norm, and there is little or no thinking in the new one. Again, the point is that after the "revolution," new conventions prevail, but once entrenched, they are conformed to as thoughtlessly as their predecessors. Does our simple model capture that basic phenomenon?

In Run 3, we begin as before, with randomly distributed initial norms and zero noise. We let the system "equilibrate," locking into neighborhood norms (as before, these appear as vertical stripes over time). Then, at $t = 130$, we shock the system, boosting the level of noise to 1.0, and holding it there for ten periods. Then we turn the noise off and watch the system re-equilibrate. Figure 10.3 chronicles the experiment.

After the shock, an entirely new pattern of norms is evident on the left-hand page. But, looking at the right-hand radius page, we see that many agents who were thoughtlessly in the L norm (black) before the shock are *equally thoughtlessly* in the R norm (white) after.

A time series plot of average radius over the course of this experiment is also revealing. See figure 10.4. Following an initial transient phase, the mean radius attains a steady state value of roughly 2.25. During the brief "shock" period of maximum noise, the average radius rises sharply,

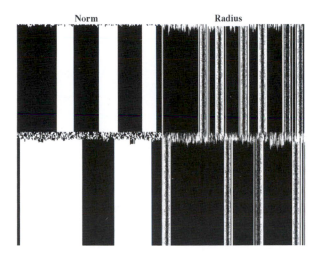

Figure 10.3. Re-equilibration after shock.

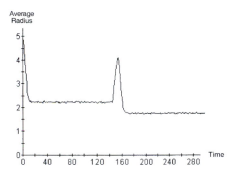

Figure 10.4. Shock experiment. Time series of average radius.

reflecting the agents' frenetic search for appropriate behavior in a period of social turmoil. One might expect that, with noise restored to zero, the average radius would relax back to its pre-shock value. In fact—as foreshadowed above—the post-shock steady state depends on the post-shock number of local norms. The lower the diversity, the lower the number of borders and, as in the present run, the lower the average radius.

Run 4. Noise of 0.15 and Endogenous Neighborhood Norms

Now, noise levels of zero and one are not especially plausible. What norm patterns, if any, emerge endogenously when initially random agents

Figure 10.5. Noise of 0.15 and endogenous norms.

play our game, but with a modest level of noise (probability of adopting a random norm)? The next four runs use the same initial conditions as Run 2, but add increasing levels of noise. With noise set at 0.15, we obtain dynamics of the sort recorded in figure 10.5.

Again, we see that individual computing is most intense at the norm borders—regions outlining the norms. We also see the emergence and disappearance of norms, the most prominent of which is the white island that comes into being and then disappears. One can think of islands as indicating *punctuated equilibria*.

For the realization depicted spatially above, the time series for average radius is given in figure 10.6. Following an initial transient phase, the average search radius clearly settles at roughly 2.0 for this realization.[5] Even at zero cost of sampling, in other words, a "stopping rule" for

---

[5]For the sake of visibility, the vertical axis ranges from zero to five. While, at this resolution, the plot may appear quite variable, the fluctuations around 2.0 are minor, given that the maximum possible radius is $(n - 1)/2$, or 95 in this case.

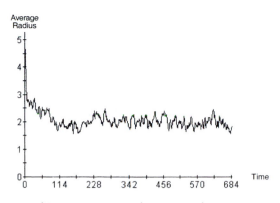

Figure 10.6. Noise of 0.15 time series of average radius.

the individual search radius emerges endogenously through local agent interactions. And this obtains at all levels of noise, as we shall see.

Now, in the cases preceding this one, there was zero noise in the agents' decision-making, and—although there would be run-to-run differences due to initial conditions and random agent call order—the point of interest was qualitative, and did not call for statistical discussion. However, in this and subsequent cases, there is noise, and quantitative matters are of interest. Hence, data from a single realization may be misleading and a statistical treatment is appropriate. The statistical analysis of simulation output has itself evolved into a large area, and highly sophisticated methods are possible. See Law and Kelton 1991 and Feldman and Valdez-Flores 1996. Our approach will be simple.

STATISTICAL RESULTS

To estimate the expected value of the long-run average search radius in this noise $= 0.15$ case, the model was rerun 30 times (so that, by standard appeal to the central limit theorem, a normal approximation is defensible) with a different random seed each time (to insure statistical independence across runs). In each run, the mean data were sampled at $t = 300$ (long after any initial transient had damped out). For a considered discussion of simulation stopping times, and all the complexities of their selection, see Judd 1999. The resulting 95% confidence interval[6] for the steady state mean search radius is [1.89, 2.03]. We double the noise level to 0.30 in Run 5.

---

[6]This is computed as $\bar{x} \pm z_{0.025} \frac{s}{\sqrt{n}}$ as in Freund (1992, 402), with $z_{0.025} = 1.96$, and $\bar{x}$ the average and s the standard deviation over our $n = 30$ runs.

Run 5. Noise of 0.30 and Endogenous Neighborhood Norms

The result, shown in figure 10.7, is a more elaborate spatial patterning than in the previous run. Again, however, we see regions of local conformity amidst a globally diverse pattern.

In this run, we see the emergence of white and black islands, indicating punctuated equilibria once more. For this realization, the mean radius time series is plotted in figure 10.8. Computed as above, the 95% confidence interval for the steady state mean radius is [2.89, 3.04].

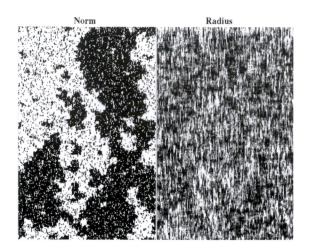

Figure 10.7. Noise of 0.30 and endogenous norms.

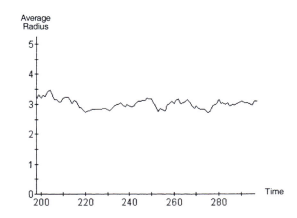

Figure 10.8. Noise of 0.30 time series of average radius.

Run 6. Noise of 0.50

Pushing the noise to 0.50 results in the patterning shown in figure 10.9, for which the average radius is plotted in figure 10.10. The 95% confidence interval for the long-run average search radius is [3.73, 3.81].

Run 7. Maximum Noise Does Not Induce Maximum Search

Finally, we fix the noise level at its maximum value of 1.0, meaning that agents are adopting the Left and Right convention totally at random.

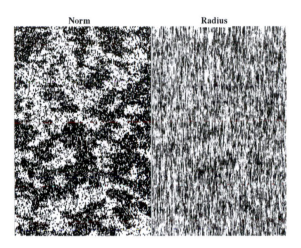

Figure 10.9. Noise of 0.50.

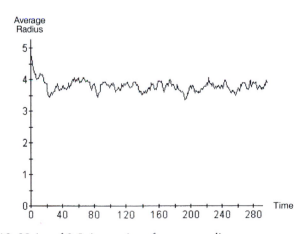

Figure 10.10. Noise of 0.5 time series of average radius.

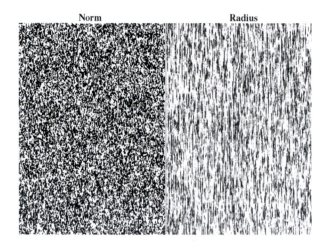

Figure 10.11. Noise of 1.0.

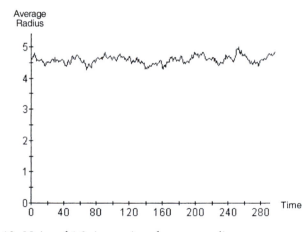

Figure 10.12. Noise of 1.0 time series of average radius.

One might assume that, in this world of maximum randomness, agents would continue to expand their search to its theoretical maximum of $(n - 1)/2$, or 95 in this case. But this is not what happens, as evident in figure 10.11. Indeed, as plotted in figure 10.12, it rises only to about 4.5. Computed as above, the 95% confidence interval is [4.53, 4.63]. Thinking—individual computing—is minimized in the monolithic world of Run 1. But, it does not attain its theoretical maximum in the totally random world of this run.

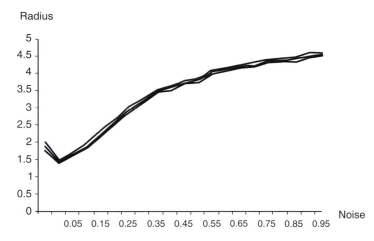

Figure 10.13. Steady state average radius and confidence intervals as function of noise at tolerance = 0.05.

Figure 10.13 gives a summary plot of the long-run average radius (middle curve) and 95% confidence intervals (outer curves) for noise levels ranging from 0 to 1, in increments of 0.05. Note that, at all noise levels, the confidence intervals are extremely narrow.

Sensitivity to the Tolerance Parameter

In all of the runs and statistical analyses given above, the tolerance parameter (see note 1) was set at 0.05, meaning that in applying the radius update rule, the agent regards $F(r)$ and $F(r+1)$ as *equal* if they are within 0.05 of one another. The agent's propensity to expand the search radius is inversely related to the tolerance. Figure 10.14 begins to explore the general relationship. For tolerances of 0.025 and 0.10, it displays the same triplet of curves as shown in figure 10.13 for $T = 0.05$ (which curve is also reproduced). All confidence intervals are again very narrow.

Even at the lowest tolerance of 0.025,[7] the average search radius does *not* attain the theoretical maximum even if the noise level does.

Finally, just to ensure that these results on the boundedness of search are not an artifact of sampling at $t = 300$, we conducted the same

---

[7] Tolerances much below this are of questionable interest. First, we detect virtually no spatial norm patterning. Second, one is imputing to agents the capacity to discern differences in relative norm frequency finer than 25 parts in a thousand, which begins to strain credulity.

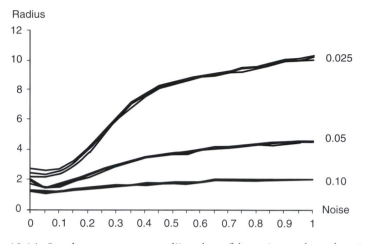

Figure 10.14. Steady state average radii and confidence intervals as function of noise at various tolerances, sampling at $t = 300$.

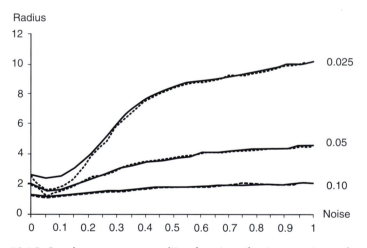

Figure 10.15. Steady state average radii as function of noise at various tolerances, sampling at $t = 300$ vs. $t = 10,000$.

analysis again, but sampling at $t = 10,000$. The results are compared in figure 10.15. The solid curves are the average search radii from the previous figure, computed at $t = 300$. The dotted curves are the corresponding data computed at $t = 10,000$. Clearly, for noise above roughly 0.20, there are no discernable differences at any of the three tolerances. (And for the low noise cases where there is some small difference, it is in fact the $t = 10,000$ curve that is lower.)[8] Search is bounded, even when noise is not.[9]

SUMMARY

My aim has been to extend the literature on the evolution of social norms with a simple agent-based computational model that generates the stylized facts regarding the evolution of norms—local conformity, global diversity, and punctuated equilibria—while capturing a central feature of norms that has been essentially ignored: that individual computing is often inversely related to the strength of a social norm. As norms become entrenched, we conform thoughtlessly. Obviously, many refinements, further sensitivity analyses, analytical treatments, and extensions are possible. But the present exposition meets these immediate and limited objectives.

Java Implementation

The model has been implemented in C++ and in Java. The Java implementation uses ASCAPE, an agent modeling environment developed at Brookings. Readers interested in running the model under their own assumptions may do so using this book's CD.

REFERENCES

Axelrod, Robert. 1984. *The Evolution of Cooperation.* New York: Basic Books.
———. 1996. An Evolutionary Approach to Norms. *American Political Science Review* 80(4): 1095–1111.
———. 1997. *The Complexity of Cooperation: Agent-Based Models of Competition and Collaboration.* Princeton: Princeton University Press.
Bicchieri, Cristina, Richard Jeffrey, and Brian Skyrms, eds. 1997. *The Dynamics of Norms.* New York: Cambridge University Press.
Canning, David. 1994. Learning and Social Equilibrium in Large Populations. In *Learning and Rationality in Economics*, ed. Alan Kirman and Mark Salmon. Oxford: Blackwell.

---

[8]The 95% confidence intervals (not shown) about the $t = 10,000$ curves are extremely narrow, as in the $t = 300$ cases.

[9]This demonstrated stability of the average search radius to $t = 10,000$ does not preclude mathematically the asymptotic approach to other values; an analytical treatment would be necessary to do that. On the other hand, even if established, the existence of asymptotic values significantly different from those that persist to (at least) $t = 10,000$ would be of debatable interest. As Keynes put it, "In the long run, we are all dead."

Feldman, David P., and James P. Crutchfield. 1997. Measures of Statistical Complexity: Why? Santa Fe Institute Working Paper 97-07-064.

Feldman, Richard M., and Ciriaco Valdez-Flores. 1996. *Applied Probability and Stochastic Processes*. Boston: PWS Publishing.

Freund, John E. 1992. *Mathematical Statistics*. 5th ed. Upper Saddle River, NJ: Prentice Hall.

Glaeser, Edward L., Bruce Sacerdote, and Jose Scheinkman. 1996. Crime and Social Interactions. *Quarterly Journal of Economics* 111:507–48.

Judd, Kenneth L. 1998. *Numerical Methods in Economics*. Cambridge: MIT Press.

Kandori, Michihiro, George J. Mailath, and Rafael Rob. 1993. Learning, Mutation, and Long-Run Equilibria in Games. *Econometrica* 61:29–56.

Kirman, Alan, and Mark Salmon, eds. 1994. *Learning and Rationality in Economics*. Oxford: Blackwell.

Law, Averill M., and W. David Kelton. 1991. *Simulation Modeling and Analysis*. New York: McGraw-Hill.

Lewis, David K. 1969. *Convention: A Philosophical Study*. Cambridge: Harvard University Press.

Ullman-Margalit, Edna. 1977. *The Emergence of Norms*. Oxford: Oxford University Press.

Vega-Redondo, Fernando. 1996. *Evolution, Games, and Economic Behaviour*. Oxford: Oxford University Press.

Young, H. Peyton. 1993a. The Evolution of Conventions. *Econometrica* 61:57–84.

———. 1993b. An Evolutionary Model of Bargaining. *Journal of Economic Theory* 59:145–68.

———. 1996. The Economics of Convention. *Journal of Economic Perspectives* 10:105–22.

———. 1998. *Individual Strategy and Social Structure*. Princeton: Princeton University Press.

GENERATING PATTERNS OF SPONTANEOUS

CIVIL VIOLENCE

THIS WORK GROWS OUT of a long-standing interest in security questions. My first three books[1] dealt exclusively with these issues, and developed mathematical models of combat, the Adaptive Dynamic Model (1985)[2] most notably. My fourth book, *Nonlinear Dynamics, Mathematical Biology and Social Science* (1997),[3] included three chapters on conflict. In one of them, I explored the use of nonlinear ordinary differential equations and nonlinear reaction diffusion systems, all drawn from mathematical epidemiology, to model the spread of revolutions specifically. That chapter concluded with the thought that, beyond these methods, "it would also be interesting to attempt the formulation of such a society using cellular automata (perhaps as a generalized Greenburg-Hastings model) or agents."[4] That thought stuck in my mind, but I didn't do much about it until instigated by John Steinbruner, who, in the wake of Rwanda, had just served on the Carnegie Commission on Preventing Deadly Conflict. Stimulated by discussions with John on the general problem, I began to accumulate sketches for a civil violence model.

Growing Artificial Societies does include resource combat between agents of different "tribes," and these arise endogenously through a cultural transmission dynamic (which is compared to Axelrod's cultural transmission model in Axtell, *et al.*).[5] However, there are no elements of central political authority (police) or political grievance against it, both of which loom large in the Civil Violence models that follow.

---

[1] Joshua M. Epstein, *Measuring Military Power* (Princeton: Princeton University Press, 1984); *Strategy and Force Planning: The Case of the Persian Gulf* (Washington, DC: Brookings Institution Press, 1987); *Conventional Force Reductions: A Dynamic Assessment* (Washington, DC: Brookings Institution Press, 1990).

[2] Joshua M. Epstein, *The Calculus of Conventional War: Dynamic Analysis without Lanchester Theory* (Washington, DC: Brookings Institution Press, 1985).

[3] Joshua M. Epstein, *Nonlinear Dynamics, Mathematical Biology, and Social Science* (Reading, MA: Addison-Wesley, 1997).

[4] Epstein, *Nonlinear Dynamics*, 86.

[5] Axtell R., Axelrod R., Epstein J. M., and Cohen M. D. 1996 Aligning Simulation Models: A Case Study and Results. *Computational and Mathematical Organization Theory* 1(2):123–41.

The aim of the modeling is to generate certain stylized facts and core dynamics of decentralized rebellions and spontaneous ethnic conflicts. It illustrates a number of themes of the Generative chapter. Autonomous agents interact locally on an explicit space. They exhibit bounded rationality in deciding whether or not to rebel, adapt to ever-changing local information, and are heterogeneous by hardship, level of political grievance, propensity to take risks, and ethnic identity. Local conformity, global diversity, and punctuated equilibrium—hallmarks of complex systems—are clearly evident here, as is the potential for empirical reconstruction (e.g., of Rwanda or Bosnia) along the lines of the Anasazi work, and policy applications in the area of peacekeeping.

The model employs the graphical technique introduced in the preceding chapter: one screen depicts the landscape of overt agent behavior, while a second simultaneously depicts the underlying "emotionscape," in which individuals are colored by their unobservable level of political grievance. My main motivation in adopting this display is to capture the "revolutionary situation" in which no one is rebelling (left screen all blue), despite the fact that political grievance is extremely high (right screen all red). As Lenin put it, "a revolution is impossible without a revolutionary situation; furthermore, not every revolutionary situation leads to revolution."

Chapter 11

MODELING CIVIL VIOLENCE: AN AGENT-BASED

COMPUTATIONAL APPROACH

JOSHUA M. EPSTEIN*

THIS ARTICLE PRESENTS an agent-based computational model of civil violence. Two variants of the civil violence model are presented. In the first a central authority seeks to suppress decentralized rebellion. In the second a central authority seeks to suppress communal violence between two warring ethnic groups.

This article presents an agent-based computational model of civil violence. For an introduction to the agent-based modeling technique, see Epstein and Axtell 1996. I present two variants of the civil violence model. In the first a central authority seeks to suppress decentralized rebellion. Where I use the term "revolution," I do so advisedly, recognizing that no political or social order is represented in the model. Perforce, neither is the overthrow of an existing order, the latter being widely seen as definitive of revolutions properly speaking. The dynamics of decentralized upheaval, rather than its political substance, is the focus here.[1] In the second model a central authority seeks to suppress communal violence between two warring ethnic groups.

*This paper results from the Arthur M. Sackler Colloquium of the National Academy of Sciences, "Adaptive Agents, Intelligence, and Emergent Human Organization: Capturing Complexity through Agent-Based Modeling," held October 4–6, 2001, at the Arnold and Mabel Beckman Center of the National Academies of Science and Engineering in Irvine, CA.

I extend special thanks to John D. Steinbruner for his long-term involvement and close collaboration. Miles T. Parker deserves particular credit for implementation of the models in ASCAPE, extensive statistical and sensitivity analysis, and valuable discussions and insights. I thank Robert Axtell, Robert Axelrod, Paul Collier, William Dickens, Michael Doyle, Steven Durlauf, Carol Graham, John Miller, Scott Page, Nicholas Sambanis, Elizabeth Wood, and Peyton Young for useful comments, criticism, and advice; Ross Hammond for research assistance; and Kelly Landis and Shubha Chakravarty for assistance in preparation of the manuscript and its graphics. The National Science Foundation and the John D. and Catherine T. MacArthur Foundation provided financial support.

This essay was published previously in *Proceedings of The National Academy of Sciences, Colloquium* 99(3): 7243–7250.

[1]The term "decentralized" deserves emphasis. Revolutionary organizations proper are not modeled. On their economic and other imperatives in important cases, see Collier 2000 and Collier and Hoeffler 1998, 1999, 2001.

And, as in model I, I am interested in generating certain characteristic phenomena and core dynamics; I do not purport to reconstruct any particular case in detail, although, as discussed in Epstein, *et al.* 2001, that is an obvious long-term objective.

CIVIL VIOLENCE MODEL I: GENERALIZED REBELLION AGAINST CENTRAL AUTHORITY

This model involves two categories of actors. "Agents" are members of the general population and may be actively rebellious or not. "Cops" are the forces of the central authority, who seek out and arrest actively rebellious agents. Let me describe the agents first. As in all agent-based models, they are heterogeneous in a number of respects. The attributes and behavioral rules of the agents are as follows.

The Agent Specification

First, in any model of rebellion there must be some representation of political grievance. My treatment of grievance will be extremely simple and will involve only two highly idealized components, which, for lack of better terminology, will be called hardship (H) and legitimacy (L). Their definitions are as follows:

H is the agent's perceived hardship (i.e., physical or economic privation). In the current model, this is exogenous. It is assumed to be heterogeneous across agents. Lacking further data, each individual's value is simply drawn from $U(0,1)$, the uniform distribution on the interval $(0,1)$. Of course, perceived hardship alone does not a revolution make. As noted in the Russian revolutionary journal, *Narodnaya Volya*, "No village ever revolted merely because it was hungry" (quoted in Kuran 1989 and deNardo 1985). Another crucial factor is:

L, the perceived legitimacy of the regime, or central authority. In the current model, this is exogenous and is equal across agents, and in the runs discussed below, will be varied over its arbitrarily defined range of 0 to 1.

The level of grievance any agent feels toward the regime is assumed to be based on these variables. Of the many functional relationships one might posit, we will assume:

$$G = H(1 - L).$$

Grievance is the product of perceived hardship (H) and perceived "illegitimacy," if you will $(1 - L)$.[2] The intuition behind this functional form is simple. If legitimacy is high, then hardship does not induce political grievance. For example, the British government enjoyed unchallenged legitimacy ($L = 1$) during World War II. Hence, the extreme hardship produced by the blitz of London did not produce grievance toward the government. By the same token, if people are suffering (high H), then the revelation of government corruption (low L) may be expected to produce increased levels of grievance.

Of course, the decision to rebel depends on more than one's grievance. For example, some agents are simply more inclined to take risks than others. Accordingly, I define R as the agent's level of risk aversion. Heterogeneous across agents, this (like H) is assumed to be uniformly distributed. Each individual's level is drawn from U(0,1) and is fixed for the agent's lifetime.

All but the literally risk neutral will estimate the likelihood of arrest before actively joining a rebellion. This estimate is assumed to increase with the ratio of cops to already rebellious—so-called "active"—agents within the prospective rebel's vision. To model this, I define v as the agent's vision. This is the number of lattice positions (north, south, east, and west of the agent's current position) that the agent is able to inspect. It is exogenous and equal across agents. As in most agent-based models, vision is limited; information is local. Letting $(C/A)_v$ denote the cop-to-active ratio within vision v, I assume the agent's estimated arrest probability P to be given by

$$P = 1 - exp[-k(C/A)_v].$$

The constant k is set to ensure a plausible estimate (of $P = 0.9$) when $C = 1$ and $A = 1$. Notice that A is always at least 1, because the agent always counts himself as active when computing P. He is asking, "How likely am I to be arrested if I go active?" Again, the intuition behind this functional form is very simple. Imagine being a deeply aggrieved agent considering throwing a rock through a bank window. If there are 10 cops at the bank window, you are much more likely to be arrested if you are the first to throw a rock ($C/A = 10$) than if you show up when there are already 29 rock-throwing agents ($C/A = 1/3$). For a fixed level of cops, the agent's estimated arrest probability falls the more actives there are. This simple idea will play an important role in the analysis.

---

[2]This is not proffered as the unique equation of human grievance. It is a simple algebraic relation intended as an analytical starting point.

TABLE 11.1
Agent State Transition

State	$(G - N)$	State Transition
Q	$> T$	$Q \Rightarrow A$
Q	$\leq T$	$Q \Rightarrow Q$
A	$> T$	$A \Rightarrow A$
A	$\leq T$	$A \Rightarrow Q$

Clearly, in considering whether or not to rebel, a risk-neutral agent won't care what the estimated arrest probability is, whereas a risk-averse agent will. It will therefore prove useful to define $N = RP$, the agent's net risk—the product of his risk aversion and estimated arrest probability. (This can be considered the special, $\alpha = 0$, case of $N = RPJ^{\alpha}$, where J is the jail term, as discussed in Epstein *et al.* 2001.) These ingredients in hand, the agent's behavioral rule is summarized in table 11.1.

If, for an agent in state Q, the difference $G - N$ exceeds some non-negative threshold T, which could be zero, then that quiescent agent goes active. Otherwise, he stays quiescent. If, for an agent in state A, the difference exceeds T, then that active agent stays active. Otherwise, he goes quiescent. In summary, the agent's simple local rule is:

Agent rule **A**: If $G - N > T$ be active; otherwise, be quiet.

This completes the agent specification.

BOUNDED RATIONALITY

It is natural to interpret this rule as stipulating that the agent take whichever binary action (active or quiescent) maximizes expected utility where, in the spirit of Kuran (1989), $G - N$ is the expected utility of publicly expressing one's private grievance, and T is the expected utility of not expressing it (i.e., of preference falsification, in Kuran's terminology). Typically, T is set at some small positive value. Notice, however, that if it takes negative values, like $-G$ (i.e., the frustration level associated with preference falsification equals the grievance level itself), agents may find it rational to rebel knowing that they will suffer negative utility. It's simply worse to "sit and take it anymore." Agents weigh expected costs and benefits, but they are not hyperrational. One might say (with all due respect to Olson 1971) that individual rationality is "local" also, in the sense that the agent's expected utility calculation excludes any estimate of how his isolated act of rebellion may affect the social order. Notice, very importantly, that deterrence is local in this model and depends on the local (individually visible)—not the global—ratio of cops to actives, which is highly dynamic in this spatial model with movement.

The Cop Specification

The cops are much simpler than prospective rebels. Their attributes are as follows:

v^*, the cop vision, is the number of lattice positions (north, south, east, and west of the cop's current position) that the cop is able to inspect. It is exogenous and equal across cops. The cops' v^* need not equal the agents' v, but will typically be small relative to the lattice size: cop vision is local also. The cops, like the other agents, have one simple rule of behavior:

> Cop rule **C**: Inspect all sites within v^* and arrest a random active agent.

Cops never defect to the revolution in this model.

Movement and Jail Terms

Although the range of motion will vary depending on the numerical values selected for v and v^*, the syntax of the movement rule is the same for both agents and cops:

> Movement rule **M**: Move to a random site within your vision.

Although the range of vision (v) is fixed, agent information (the number of cops and actives they see) is heterogeneous because of movement.

Regarding jail terms for arrested actives, these are exogenous and set by the user. Specifically, the user selects a value for the maximum jail term, J_max. Then, any arrested active is assigned a jail term drawn randomly from U(0, J_max). J_max will affect the dynamics in important ways by removing actives from circulation for various durations. However, for the present version of the model with alpha implicitly set to zero (see remark above), there is no deterrent effect of increasing the jail term. Setting alpha to a positive value would produce a deterrent effect. In addition to having no deterrent effect, it is assumed that agents leave jail exactly as aggrieved as when they entered.

MEASUREMENT

It is important to state forthrightly that I make no pretense to measuring model variables such as perceived hardship (Graham and Pettinato 2002) or legitimacy. The immediate question is whether this highly idealized model is sufficient to generate recognizable macroscopic revolutionary dynamics of fundamental interest. If not, then issues of measurement are

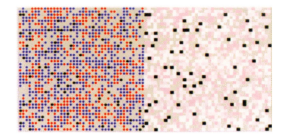

Figure 11.1. Action and grievance screens.

moot. So, the first issue is whether the model produces interesting output. In addition to data generated by the model, run-time visualization of output is very useful. My graphical strategy is as follows.

GRAPHICS

Events transpire on a lattice. Agents and cops move around this space and interact. I am interested in the dynamics of grievance and—quite separately—in the dynamics of revolutionary action. The point of separating these private and public spheres is to permit illustration of a core point in all research on this topic: public order may prevail despite tremendous private opposition to—feelings of grievance toward—a regime. Given this important distinction between private grievance and public action, two screens are shown (see fig. 11.1).

On the right screen, agents are colored by their private level of grievance. The darker the red, the higher the level of grievance. On the left screen, agents are colored by their public action: blue if quiescent; red if active. Cops are colored black on both screens. Simply to reduce visual clutter, all agents and cops are represented as circles on the left screen and squares on the right. Unoccupied sites are sand-colored on both screens.

RUNS

To begin each run of the model, the user sets L, J, v, $v*$, and the initial cop and agent densities. To ensure replicability of the results, input assumptions for all runs are provided in table 11.2. Agents are assigned random values for H and R, and cops and (initially) quiescent agents are situated in random positions on the lattice. The model then simply spins forward under the rule set: {**A, C, M**}. An agent or cop is selected at random (asynchronous activation) and, under rule **M**, moves to a random site within his vision, where he acts in accord with rule **C** (if a cop) or **A** (if an agent). The model simply iterates this procedure until the user quits or some stipulated state is attained. What can one generate in this extremely simple model?

TABLE 11.2
Input Assumptions for Runs

Variable Name	Model One				Model Two			
	Run 1	Run 2	Runs 3&4	Run 5	Run 6	Run 7	Run 8	
Cop vision	1.7	7	7	7	1.7	1.7	1.7
Agent vision	1.7	7	7	7	1.7	1.7	1.7
Legitimacy	0.89	0.82	0.9	0.8	0.9	0.8	0.8
Max. jail term	15	30	Infinite	Infinite	15	15	15
Movement	None	Random site in vision	Random site in vision	Random site in vision	Random site in vision	Random site in vision	Random site in vision
Initial cop density	0.04	0.04	0.074	0.074	0	0	0.04

All models: lattice dimensions (40 × 40); topology (tours); cloning probability (0.05); arrest probability constant, k (2.3); max age (200); agent "active" threshold, for $G - N$ (0.1); initial population density (0.7) agent updating (asynchronous); agent activation (once per period, random order). Departures noted in text.

Individual Deceptive Behavior

Despite their manifest simplicity, the agents exhibit unexpected deceptive behavior: privately aggrieved agents turn blue (as if they were non-rebellious) when cops are near, but then turn red (actively rebellious) when cops move away. They are reminiscent of Mao's directive that revolutionaries should "swim like fish in the sea," making themselves indistinguishable from the surrounding population. *Ex post facto*, the behavior is easily understood: the cop's departure reduces the C/A ratio within the agent's vision, reducing his estimated arrest probability, and with it his net risk, N, all of which pushes $G - N$ over the agent's activation threshold, and he turns red. But it was not anticipated. Moreover, it would probably not have been detected without a spatial visualization (see Epstein, *et al.* 2001); individual deception would not be evident in a time series of total rebels, for example.

Free Assembly Catalyzes Rebellious Outbursts

With both agents and cops in random motion, it may happen that high concentrations of actives arise endogenously in zones of low cop density. This can depress local C/A ratios to such low levels that even the mildly aggrieved find it rational to join. This catalytic mechanism is illustrated

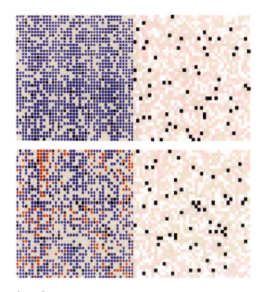

Figure 11.2. Local outbursts.

in figure 11.2.[3] Random spatial correlations of activists catalyze local outbursts.

This is why freedom of assembly is the first casualty of repressive regimes. Relatedly, it is also the rationale for curfews. The mechanism is that local activist concentrations reduce local C/A (cop to active) ratios, reducing (via the equation for P above) the risk of joining the rebellion. To be the first rioter, one must be either very angry or very risk-neutral, or both. But to be the 4,000th—if the mob is already big, relative to the cops—the level of grievance and risk-taking required to join the riot is far lower. This is how, as Mao Tse-tung liked to say, "a single spark can cause a prairie fire" (quoted in Kuran 1989). Coincidentally, the Bolshevik newspaper founded by Lenin was called *Iskra*, the spark! The Russian revolution itself provides a beautiful example of the chance spatial correlation of aggrieved agents. As Kuran (1989) recounts, "On February 23, the day before the uprising, many residents of Petrograd were standing in food queues, because of rumors that food was in short supply. Twenty thousand workers were in the streets after being locked out of a large industrial complex. Hundreds of off-duty soldiers were outdoors looking for distraction. And, as the day went on, multitudes of women workers left their factories early to march in celebration of

---

[3] All animations screen-captured in this article can be viewed in full as QuickTime movies on this book's CD.

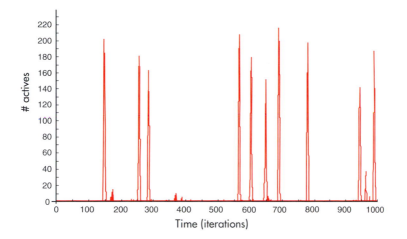

Figure 11.3. Punctuated equilibrium.

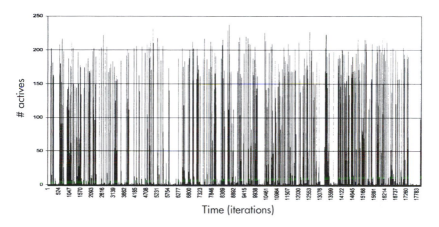

Figure 11.4. Punctuated equilibrium persists.

Women's Day. The combined crowd quickly turned into a self-reinforcing mob. It managed to topple the Romanov dynasty within 4 days." A random coalescence of aggrieved agents depresses the local C/A ratio, quickly emboldening all present to openly express their discontent. A time series of total rebels is also revealing. It displays one of the hallmarks (Young 1998) of complex systems: punctuated equilibrium (fig. 11.3). Long periods of relative stability are punctuated by outbursts of rebellious activity. And indeed, many major revolutions (e.g., East German) are episodic in fact. The same qualitative pattern of behavior— punctuated equilibrium—persists indefinitely, as shown in figure 11.4, which plots the data over some 20,000 iterations of the model.

WAITING TIME DISTRIBUTION

Is there any underlying regularity to these complex dynamics? For many complex systems, it turns out to be of considerable interest to study the distribution of waiting times between outbursts above some threshold. In this analysis we set the threshold at 50 actives. An outburst begins when the number of actives exceeds 50 and ends when it falls below 50. I am interested in the time between the end of one outburst and the start of the next. Sometimes, one must wait a long time (e.g., 100 periods) until the next outburst. Sometimes, the next outburst is nearly immediate (e.g., a gap of only two periods). The frequency distribution of these inter-outburst waiting times, for 100,000 iterations of the model, is shown in figure 11.5.

In the complexity literature, one often encounters the notion of an "emergent phenomenon." I have argued elsewhere (Epstein 1999) that substantial confusion surrounds this term. However, if one defines emergent phenomena simply as "macroscopic regularities arising from the purely local interaction of the agents" (Epstein and Axtell 1996), then this waiting time distribution surely qualifies. It was entirely unexpected and would have been quite hard to predict from the underlying rules of agent behavior. For instance, figure 11.5 suggests a Weibull or perhaps Lognormal distribution. Although rigorous identification is a suitable topic for future research, these data are clearly not uniformly distributed. But all distributions used in defining the agent population—the distribution of hardship and risk aversion—are uniform. In a uniform waiting time distribution, one is just as likely to wait 100 cycles as 50;

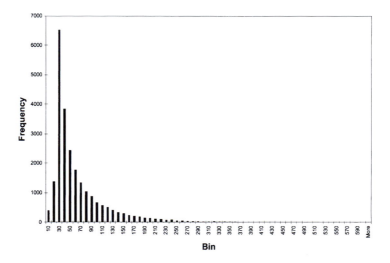

Figure 11.5. Waiting time distribution.

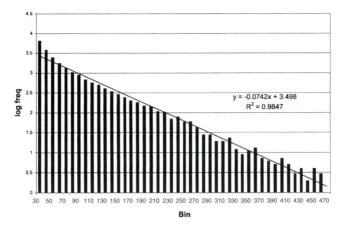

Figure 11.6. Logged data (truncated).

that is not the case in my model, at least for these parameters. The mean of these data—the average duration between outbursts—is 60. (The SD is 55.) Clearly, most of the probability density is concentrated around this value: this means that one is much more likely to see successive outbursts within 60 cycles than within 100.

If one is willing to truncate this distribution—throwing out the most high-frequency events (waiting times less than 30 cycles), the remaining distribution is well fit by the negative exponential. In particular, the logged (truncated) data are shown in figure 11.6.

Ordinary least-squares regression yields an R-squared of 0.98 with slope -0.07 and intercept 3.5. The negative exponential distribution is ubiquitous in the analysis of failure rates—the rates at which electrical and mechanical systems break down. It would be interesting if "social breakdowns" followed a similar distribution. Another obvious issue is the sensitivity of this distribution to variations in key parameters. For example, how would an increase in the jail term deform the distribution? One might conjecture that, by removing rebellious agents from circulation for a longer period, increasing the jail term would "flatten" the distribution and raise its mean. All of these issues could be fruitfully explored in the future. For the moment, the core point is that a powerful statistical regularity underlies the model's punctuated equilibrium dynamics.

OUTBURST SIZE DISTRIBUTION

A second natural statistical topic is the size distribution of rebellious outbursts. To study this, I use the same parameterization as above and adopt the same threshold: 50 actives. But there are numerous ways to

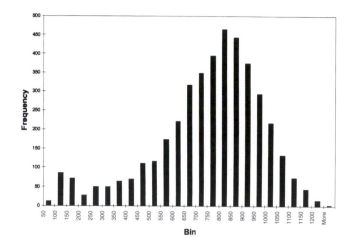

Figure 11.7. Total activation distribution.

measure outburst size for statistical purposes. For example, imagine a flare-up lasting 5 days with the following number of actives per day: 60, 100, 120, 95, and 80. This outburst has a peak active level of 120. It has average (daily) activation of 100 and total activation (the sum) of 500. The size distribution of outbursts using the total activation measure is shown in figure 11.7.

The mean and SD are, respectively, 708 and 230. The distributions using the peak and average data are qualitatively similar. As in the case of the waiting time distribution, one could conduct further analysis to identify the best fit to figure 11.7. The point to emphasize here is not which distribution is best, but that some macroscopic regularity emerges. A strength of agent models is that they generate a wealth of data amenable to statistical treatment.

A RIPENESS INDEX

Turning to another topic, we often speak of a society as being "ripe for revolution." In using this terminology, I have in mind a high level of tension or private frustration. Does the present model allow me to quantify this in an illuminating way? As a first cut, I noted earlier that society can be bright red on the right screen (indicating a high level of grievance) while being entirely blue on the left (indicating that no one is expressing, or "venting," their grievance). So, if this combination of high average grievance $\overline{G}$ on the right and high frequency of blues $\overline{B}$ on the left were the best indicator of high tension, a reasonable "ripeness" index would be simply their product: $\overline{G}\,\overline{B}$. This, however, ignores the crucial question, why are agents blue? If they are inactive simply because

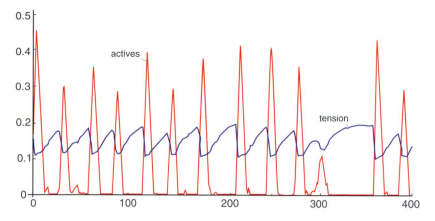

Figure 11.8. Tension (blue) and actives (red).

they are risk averse and have no inclination to go active, then they are not truly frustrated in the inactive blue state. So, for fixed $\overline{G}$ and $\overline{B}$ a good tension index should increase as average risk aversion falls (more agents want to act out, but are nonetheless staying blue). Hence, a better simple measure is: $\overline{G}\,\overline{B}/\overline{R}$, where $\overline{R}$ is average risk aversion. In figure 11.8, I plot this against a curve of actives designed to exhibit high volatility. It is clear that a buildup of tension precedes each outburst and might be the basis of a warning indicator.

I turn now to a comparison of two runs involving reductions in legitimacy.[4] Both runs begin with legitimacy at a high level. In the first, I execute a large absolute reduction (from $L = 0.9$ to $L = 0.2$) in legitimacy, but in small increments (of a percent per cycle). In the second, I reduce it far less in absolute terms (from $L = 0.9$ to $L = 0.7$), but do so in one jump. Which produces the more volatile social dynamics, and why?

Salami Tactics of Corruption

Figure 11.9 shows the results when I reduce legitimacy in small increments. It displays three curves. The downward sloping upper curve plots the steady incremental decline in legitimacy over time. (To make these graphs clear, I actually plot 1,000 L.) The horizontal red curve just above the time axis shows the number of actives in each time period. Even though legitimacy declines to zero, there is no red spike, no explosion, because—as discussed earlier—each new active is being picked off in isolation, before he can catalyze a wider rebellion. And this is why the

---

[4] I thank Miles T. Parker for this comparison.

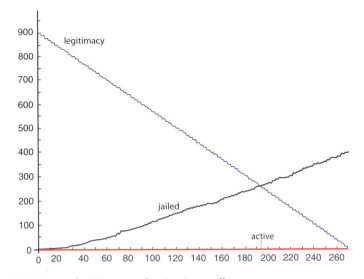

Figure 11.9. Large legitimacy reduction in small increments.

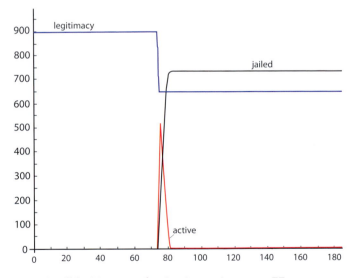

Figure 11.10. Small legitimacy reduction in one jump, $t = 77$.

middle curve—representing the total jailed population—rises smoothly over time.

The same variables are plotted in figure 11.10. However, the scenario is different. I hold legitimacy at its initially high level (of 0.90) for 77 periods. Then, in one jump, I reduce it to 0.70, where it stays. The upper legitimacy curve is a step function. Even though the absolute legitimacy reduction (of 0.30) is far smaller than before, there is an explosion of actives, shown by the red spike. And, in turn, there is a sharp rise in the jailed population, whose absolute size exceeds that of the previous run.

Now, why the difference? In the incremental legitimacy reduction scenario, the potentially catalytic agents at the tail of the grievance distribution are being picked off in isolation, before they can stimulate a local contagion. The sparks, as it were, are doused before the fire can take off. In the second—one-shot reduction—case, even though the absolute legitimacy decline is far smaller, multiple highly aggrieved agents go active at once. And by the same mechanism as discussed earlier, this depresses local C/A ratios enough that less aggrieved agents jump in. Hence, the rebellious episode is greater, even though the absolute legitimacy reduction is smaller. It is the rate of change—the derivative—of legitimacy that emerges as salient.

This result would appear to have important implications for the tactics of revolutionary leadership. Rather than chip away at the regime's legitimacy over a long period with daily exposés of petty corruption, it is far more effective to be silent for long periods and accumulate one punchy exposé. Indeed, the single punch need not be as "weighty," if you will, as the "sum" of the daily particulars. (The one-shot legitimacy reduction need not be as great as the sum of all of the incremental deltas.) Perhaps this is why Mao would regularly seclude himself in the mountains in preparation for a dramatic reappearance, and why the return of exiled revolutionary leaders—like Lenin and Khomeini—are attended with such trepidation by authorities. Perhaps this is also why dramatic "triggering events" (e.g., assassinations) loom so large in the literature on this topic; often, they are instances in which the legitimacy of the regime suddenly takes a dive. By the same token, the earlier run (incremental legitimacy reductions) explains the counterrevolutionary value of agent provocateurs: they incite the most aggrieved agents to go active prematurely, allowing them to be arrested before they can catalyze the wider rebellion.[5] While often sufficient, sharp legitimacy reductions are not the only inflammatory mechanism.

Cop Reductions

Indeed, "it is not always when things are going from bad to worse that revolutions break out," wrote Tocqueville (1955). "On the contrary, it oftener happens that when a people that has put up with an oppressive rule over a long period without protest suddenly finds the government relaxing its pressure, it takes up arms against it." According to Kuran

---

[5] I thank Robert Axelrod for this point.

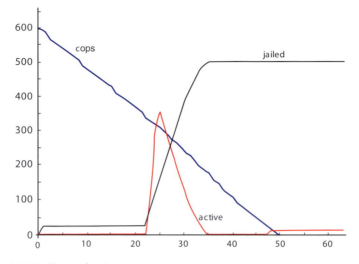

Figure 11.11. Cop reductions.

(1989), in the cases of the French, Russian, and Iranian revolutions, "substantial numbers of people were privately opposed to the regime. At the same time, the regime appeared strong, which ensured that public opposition was, in fact, unalarming. What, then, happened to break the appearance of the invincibility of the regime and to start a revolutionary bandwagon rolling? In the cases of France and Iran, the answer seems to lie, in large measure, in a lessening of government oppression." Indeed, Tocqueville wrote that "liberalization is the most difficult of political arts." Here I interpret liberalization as cop reductions. Beginning at a high level, I walk the level of cops down. Figure 11.11 shows the typical result.

Unlike the case of incremental legitimacy reductions above (salami tactics), there comes a point at which a marginal reduction in central authority does "tip" society into rebellion. The dynamics under reductions in repressive potential (cops) are fundamentally different from the dynamics under legitimacy reduction in this model—and perhaps in societies. Because both types of reduction are emboldening to revolutionaries, one might have imagined that reductions in the regime's repressive power—the cop level—would produce dynamics equivalent to those under legitimacy reduction. As we see, however, the dynamics are fundamentally different.

Stylized Facts Generated in Model I

Although model I is exploratory and preliminary, it does produce noteworthy phenomena with some qualitative fidelity and therefore seems, at the very least, promising. It showed, first, the unexpected emergence of individually deceptive behavior, in which privately aggrieved agents hide their feelings when cops are near, but engage in openly rebellious activity when the cops move away. In general, the model naturally represents political "tipping points"—revolutionary situations in which the screen is blue on the left (all agents quiescent) but red on the right. Surface stability prevails despite deep and widespread hostility to the regime. When pushed beyond these tipping points, the model produces endogenous outbursts of violence and punctuated equilibria characteristic of many complex systems. For some parameter settings, the size distribution of rebellious outbursts and the distribution of waiting times between outbursts exhibit remarkable regularities. The model explains standard repressive tactics like restrictions on freedom of assembly and the imposition of curfews. Such policies function to prevent the random spatial clustering of highly aggrieved risk-takers, whose activation reduces the local cop-to-active ratio, permitting other less aggrieved and more timid agents to join in. This same catalytic dynamic underlies the intriguing "salami tactic" result: Legitimacy can fall much farther incrementally than it can in one jump, without stimulating large-scale rebellion. The reason is that, in the former (salami) case, the tails of the radical distribution— the sparks—are being picked off before they can catalyze joining by the less aggrieved, and this had implications for both revolutionary and counterrevolutionary tactics. The model bears out Tocqueville's famous adage that "liberalization is the most difficult of political arts," showing that (quite unlike legitimacy reductions) incremental reductions in repressive potential (cops) can produce large-scale tipping events.

It should be added that the individual-level specification is quite minimal, imposing bounded demands on the agent's (and cops) information and computing capacity, while still insisting that the agent crudely weigh expected benefits against expected costs in deciding how to act. Agents are boundedly rational and locally interacting; yet interesting macroscopic phenomena emerge. As noted at the outset, the model seems most promising for cases of decentralized upheaval. Although one could argue that certain effects of revolutionary leadership—reductions in perceived legitimacy through rousing speeches or writings that expose regime corruption—are captured, explicit leadership as such is really not modeled. That could be a weakness in cases—for example, the communist Chinese revolution—where central leadership was important, although some would argue that, even there, the main issue was not the individual leader, but society's "ripeness for revolution." As Engels wrote, "in default of Napoleon, another would have been found" (cited in Kuran 1989). My tension index might be a crude measure of this ripeness. Let me turn now to situations of interethnic violence.

CIVIL VIOLENCE MODEL II: INTER-GROUP VIOLENCE

Although distinct cultural groups have been generated in agent-based computational models (Epstein and Axtell 1996; Axelrod 1997), here, I will posit two ethnic groups: blue and green. Agents are as in model I and turn red when active. But now, "going active" means killing an agent of the other ethnic group. The killing is not confined to agents of the out-group known to have killed. It is indiscriminate. In this variant of the model, legitimacy is interpreted to mean each group's assessment of the other's right to exist, and for the moment, L is exogenous and the same for each group. For model II, I also introduce some simple population dynamics. Specifically, agents clone offspring onto unoccupied neighboring sites with probability p each period. Offspring inherit the parent's ethnic identity and grievance. Because there is birth, there must be death to prevent saturation. Accordingly, agents are assigned a random death age from U(0, max_age). Here, max_age = 200. Cops are as before, and arrest—evenhandedly—red agents within their vision. (This assumption of even-handedness can, of course, be relaxed.) There is no in-group policing in this version of the model, although Fearon and Laitin (1996) argue convincingly that this may be important in many cases.

Peaceful Coexistence

For the first run of model II, I set legitimacy to a high number, just to check whether peaceful coexistence prevails with no cops. Figure 11.12 depicts a typical situation. The left screen clearly shows spatial heterogeneity and peaceful mixing of groups with no red agents. On the right, only the palest of pink shades, indicating low levels of grievance, are seen. Harmonious diversity prevails. However, with no cops to regulate the competition, if L falls, even to 0.8, the picture darkens substantially. Indeed, ethnic cleansing results.

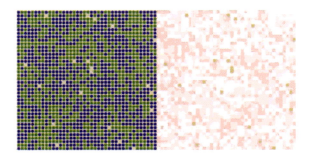

Figure 11.12. Peaceful coexistence.

Ethnic Cleansing

The sequence of five panels in figure 11.13 clearly shows local episodes of ethnic cleansing, leading ultimately to the annihilation of one group by the other: genocide.

Over a large number of runs ($n = 30$), genocide is always observed. The victor is random. The phenomenon is strongly reminiscent of

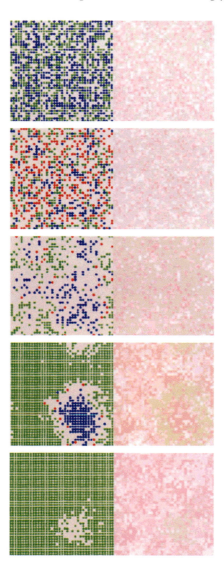

Figure 11.13. Local ethnic cleansing to genocide.

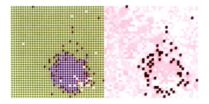

Figure 11.14. Safe havens emerge under peacekeeping.

"competitive exclusion" in population biology (see May 1981). When two closely related species compete in a confined space for the same resource, one will gain an edge and wipe the other out. If, however, the inter-species competition is regulated by a predator that feeds evenhandedly on the competitors, then both can survive. Peacekeepers are analogous to such predators. I introduce them presently.

Safe Havens

I begin this run exactly as in the previous genocide case. But, at $t = 50$, I deploy a force of peacekeepers. They go to random unoccupied sites on the lattice. And this typically produces safe havens. A representative result is shown in figure 11.14. Rather than begin with no cops initially, as in the previous run, a case with high initial cop density was also examined. Once a stable pattern had emerged, the cops were withdrawn. Here, with heavy authority from the start (a high cop density of 0.04), a stable, but nasty, regime emerges. The presence of cops prevents either side from wiping the other out, but their coexistence is not peaceful: ethnic hostility is widespread at all times. When (*ceteris paribus*) all cops are withdrawn—the peacekeepers suddenly go home—there is reversion to competitive exclusion and, eventually, one side wipes out the other.

Clearly, peacekeeping forces can avert genocide. But what is the overall relationship between the size of a peacekeeping presence and the incidence of genocide? As an initial exploration of this complex matter, a sensitivity analysis on cop density is conducted.

Cop Density and Extinction Times

For purposes of this analysis, all cops are in place at time 0. But there will be random run-to-run variations in their initial positions and, of course, in their subsequent movements. These initial cop densities are systematically varied from 0.0 up to 0.1, in increments of 0.002. For each such value, the model was run 50 times until the monochrome—genocide—state was reached. (If it was not reached

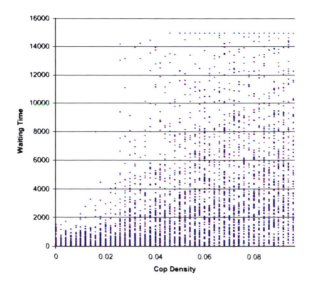

Figure 11.15. Waiting time and initial cop density data.

after 15,000 cycles, the run was terminated.) The data appear in figure 11.15.

There are three things to notice. First, at low force densities (0 to 0.02), convergence to genocide is rapid. Second, the same rapid convergence to genocide is observed at all force densities. Third, reading vertically, at high force densities (0.08 and above), there is high variance. One can have high effectiveness (delays of over 15,000 cycles) or extremely low effectiveness (convergence in tens of cycles). The devil would appear to be in the details in peacekeeping operations.

As noted earlier, the model was run 50 times at each density (with a different random seed each time). So, at each density, there is a sample distribution of waiting times over the 50 runs. I plot the means of these distributions at each initial density in figure 11.16, along with the best linear fit to the same data. On average, the larger the initial force of peacekeepers, the more time one buys. At the same time, however, the SD is also rising, as shown in figure 11.17. Hence, the confidence interval about the mean is expanding. Thus, as the mean waiting time to genocide grows, we have decreasing confidence in it as a point estimate of the outcome, all of which is to say, perhaps, that the peacekeeping process is highly path dependent and uncertain. (Giving cops the capacity to communicate could affect these results.)

This entire analysis proceeds from the assumption that all forces are in place at time 0. The same analysis could be conducted for different

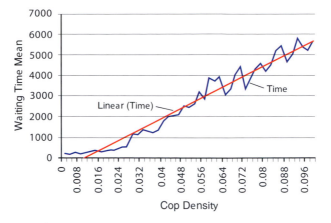

Figure 11.16. Waiting time mean and initial cop density.

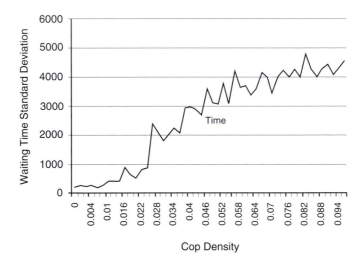

Figure 11.17. Waiting time standard deviation and initial cop density.

arrival schedules. At the moment, the claim is simply that the agent-based methodology permits one to study the effects of early and late interventions of different scales, which is obviously crucial in deciding how to size, design, and operate peacekeeping forces.

Summary of Model II Results

With high legitimacy (mutual perception by each ethnic group of the other's right to exist), peaceful coexistence between ethnic groups is observed; no peacekeepers are needed. However, if the force density is held at zero, and legitimacy is reduced (to 0.8), local episodes of ethnic cleansing are seen, leading to surrounded enclaves of victims, and ultimately to the annihilation of one group by the other. With early intervention on a sufficient scale, this process can be stopped. Safe havens emerge. With high cop density from the outset, the same level of legitimacy (0.8) produces a stable society plagued by endemic ethnic violence. If cops are suddenly removed, there is reversion to competitive exclusion and genocide. The statistical relationship between initial cop densities and the waiting time to genocide was studied. Although the mean relationship was positive, quick convergence to genocide at extremely high force levels, it was shown, is not precluded, because of the path-dependent and highly variable dynamics of interethnic civil violence.

CONCLUSION

Agent-based methods offer a novel and, I believe, promising approach to understanding the complex dynamics of decentralized rebellion and interethnic civil violence, and, in turn, to fashioning more effective and efficient policies to anticipate and deal with them.

REFERENCES

Axelrod, Robert. 1997. The Dissemination of Culture: A Model with Local Convergence and Global Polarization. *Journal of Conflict Resolution* 41: 203–26.

Collier, Paul. 2000. Rebellion as a Quasi-Criminal Activity. *Journal of Conflict Resolution* 44:839–50.

Collier, Paul, and Anke Hoeffler. 1998. On Economic Causes of Civil War. *Oxford Economic Papers* 50:563–73.

———. 1999. Justice-Seeking and Loot-Seeking in Civil War. Working paper, World Bank, Washington, DC, April.

———. 2001. Greed and Grievance in Civil War. Policy Research Working Paper 2355, World Bank, Washington, DC.

Dean, Jeffrey S., George J. Gumerman, Joshua M. Epstein, Robert L. Axtell, Alan C. Swedlund, Miles T. Parker, and Stephen McCarroll. 2000. Understanding Anasazi Culture Change through Agent-Based Modeling. In *Dynamics in Human and Primate Societies: Agent-Based Modeling of Social and Spatial Processes*, ed. Timothy A. Kohler and George J. Gumerman. New York: Oxford University Press.

DeNardo, James. 1985. *Power in Numbers: The Political Strategy of Protest and Rebellion*. Princeton: Princeton University Press.

Epstein, Joshua. M. 1999. Agent-Based Computational Models and Generative Social Science. *Complexity* 4(5): 41–60.

Epstein, Joshua M., and Robert L. Axtell. 1996. *Growing Artificial Societies: Social Science from the Bottom Up*. Washington, DC: Brookings Institution Press; Cambridge: MIT Press.

Epstein, Joshua M., John D. Steinbruner, and Miles T. Parker. 2001. Modeling Civil Violence: An Agent-Based Computational Approach. Working Paper 20, Center on Social and Economic Dynamics, Washington, DC.

Fearon, James D., and David D. Laitin. 1996. "Explaining Interethnic Cooperation." *American Political Science Review* 90:715–35.

Graham, Carol, and Stefan Pettinato. 2002. *Happiness and Hardship: Opportunity and Insecurity in New Market Economies*. Washington, DC: Brookings Institution Press.

Kuran, Timur. 1989. Sparks and Prairie Fires: A Theory of Unanticipated Political Revolution. *Public Choice* 61:41–74.

May, Robert M., ed. 1981. *Theoretical Ecology: Principles and Applications*. 2nd ed. Oxford: Blackwell.

Olson, Mancur. 1971. *The Logic of Collective Action: Public Goods and the Theory of Groups*. Rev. ed. Cambridge: Harvard University Press.

Tocqueville, Alexis de. 1955. *The Old Regime and the French Revolution*. Trans. Stuart Gilbert. New York: Doubleday.

Young, H. Peyton. 1998. *Individual Strategy and Social Structure*. Princeton: Princeton University Press.

GENERATING EPIDEMIC DYNAMICS

I LOVE MATHEMATICS. I also believe that mathematical theories can offer fundamental insights. In areas where there is a well-developed and powerful mathematical theory, one should master the theory before building an agent model. Moreover, it is often of great value to "dock" the agent model—or some special case of it—to the classical mathematical one.

In the field of epidemiology, there is a beautiful underlying mathematical theory that applies powerfully to an important class of cases. While we offer an agent-based epidemic model in chapter 12, the effort began with a "docking exercise" worth recounting as a prelude.

The Kermack-McKendrick Equations

Perhaps the most famous equations of mathematical epidemiology are the Kermack-McKendrick equations, published in 1927.[1] In the basic model, the total population is constant and is comprised of three homogeneous pools: Susceptibles $S(t)$, Infectives $I(t)$, and Removeds $R(t)$. Conceptually, susceptibles contract the disease through contact with infectives; the infected are then removed from circulation through death or permanent immunity. Since the flow is from susceptible (S) to infective (I) to removed (R), it is termed an *SIR* epidemic model.[2] The flow from state to state is governed by three differential equations. Letting r denote the infection rate (the per contact transmission probability) and p the rate at which infectives are removed from circulation (e.g., by death), the Kermack-McKendrick model is

$$dS/dt = -rSI \tag{1}$$

$$dI/dt = rSI - pI \tag{2}$$

$$dR/dt = pI. \tag{3}$$

---

[1] W. O. Kermack and A. G. McKendrick, "Contributions to the Mathematical Theory of Epidemics," *Proceedings of the Royal Statistical Society*, ser. A, 115:700–721.

[2] For a thorough modern mathematical exposition, see Paul Waltman, *Deterministic Threshold Models in the Theory of Epidemics* (New York: Springer, 1974).

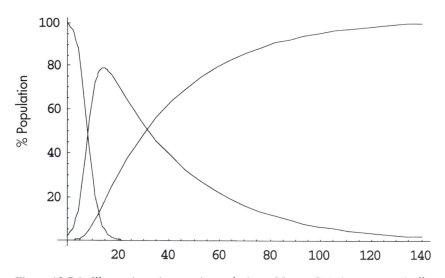

Figure 12.P.1. Illustrative time series solution. Note: $S(t)$ is monotonically decreasing, $R(t)$ monotonically increasing, and $I(t)$ rises and then falls.

A typical solution of this system is graphed in figure 12.P.1.

The susceptible curve falls, as individuals contract the disease and transfer into the infective pool. Accordingly, the infective curve rises. But, after a period of growth, this curve begins to fall also, as agents are removed from circulation, through death or permanent immunity, for example.

Now these equations assume a type of *perfect mixing* with which we will dispense in the agent-based smallpox model proper. In particular, the equations posit that the rate of flow from the susceptible to the infective pool dS/dt is proportional to *the product of the pool sizes*, $S(t)$ and $I(t)$. Implicitly, the model lines up all S susceptibles in a row. Then each of the I infectives marches down the entire line of susceptibles and sneezes in each one's face. That produces SI contacts. That's perfect mixing, and the model is very unrealistic in many cases. So what? Realism per se is overrated. It is a very revealing model! Two important and counterintuitive insights are immediate, and they apply even in the agent model we will ultimately build. Indeed, an agent model incapable of capturing them will be of dubious value.

First Counterintuitive Insight: Herd Immunity

Imagine a contagious disease of some sort. And imagine that you possess a perfectly effective vaccine for that disease. Intuition would surely suggest that, to prevent epidemics of the disease, it is necessary to vaccinate the entire population. The Kermack-McKendrick equations say otherwise! How?

To say that an epidemic occurs is to say that the infectious pool, $I(t)$, is growing. In other words, $dI/dt > 0$. But, by (2), this occurs if and only if $rSI > pI$, which is to say

$$S > p/r. \tag{4}$$

The ratio on the right—the removal rate over the infection rate—is a threshold. If the susceptible pool exceeds it, the infection spreads. If not, the epidemic fizzles. The very important implication of this threshold result is that *less than universal immunization is required to prevent epidemics*. Indeed, by (4), the fraction immunized need only be big enough that the unimmunized fraction—the remaining susceptible pool, S—be below the threshold p/r. *Not every cow needs to be immune for the herd to be immune*. For instance, diphtheria and scarlet fever require 80 percent immunization to achieve "herd immunity,"[3] as the effect is known. Hethcote and Yorke argue that "a vaccine could be very effective in controlling gonorrhea...for a vaccine that gives an average immunity of 6 months, the calculations suggest that random immunization of $1/2$ of the general population each year would cause gonorrhea to disappear."[4] While our smallpox model will differ profoundly from the Kermack-McKendrick picture, the goal of herd immunity is at the core of our vaccination strategy.

Second Counterintuitive Insight

As for the second insight, consider the following proposition: The more deadly a disease is to the individual, the more widespread is the epidemic it produces. Right? Surprisingly, the answer is no, and this, too, is evident from the threshold equation (4). The equation says that if we vaccinate until the remaining susceptible (i.e., unvaccinated) pool falls below p/r, the epidemic will fizzle. So, for the sake of argument, fix r (contagiousness per contact). Now, as p (individual deadliness) gets higher and higher, so does p/r, and we need to vaccinate fewer and fewer to achieve herd immunity. In the limit of perfect deadliness, we don't need to vaccinate anyone: the disease kills off its hosts before they can spread

---

[3] Leah Edelstein-Keshet, *Mathematical Models in Biology* (New York: Random House, 1988), 255.

[4] H. W. Hethcote and J. A. Yorke, *Gonorrhea Transmission Dynamics and Control* (New York: Springer, 1980).

the disease to others, precluding any epidemic! The original proposition is not right. A disease can be so deadly to the individual that it can't spread efficiently. (Smallpox, as we shall see, is more clever than that.)

The Kermack-McKendrick equations, highly idealized as they may be, thus offer two fundamental qualitative insights. And, as a first activity, our team built an extremely simple agent model to "dock" to that classic model.

An Agent Model

In this minimal agent model, events transpire on a green "playground," if you like. Healthy susceptible agents are colored blue, while a few initial infective agents are colored red. As an approximation to Kermack-McKendrick perfect mixing, agents move to neighboring sites randomly (they execute a 2D random walk in their von Neumann neighborhoods), bumping into other agents. Any time an infective (red) bumps into a healthy (blue), the disease is transmitted with fixed probability (0.2), just as in Kermack-McKendrick. And, as in that model, there is no heterogeneity within the susceptible or infective pools, and infectives are removed (die) with probability 0.4 per period. A movie of the typical epidemic is on the CD. Snapshots are offered in run 1.

Over time, the blue population turns red, as susceptibles contract the disease, and reds disappear, as they are removed from circulation. The time series are exactly as in the figure 12.P.1 solution curves for the Kermack-McKendrick model, as shown in figure 12.P.2. So this is a

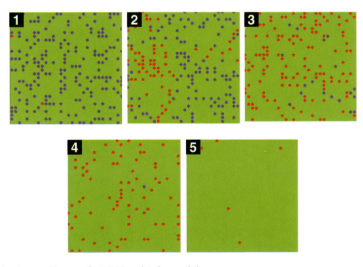

Run 1. Agent Kermack-McKendrick model.

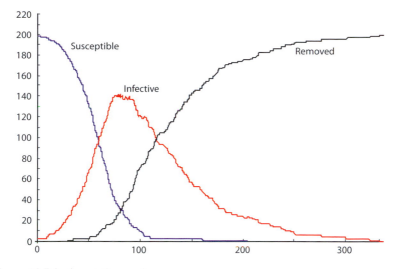

Figure 12.P.2. Agent time series.

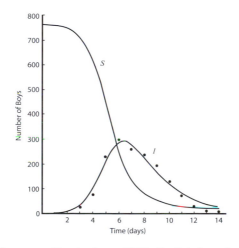

Figure 12.P.3. Influenza epidemic data, 1978, English boarding school. Of 763 boys, 512 were confined to bed, 22 January–February 1978. (Source: *British Medical Journal*, 4 March 1978.)

reasonable "docking" of an agent model to the classic one. Moreover, they both do very nicely in cases where the well-mixed assumption holds, as in the data from a 1978 influenza epidemic in an English boys' school shown in figure 12.P.3. The epidemic took place in winter on a small campus where dining commons and dormitories ensured high contact rates, permitting the well-mixed model to apply nicely. Notice the close fit between theoretical and observed infective curves.

For modeling the spread of smallpox in modern urban settings, however, this well-mixed assumption can be dangerously misleading, and we will depart from it. Some degree of spatial realism is fundamentally important. Agents, moreover, will need to be heterogeneous in a variety of important ways not reflected in the Kermack-McKendrick model.

However, that model, and explicit mathematical models in general, are very important tools. We are all prisoners of our tools. But if one must live in a prison cell, one would like it to be as spacious as possible. So one should try to master all the tools one can.

Chapter 12

TOWARD A CONTAINMENT STRATEGY FOR

SMALLPOX BIOTERROR: AN INDIVIDUAL-BASED

COMPUTATIONAL APPROACH

JOSHUA M. EPSTEIN, DEREK A. T. CUMMINGS,
SHUBHA CHAKRAVARTY, RAMESH M. SINGHA,
AND DONALD S. BURKE*

INTRODUCTION

SINCE the September 11, 2001, terrorist attacks in New York and Washington and the subsequent anthrax outbreaks on the east coast of the United States, bioterror concerns have focused on smallpox. Routine smallpox vaccinations in the United States ended in 1972. The level of immunity remaining from these earlier vaccinations is uncertain but is assumed to be degraded substantially. For present modeling purposes, we assume it to be nil.

As a weapon, smallpox would be very different from anthrax. Anthrax is not a communicable disease. Smallpox is highly communicable. With a case fatality rate of roughly 30 percent (meaning that 30 percent of infected individuals die), it is also very deadly. Many of those who survive the disease, furthermore, are permanently disfigured, their well-being compromised for life.

There is now heated debate on the appropriate national strategy for smallpox bioterror.[1] Who should be vaccinated? Everyone who volunteers? Targeted subpopulations? When should immunization begin?

*This research was funded by a grant from the Alfred P. Sloan Foundation. For helpful discussion, the authors thank Ellis McKenzie, Edward Kaplan, James Koopman, Jon Parker, Nancy Gallagher, Elisa Harris, and John Steinbruner.

This essay was published previously as a monograph by The Brookings Institution Press, Washington, D.C. 2004.

[1] In the summer of 2001, researchers at the Johns Hopkins Center for Civilian Biodefense Strategies, in collaboration with several other organizations, formulated a policy exercise known as Dark Winter, which raised many important questions for bioterror attack response; see O'Toole, Mair and Inglesby 2002.

Immediately? Only after a confirmed attack? What is the role of quarantine?

In this monograph, we present a county-level individual-based computational model of a smallpox epidemic.[2] We review and criticize the two main vaccination strategies currently under discussion: trace and mass vaccination. Based on the model, we then develop a distinct "hybrid" strategy that differs sharply from both, while combining useful aspects of each. It involves both preemptive (that is, pre-release) and reactive measures. As the basis for a national smallpox containment strategy, we believe it offers important advantages over the alternatives.

Models

In gauging the scale of a smallpox bioterror threat, and in designing an effective policy response, it is crucial to have *epidemic models* depicting the spatial spread of the disease in a relevant setting. Without the use of explicit models, there is no systematic way to gauge uncertainty or to evaluate competing intervention strategies. Building on previous work, we have developed an individual-based computational modeling environment for the study of epidemic dynamics in general (see appendix A).[3] This can be applied to an indefinite variety of pathogens and social structures. Here, we develop an individual-based model of smallpox at the county level (an application to genetically modified smallpox is also noted).[4]

In contrast to compartmental epidemic models, which assume perfect homogeneous mixing and mass action kinetics (Anderson and May 1991; Kaplan, Craft, and Wein 2002), the individual-based approach explicitly tracks the progression of the disease through *each individual* (thus populations become highly heterogeneous by health status during simulations) and tracks the contacts of each individual with others in relevant social networks and geographical areas (for example, family members, coworkers, schoolmates). All rules for individual agent movement (for example, to and from workplace, school, and hospital) and for contacts with and transmissions to other people are explicit, as is stochasticity (for example, in contacts). No homogeneous mixing assumptions are

---

[2] Individual-based modeling is also called agent-based modeling. To avoid confusion between our agents (individual people) and infectious disease agents, we use the term individual-based modeling predominantly.

[3] See Burke 1998; Grefenstette *et al.* 1997; Burke *et al.* 1998; Epstein and Axtell 1996; Epstein 1997.

[4] For an introduction to the individual-based modeling technique, see Epstein and Axtell 1996. For diverse applications of the methodology, see Brian *et al.* 2002.

employed at any level. The prime social units that loom largest in the smallpox data (Mack 1972), such as hospitals and families, are explicitly represented, and our vaccination (and isolation) strategy is focused on these units of social structure. Calibration of our model to these data, and statistical analysis of core model runs, are discussed below.

Our model differs from the primary (and valuable) competing approaches, in a number of ways. For example, it differs from that of Halloran *et al.* (2002) in its explicit inclusion of hospitals. Most fundamentally, as a "pure" individual-based model, it eschews all homogeneous mixing assumptions at any level, in contrast to the models of both Halloran *et al.* (2002) and Kaplan, Craft, and Wein (2002).[5]

THE COUNTY-LEVEL MODEL

The software we have developed permits generalization to multiple levels of social structure. For present purposes, we model a county composed of towns, hospitals, households, schools, and workplaces.

Structure and Calibration

The model structure was chosen for comparability to historical data describing the relationship between smallpox cases and the individuals who transmitted the disease to those cases in forty-nine outbreaks of smallpox in Europe from 1950 to 1971. This data set reveals the crucial role of hospital and household transmission in smallpox outbreaks. We wished to build the simplest model that captured the heterogeneity of transmission in the different settings of hospitals, families, workplaces, and schools. To ensure the replicability of our results, numerical assumptions and technical details are given in appendix A. Selected salient assumptions are noted in the text.

The model parameters governing the probability of transmission per contact and the contact rates in different social settings were chosen through a calibration of simulated epidemics with the historical data. As noted above, these data describe outbreaks resulting from forty-nine importations of a single case of smallpox into nonendemic Europe during the period 1950–71 (Mack 1972). Two distributions from these data were used for the calibration: the distribution of the number of cases

---

[5]There are further differences, including parametric ones. For useful remarks comparing continuous and discrete individual approaches in the present connection, see Koopman (2002).

resulting from each of these importations and the distribution describing the proportion of cases resulting from exposure in a hospital setting, in a workplace or school setting, and in the home. A combinatorial sweep of the core parameters—the per contact transmission probability and the contact rates in the hospital, the home, and the workplace or school—was performed and the distribution of these results over many simulation runs was compared to the historical data. In all, approximately 10,000 runs were performed. The parameters that minimized the sum of squared deviations from the two historical distributions were chosen. In the present version of the general model, each town is assumed to contain 100 family households, each with two working adults and two school-aged children—400 individuals in total. With two towns, the county population is thus 800.

Each town has one school and one workplace. All children attend their own town's school (there is no intertown busing). A small fraction of adults, by contrast, do commute to work in the other town. In our base runs, we assume that 10 percent of adults commute. There is a single county hospital, used by both towns. A small number of adults (in the present version, five) from each town work in the common hospital. Finally, there is a single morgue housing all individuals who have died.

Time and Contacts

Each modeling day is equally divided between a "daytime," when adults work and children attend school, and a "nighttime," when family members (exclusively) interact at home. Each of these phases of the day is composed of several rounds, in which each individual is processed, or "activated," once. The essential event that occurs when an individual is active is contact with other individuals. In our model, the active individual is contacted by randomly selected individuals from the relevant pool (family members or immediate neighbors at work or school). Numerical assumptions regarding contacts and transmissions per contact are given in appendix A. Note that the per contact probability of contracting the infection depends on the stage of illness the contacting individual is in.

Graphics

The graphical setup is depicted in figure 12.1, which shows two snap-shots of the county, labeled nighttime and daytime. Two towns can be seen, Circletown and Squaretown, inhabited respectively by circle individuals and square individuals. (Circles and squares are used simply

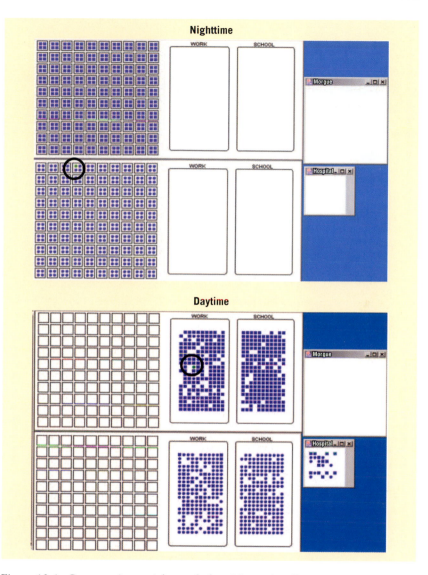

Figure 12.1. County view: night and day. Note: An illustrative commuter is colored green and circled.

to make commuting individuals discernible and to depict the hospital workers' hometowns.) As runs progress, individuals return home at night and go to work and school during the day, a process that iterates indefinitely. That summarizes the social contact process. Meanwhile, the epidemic is running its course.

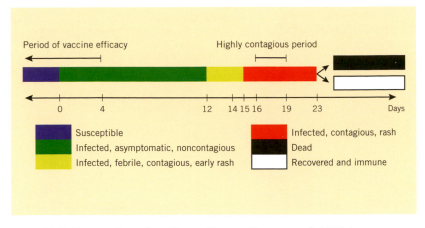

Figure 12.2. Progression of smallpox. (Source: Fenner *et al.* 1988.)

Smallpox Assumptions

Our assumptions about the natural history of smallpox in the individual are illustrated by the timeline in figure 12.2, which also describes the color coding used in the model graphics. Before we release our index case—the first infective individual—into the population, we assume all individuals to be susceptible; that is, we assume no background of immunity (for example, from previous vaccinations). Susceptible individuals are colored blue. Referring to the timeline, let us assume an individual contracts the infection at day 0. At that point, she/he is colored green. Although the person is infected with smallpox, she/he is asymptomatic and noncontagious for twelve days. However, unless the infected individual is vaccinated within four days of exposure, the vaccine is ineffective. As a policy matter, this is a critical point: it will be necessary to vaccinate infected people *before* they manifest any symptoms. From day 12 to day 15, infected individuals are assumed to be febrile and contagious (infectious) with smallpox, but do not yet exhibit the smallpox rash. They feel sick, but precise diagnosis is not yet possible.

At the end of day 15, smallpox rash is finally evident. After twelve hours in this state, individuals are assumed to be hospitalized. After eight more days (day 23 of illness), during which they have a cumulative 30 percent probability of mortality, surviving individuals recover and return to circulation permanently immune to further infection. Dead individuals are colored black and placed in the morgue.[6] Immune individuals are colored white. Contagiousness varies in the course of the infection.

---

[6]The morgue is a closed system, so no transmission occurs there.

Individuals are assumed to be 2 times as infectious during days 16 through 19 as during days 12 through 15. In the final phases of the rash, infectivity returns to the day 12 value, as indicated in figure 12.2. In the simulated epidemics below, individuals will be colored by their state: healthy (blue), infected (green), contagious early rash (yellow), rash (red), dead (black), or immune (white). At any time, the population will be heterogeneous by health status.

SIMULATED EPIDEMICS

We present a number of runs and statistical analyses. All simulations in this paper assume a single initial infective individual (for example, a bioterrorist or bioterror victim), who is an adult commuter.[7]

We present snapshots from our computer simulation. It is important to note, when presenting a simulated epidemic, that any such realization is but one sample path of a stochastic process. There are run-to-run differences due to random effects, even when all parameters are fixed across runs. Indeed, as we shall see, these random effects can be dramatic, spelling the difference between large-scale epidemics and abortive ones. Hence, a statistical treatment is necessary and is offered below. To begin, however, it builds intuition to "watch" the base case epidemic unfold in our county over time. Again, we imagine a smallpox bioterrorist initiating the epidemic by infecting (or by being) a commuting adult.

Base Case: No Intervention

The base case scenario is obtained by setting model parameters to values found by calibration to the European data set and by assuming no preexisting immunity in the population. The epidemic is allowed to simply run its course, without any vaccination or isolation strategy. Figure 12.3 presents nine frames (snapshots) from the full simulation, which can be viewed as a movie on this book's CD.

Frame 1 simply shows our index case, a Circletown commuter, at home at night. He is green, indicating that he is infected but asymptomatic.

---

[7]On reflection, the assumption of a single index case is quite conservative. In a city of 12 million (such as Manhattan), one initial infective per 800 would translate into 15,000 initial infectives. Of course, this assumes a *linear* scale-up, which may well be unrealistic. Our point is simply that, scaled in any plausible way, one in 800 will translate into an enormous attack force in the bioterror interpretation. Although we do not believe we are artificially simplifying the problem, our software allows for expansion by orders of magnitude, as discussed below.

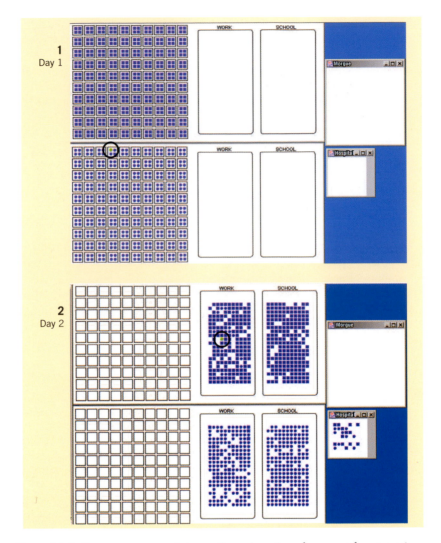

Figure 12.3. Base case run: no interventions. (*continued on next four pages*)

Frame 2 shows the index case at his Squaretown workplace the next day. Frame 3 depicts the situation on day 15, at which point our index case has developed the full smallpox rash. Notice that by this point, he has spread the disease to others (colored green), but none of them are aware that they are ill, a situation that persists into day 16 (frame 4), when the index reports to the hospital. In this particular run, he dies eight days later and is taken to the morgue on day 24, as shown in frame 5. No one else in the county yet realizes that he or she is sick.

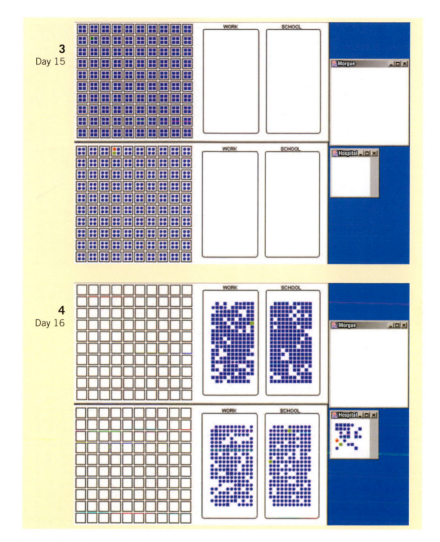

Figure 12.3. (*continued*)

Frame 6 depicts the situation at day 42. Notice that the epidemic is now far worse in Squaretown than in Circletown, despite the fact that it began in Circletown. So, seemingly sensible strategies like "concentrate vaccination on the town where the outbreak begins" may do poorly. By the time one vaccinates there, the epidemic may well have spread beyond. Frame 7 (day 52) and frame 8 (day 62) show the hospital filling to capacity and the morgue filling up. They also show that many people recover (colored white). Finally, the epidemic's end state is shown in

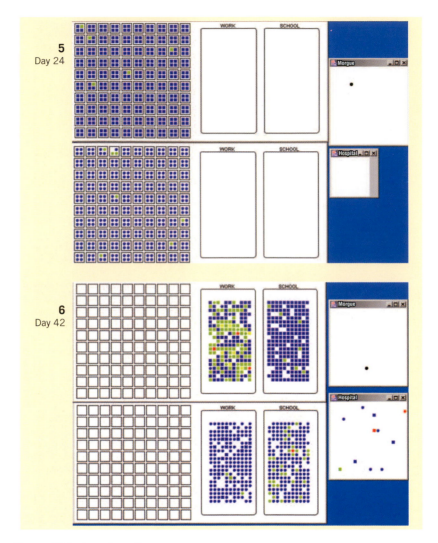

Figure 12.3. (*continued*)

frame 9 (day 82). With no intervention, everyone in the county eventually contracts smallpox, and roughly 30 percent die of the disease.[8] It is noteworthy that the base case assumes no background of immunity. It may represent well the dynamics when European smallpox was first introduced into virgin indigenous populations.

---

[8] Here, we assume that agents continue to go to work and school, that hospitalized individuals are not isolated, and that no agents flee the county.

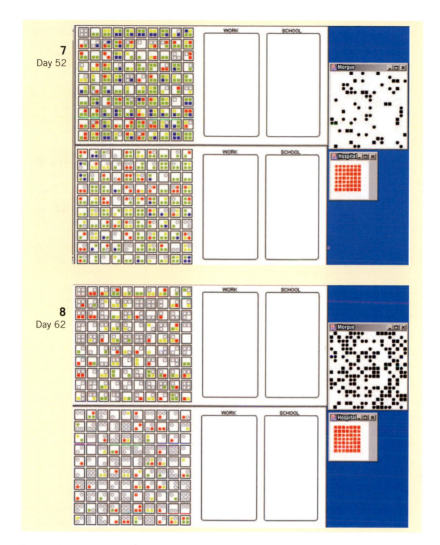

Figure 12.3. (*continued*)

Figure 12.4 shows typical time series of incidence (top panel) and number of infected individuals (bottom panel) for a representative simulation in which there is no intervention.

This, then, is the problem we wish to address. What is the appropriate policy response? We begin with a review of traditional vaccination strategies and their problems. We then offer a hybrid vaccination strategy of our own. The substantial role of voluntarism is noted.

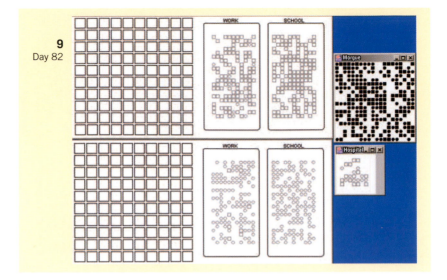

Figure 12.3. (*continued*)

VACCINATION STRATEGIES

The vaccination strategies that have loomed largest in the policy debate thus far are trace vaccination and mass vaccination. Each has advantages and disadvantages.

Trace Vaccination

Trace vaccination is an elegant idea. Given a confirmed smallpox case, one traces every contact the individual has had and vaccinates that group. The Centers for Disease Control has adopted priority-based trace vaccination in its Smallpox Response Plan and Guidelines,[9] discussed more fully in appendix B. Contacts are technically defined as "persons who had...close proximity contact (<2 meters = 6.5 feet) with a confirmed or suspected smallpox patient after the onset of the smallpox patient's fever."

This approach was effectively used in the worldwide smallpox eradication effort (Fenner *et al.* 1988). However, there is great concern that in advanced industrial settings, an individual's network of contacts is huge. It will include persons who rode the same urban metro system

---

[9]U.S. Centers for Disease Control, "Smallpox Response Plan and Guidelines" (version 3.0), ww.bt.cdc.gov/agents/smallpox/response-plan/ [accessed December 5, 2002].

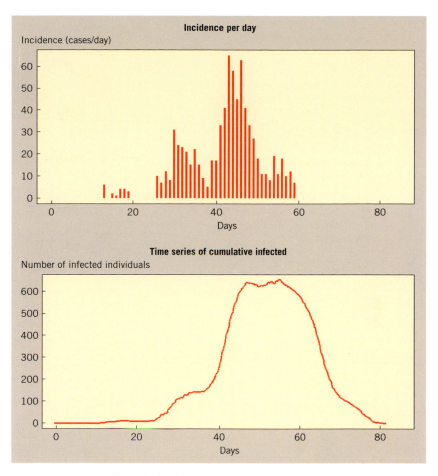

Figure 12.4. Typical results for base case run.

or flew out of the same airport, and thus contacts will be dispersed all over the city or country. The resource demands for full trace vaccination quickly become daunting. The value of incomplete, or imperfect, trace vaccination has not received sufficient attention. Below, we present a new strategy involving such an approach.

Mass Vaccination

Indiscriminate mass vaccination poses a distinct set of problems. The first is that administration of the smallpox vaccine is not without risk. Complications from the vaccine include post-vaccinal encephalitis, progressive vaccinia, eczema vaccinatum, generalized vaccinia, and accidental infection. It is estimated that forty complications would result from every 1 million doses given. Of these, an estimated one in every 1 million persons vaccinated would die from complications (Lane *et al*. 1969).

Second, vaccination is not recommended for a significant proportion of the population—groups at special risk of vaccine complications. These groups include persons with eczema; patients undergoing chemotherapy for leukemia, lymphoma, or generalized malignancy; patients with HIV; persons with hereditary immune deficiencies; and pregnant women (Henderson *et al*. 1999). Vaccination of these persons, or even inadvertent inoculation with the vaccine strain, could lead to serious disease or death.

In summary, while perfect trace vaccination is infeasible from a practical standpoint, mass vaccination carries relatively greater risks of vaccine-related morbidity and mortality.

The Policy Challenge

The challenge for government is therefore as follows: *Design a policy that is more feasible than trace vaccination, less risky than mass vaccination, and highly effective in containing a smallpox epidemic.* In designing such a policy, we exploit the essential feature of epidemics as dynamic processes: they are nonlinear stochastic phenomena.

BIFURCATION AND EPIDEMIC QUENCHING

Epidemiologists have long known that introductions of disease into populations with some background level of immunity can yield large outbreaks or outbreaks of just a handful of cases, with no outbreaks of sizes between the very small and the very large. This bifurcation phenomenon is described in the literature by the results of stochastic compartmental models, using the terms *stochastic extinction* or *fade-out* (Anderson and May 1991; Bailey 1953; Whittle 1955). We introduce the term *epidemic quenching* to denote dynamics in which the stochastic extinction occurs at the scale of discrete social units. Thus, introductions can be quenched at the level of the family, the workplace, or the town. We believe this approach accurately captures the local stochastic nature of real epidemics. The best vaccination strategies may well be those that take advantage of the importance of social structure to real epidemics. What strategies might give the public a reasonable chance that an epidemic will be "quenched"?

The actual data on European introductions of smallpox from 1950 to 1971 are shown in figure 12.5. This data set is focused on the

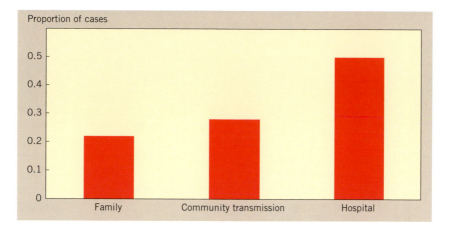

Figure 12.5. Smallpox cases, by relationship to transmitting case, Europe, 1950–71. Note: Sample includes 680 cases. (Source: Mack 1972.)

question, where did infected individuals contract smallpox? Importantly, 50 percent contracted it at the hospital and 22 percent contracted it from family. The remaining community transmission (28 percent) resulted primarily from workplace, school, and casual contacts.[10] Our model was built with these units of social structure in mind, and includes hospitals, families, work, and other venues precisely to allow calibration to these data. As discussed above, we have fit the model to these data. A strength of the agent-based approach is that it facilitates a focus on heterogeneous social units with distinctive internal dynamics, in contrast to models with homogeneous compartments. These European smallpox epidemic data clearly suggest that vaccinating hospital workers preemptively (before an attack) and vaccinating family contacts reactively (as soon as possible afterward) would be a powerful defense. Figure 12.6 indicates that this is, in fact, the case. The red curve is the time series for a typical run with no interventions. The black curve is the time series for the strategy just stated: preemptive hospital vaccination, isolation of cases in the hospital, and reactive contact tracing of household members of cases. Notice that none of these measures involves elaborate contact tracing. If, to these measures, we begin to add moderate levels of mass preemptive vaccination, a significant fraction of epidemics are quenched in our model.

---

[10]These data exclude fomite transmissions and include some undetermined and un-reported transmissions.

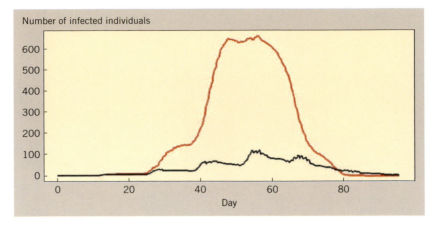

Figure 12.6. Results of interventions. Note: The black time series shows a typical run that implements our suggested interventions. The red time series is the original curve from figure 12.4, which shows the no-intervention case.

Quenching through Combined Vaccination Efforts

Combining vaccination efforts—preemptive mass vaccination, preemptive vaccination of hospital workers, and reactive household trace vaccination—has dramatic effects on both the quenching (confinement of the disease to discrete social units) and the extent (absolute number of cases) of the epidemic. Since epidemics are stochastic, single runs can be misleading. Therefore, we conducted a statistical analysis. We assume that all hospital workers in the model are vaccinated preemptively. Then, for five distinct levels of reactive family trace vaccination (0, 25, 50, 75, and 100 percent), we study how the course of the epidemic varies as we increase the level of preemptive mass vaccination. At each level of mass vaccination, the model was run 100 times, with a different random seed each run. The number infected in each run is plotted in red.

The entire analysis is shown in figure 12.7. The main observation is that with increasing levels of family contact tracing, the distribution of the number infected in each epidemic shifts downward at every level of preemptive mass vaccination.

While the best policy results are obtained at 100 percent family contact tracing (panel 7e), the most illuminating scientific results are evident in panel 7d, which displays 75 percent family contact tracing. Here, we see clear bifurcations: particularly at lower levels of preemptive mass vaccination (50 percent or less) the addition of reactive family contact tracing produces a trimodal distribution of epidemics. We examine this 75 percent family contact tracing case in much greater detail in

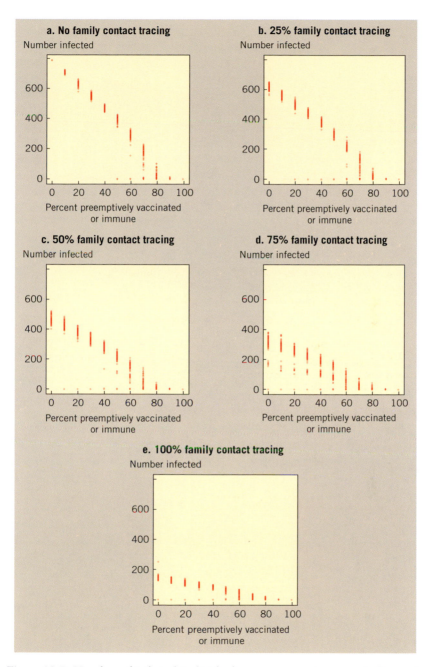

Figure 12.7. Number of infected individuals versus percent preemptively vacci-nated, by level of family contact tracing.

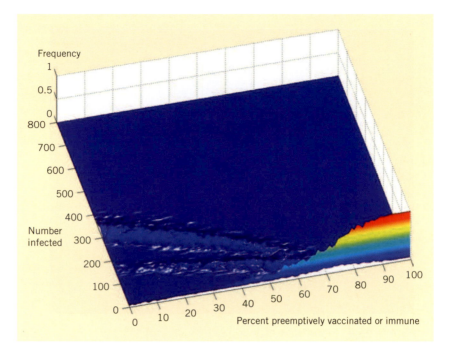

Figure 12.8. Probability surface of the 75 percent family contact tracing case.

figure 12.8. This offers a higher resolution version of panel 7d, with the frequency explicitly plotted above each point.

We believe that the bifurcations seen in figures 12.7 and 12.8 are clear evidence of quenching at the level of structural social units. We are studying this phenomenon in greater depth.

A BALANCED POLICY

As noted, the best *policy* results are obtained at 100 percent reactive family contact tracing, shown in figure 12.7, panel e. Figure 12.9 shows the cumulative distribution of infection for the 60 percent preemptive mass vaccination level in that panel (7e). We cite this level of preemptive vaccination because it is obtainable at minimum risk, by revaccinating those individuals successfully vaccinated in the past.[11] This group is highly unlikely to suffer any of the side effects emphasized above.

---

[11]U.S. Bureau of the Census, 2000 Summary File, www.census.gov/census2000/states/us.html [accessed December 5, 2002].

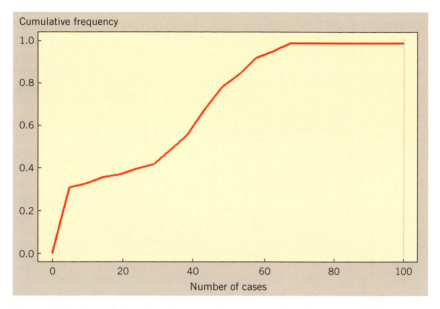

Figure 12.9. Cumulative distribution of outcomes for the case of 100 percent family contact tracing and preemptive hospital vaccination and 60 percent preemptive mass vaccination.

The vertical axis of figure 12.9 is the frequency of simulated outbreaks (for $n = 100$) that result in fewer than the number of infections indicted on the x-axis. So, for example, 100 percent of the simulation runs result in fewer than seventy infections (twenty-one deaths).

Now, what one sees reported in the media is, in a sense, the easy part of the policy problem. If there is a confirmed bioterror release of smallpox, the government must, of course, provide vaccine. Politically, there is no alternative. Hence the U.S. government is stockpiling 286 million doses (Kaplan, Craft, and Wein 2002). But the deeper and politically tougher question is what to do *before* any release to contain the epidemic and ease the burden of further vaccination if necessary (and the attendant risks of indiscriminate immunizations). In our two-town county model, the following mix of preemptive and reactive policy measures achieves these goals:

PREEMPTIVE

1. Vaccination of 100 percent of hospital workers
2. Voluntary revaccination of healthy vaccinees (individuals successfully vaccinated in the past)

3. Isolation in hospital of confirmed cases
4. Vaccination of household members of confirmed cases

Referring to figure 12.9, under this package, 100 percent of the simulated outbreaks result in fewer than seventy cases (twenty-one deaths), 75 percent of outbreaks yield fewer than forty-five cases (fourteen deaths), and 50 percent of outbreaks yield fewer than thirty-five cases (eleven deaths). This certainly qualifies as *containment* compared to the no-intervention base case, in which the entire population of 800 individuals becomes infected and roughly 240 die in virtually all runs.

In our model, this package of measures offers the public an excellent chance that a bioterror smallpox attack will be quenched and limited in its severity and sharply reduces the logistical burden and public health risk of further vaccination if necessary. In particular, it minimizes the risks of indiscriminate mass vaccination and, in contrast to complete trace vaccination, is entirely feasible. Given a credible bioterrorist threat,[12] this combination of measures can serve as the basis for a smallpox containment strategy.

RESEARCH CONDUCTED UNDER THE AUSPICES OF THE SMALLPOX MODELING WORKING GROUP

Since the completion of the above research, the Brookings-Hopkins team has joined the Smallpox Modeling Working Group of the Secretary's Advisory Council on Public Health Preparedness of the Department of Health and Human Services. This working group was established and is chaired by D. A. Henderson, who also chairs the advisory council. One of the working group's major undertakings was to collectively agree on core model parameters, distributions, and behavioral assumptions. The following major model extensions were then assigned and have since been made.

- Expand the population from 800 to 6,000 and 50,000 agents
- Expand the range of bioterror attacks from the single initial infective to 10 initial infectives in the 6,000 agent case, and to 500 initial infectives (for example, an aerosol release in a movie theater) in the 50,000 agent case

---

[12] The credibility of such a threat at this time is a topic that lies outside the bounds of this research.

- Explore social structures beyond the original two-town setup, including a ring of towns and a hub-and-spoke arrangement of towns
- Beyond ordinary smallpox, model modified (by background immunity) and hemorrhagic smallpox
- Use full distributions for the incubation periods and infectiousness of smallpox, rather than the step functions of the original model (arriving at agreed assumptions for the natural history of the disease in humans consumed a number of working group meetings)
- Use explicit assumptions about when smallpox cases would be recognized in the hospital (interventions start once the first case is recognized)
- Vary the care-seeking behavior of individuals based upon the type of smallpox they contract (for example, hemorrhagic cases seek care almost immediately after the onset of symptoms, whereas modified cases have some chance of not reporting to the hospital for three days after the onset of symptoms)
- Model ten different intervention scenarios, from preemptive vaccination of hospital workers only (various percentage levels) to surveillance and containment to varying levels of post-attack voluntary mass vaccination, and combinations of these.

All of these extensions and results are presented in the report "Individual-based Computational Modeling of Smallpox Epidemic Control Strategies" (Burke *et al.* 2004). While the scenarios that were assigned the working group for analysis differed from the one treated above, none of the analyses in that report challenge the fundamental soundness of the containment approach developed in the present monograph.

FURTHER RESEARCH

We plan to deepen our analysis of smallpox proper, extend our study of interventions, and examine a number of further topics.

Expanding Scale

Regarding smallpox proper, further sensitivity analyses will be worthwhile. First among them, perhaps, is the question of scale-up. Do our fundamental results change when we expand the model to populations orders of magnitude larger? We plan to scale the model up to 1.7 million individuals. Other worthwhile sensitivity analyses would further vary the number of initial cases and the patterns and levels of commuting, for example.

Vaccinating Contacts Of Contacts

Current modeling efforts (including ours) assess trace vaccination efforts assuming that these entail only the identification and vaccination of contacts of confirmed infected individuals. They have not assessed the effects of vaccinating contacts of contacts of confirmed infected individuals. The Centers for Disease Control's Smallpox Response and Guidelines and the worldwide smallpox eradication effort both cite the importance of vaccinating contacts of contacts in containing smallpox outbreaks.[13] Our modeling effort has built the capability to quantitatively assess the impact of such an approach, and this will be the subject of further inquiry.

Seasonality

Smallpox epidemic dynamics are known to vary with the seasons. Spread efficiency increases in winter relative to summer. These seasonal variations could affect the appropriate mix of intervention strategies. Seasonality, then, is another promising area of further research.

Family Isolation

Beyond vaccination, isolation is another policy. One can imagine "trace isolation," in which the dendrite of a confirmed smallpox carrier's contacts is traced and isolated. But this is as intractable as trace vaccination. One can also imagine broad isolation strategies, like closing all schools and workplaces or banning cross-town traffic (essentially quarantine of entire towns).

   More discriminating than quarantine but less demanding than trace isolation is the following strategy, which we term *family isolation*: if any member of a household is diagnosed (presumably at the hospital) to have smallpox, all other household members stay home. This is surprisingly effective on its own. Indeed, as figure 12.10 suggests, with no vaccination or other additional interventions, this strategy is roughly as effective as random vaccination of half the population.

   This makes the important methodological point that epidemics involve two dynamics. The first, the course of the disease in the individual, is

---

[13]U.S. Centers for Disease Control, "Smallpox Response Plan and Guidelines" (version 3.0), www.bt.cdc.gov/agents/smallpox/response-plan/ [accessed December 5, 2002]; Fenner *et al.* 1988.

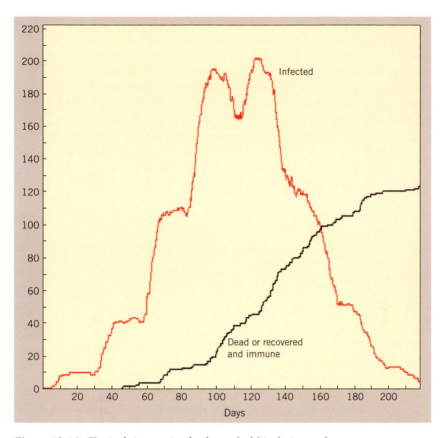

Figure 12.10. Typical time series for household isolation only.

biomedical. The second, the spatial contact process among individuals, is social. Our family isolation policy operates only on the social contact process, but would be a surprisingly powerful adjunct to the vaccination strategies articulated earlier. Isolation is particularly relevant to SARS (Severe Acute Respiratory Syndrome), for which no vaccine is available. Further voluntary measures worthy of analysis are the use of masks, gloves, and other individual protective options. Among topics beyond smallpox, the threat of novel pathogens looms large.

Novel Pathogens: IL-4 Smallpox

In the wake of the recent Australian mousepox incident, there has been concern that incorporation of the interleukin-4 gene into smallpox would produce a more deadly pathogen that we have termed IL-4 smallpox.[14] Precisely how IL-4 smallpox would behave in human hosts is uncertain, but it is known that "interleukin-4 mediates down regulation of antiviral cytokine expression and cytotoxic T-lymphocyte responses and exacerbates vaccinia virus infection in vivo" (Sharma *et al.* 1996). As a consequence, it is plausible that the pathogenicity (unvaccinated fatality rate) of IL-4 smallpox would substantially exceed that of unadulterated smallpox, that the smallpox vaccine would be considerably less effective against IL-4 smallpox than against smallpox, and that the transmissibilities of IL-4 and unmodified smallpox would be comparable. We have begun to explore how IL-4 smallpox would spread in our county-level model on plausible, albeit uncertain, numerical assumptions—for example, that IL-4 smallpox pathogenicity is twice that of smallpox, that smallpox vaccine is 50 percent as effective against IL-4 smallpox as against smallpox, and that IL-4 smallpox transmissibility equals that of smallpox. On these assumptions, containment of IL-4 smallpox is far more demanding than smallpox containment. Further research on IL-4 smallpox, and on the problem of novel pathogens in general, is planned.

It should be noted that the problem of engineered pathogens quickly raises a host of policy issues regarding the governance of scientific research in both academia and the private sector. This is another important topic for research (see Steinbruner *et al.* 2002).

Appendix A: Technical Discussion

The model was written in Java, using the Ascape modeling framework. In Ascape, models consist of a variable-sized population of individual agents (objects) who coexist on a landscape of variable size and shape. In the case of this model, the landscape chosen was a two-dimensional grid resembling an overhead map. The use of an object-oriented *class* to implement Ascape agents allows for a large degree of heterogeneity among agents. Each agent object contains and updates a range of information (such as the agent's infection status, her location on the grid, and so forth). The agent decides her own actions (for example, go to work, go to the hospital, interact with another agent). Agents can be coded to have variable actions, behaviors, and data simply by creating subclasses of the basic agent type. The Ascape library of classes also provides a wide range of methods to develop interagent (and agent-landscape) interactions. In this case, the landscape was discretized into spaces corresponding to our model's major social units: homes, schools,

---

[14] For the mousepox incident, see Jackson *et al.* 2001.

workplaces, hospital, and morgue. Each agent has memory of where she lives, who her family members are, and where she works. The model also keeps track of all those agents with whom she has interacted over a variable length of time, which allows us to model interventions such as contact tracing.

When a run of the model begins, all agents are at home, and one agent (a commuter) has already been infected with smallpox. The model is started on day 10 of that initial agent's infection. The model proceeds in rounds: each round consists of one iteration through the entire agent population. The call order is randomized each round, and agents are processed, or activated, serially (asynchronously). On each round, when an agent is activated, she identifies her immediate neighbors (she has up to eight so-called Moore neighbors, depending on her location on the landscape) for interaction. Each interaction may, depending on a random number draw, result in a *contact*. In turn, that contact results in a transmission of the infection from the contacted agent to the active agent with probability 0 if the contacted agent is not infectious, or a positive probability (see table 12.A.1 below) that varies according to the progress of the contacted agent's disease. Both agents record the contact in their memories, regardless of whether it resulted in a transmission (since, in reality, neither would know if transmission had in fact occurred). In the event the active agent contracts the disease, she turns green and her own internal clock of disease progression begins. After twelve days, she will turn yellow and begin infecting others.

This construction of agent interactions allows for enormous flexibility in modeling. The number of rounds each agent spends at work or school, the number of interactions each agent has per round, how often an interaction results in a contact, how often a contact results in transmission are all variable in our model and subject to sensitivity analysis. For the runs presented in this paper, each full day consists of twenty rounds, divided equally between daytime and nighttime. Thus a child spends ten rounds at home (night) and ten rounds at school (day). On each round, agent interactions proceed as discussed earlier. Note that the agent's neighbors are fixed at home (they are the same each day), whereas they are variable at work (the agent lands at the same workplace, but in a different random location at work each day).

We further make the number of contacts stochastic. Fewer contacts are assumed to occur at the workplace or school than at the home or hospital. This reflects the observation that transmission usually occurs as the result of direct contact between individuals, which is more likely to occur at home (Fenner 1988). The likelihood of an interaction resulting in a contact at home is 1.0 and at work is 0.3. The model records an agent's contacts during the three days before she turns red.

TABLE 12.A.1
Numerical Parameters

Parameter	Chosen Value	Possible Range
Transmission rate per contact during regular transmission period	0.2	0–1.0
Transmission rate during high-transmission period	0.4	0–1.0
Length of noncontagious period (days)	12	0–25
Length of early rash contagious period (days)	3.0	0–25
Start of high-transmission period (days)	16	12–25
End of high-transmission period (days)	19	12–25
Pathogenicity	0.3	0.2–0.6 (IL-4)
Percent initially vaccinated	0.0	0–100
Percent hospital workers initially vaccinated	0.0	0–100
Vaccine efficacy	1.0	0.5–1.0
Number of agents initially infected	1.0	1–800
Number of adults who work in hospital	10	0–50
Hospital size	10×10	$10 \times 10 – 30 \times 30$
Allow hospital visitors (Boolean)	False	True, false
Percent adults who commute	10	0–100
Family stays home when first member infected (Boolean)	True	False, true
Family contact tracing (percent)	100	0–100
Work contact tracing (percent)	20	0–100
Family contacts of contacts (percent)	0.0	0–100
Work contacts of contacts (percent)	0.0	0–100
Number of days of accumulated contacts to trace	3.0	0–15
Number of interactions per day per agent	10	10, 80
Rounds per day	10	1–50
Probability of contact per home interaction	1.0	0–100
Probability of contact per work or school interaction	0.3	0–100
Probability of contact per hospital interaction	1.0	0–100
Contact tracing maximum delay (days)	2	0–10

To summarize, a day consists of ten rounds at home followed by ten rounds at work or school. The model tracks each individual agent's disease progression on a daily basis.

Stochasticity plays an important role in the model. Our model employs the pseudorandom number generator from the Java 2 platform. The generator uses a 48-bit seed, which is modified using a linear congruential formula (see Knuth 1998, sec. 3.2.1). By recording the random seed used in each run, we can faithfully reproduce any run generated by the model. All random events occur at the agent level; that is, the agent draws a random number from a uniform distribution and, depending on the parameter value for the event in question, the agent's state changes. The following elements of the model depend on a random draw:

- Which agent is the index case
- The order in which agents are activated
- The agent's workplace (home town, other town, hospital)
- The agent's daily location in her workplace
- Whether or not an agent is traced and/or vaccinated
- Vaccine efficacy
- Whether or not a given interaction results in a contact
- Whether or not a given contact results in a transmission.

There are also various aspects of epidemics that involve delays and lags. Although we have not fully explored these areas in our model, we plan to introduce a stochastic delay in trace vaccination. The current version assumes a fixed delay of two days between the time an agent is diagnosed and when her contacts are vaccinated. In future work, this delay will be drawn from a Poisson distributed clock.

Table 12.A.1 summarizes the model's numerical parameters. For each parameter, we list the range of possible (and plausible) values and the value assigned in the runs presented. Calibration to historical data was discussed in the text. Departures from these parameter settings are also noted in the text.

Appendix B: Current Smallpox Policy

Until 1972, immunization was required for all individuals over the age of one year in the United States. In 1972, however, the government discontinued routine vaccination because the risk of serious adverse effects outweighed the rather low risk of infection, due to high vaccine coverage and minimal exposure to smallpox. In addition, practices such as ring vaccination, which were extremely successful in the global eradication effort, further encouraged the cessation of routine smallpox vaccination.[15]

However, bioterrorism concerns have renewed interest in the creation of a substantial smallpox vaccination policy in the United States. Therefore, the Centers for Disease Control (CDC) has updated the response plans used in the 1970s. The current interim Smallpox Response Plan

---

[15] The term *ring vaccination* is used variously to denote different forms of targeted (as against mass) vaccination. As in the CDC's usage, it normally involves, but need not be limited to, trace vaccination.

and Guidelines (available at: www.bt.cdc.gov/agent/smallpox/response-plan/index.asp [accessed December 10, 2002]) employs many of the methods used successfully to control outbreaks more than thirty years ago. The main concept is to control any smallpox epidemic using ring vaccination. The size of the ring of individuals may be modified according to the scale of the outbreak, the level of resources available, and the effectiveness of the method. Thus, health officials would first isolate suspected and confirmed smallpox cases. Subsequently, they would trace and vaccinate contacts of the isolated cases, as well as vaccinating the household members of the contacts.

The CDC guidelines prioritize groups for immunization as follows:

1. Face-to-face close contacts ($\leq$6.5 feet or 2 meters) or household contacts with smallpox patients after the onset of the patient's fever.
2. Persons exposed to the initial release of the virus (if the release was discovered during the first generation of cases and vaccination may still provide benefit).
3. Household members (without contraindications to vaccination) of contacts with smallpox patients (to protect household contacts should smallpox case contacts develop disease while under fever surveillance at home).
4. Persons involved in the direct medical care, public health evaluation, or transportation of confirmed or suspected smallpox patients.
5. Laboratory personnel involved in the collection and/or processing of clinical specimens from suspected or confirmed smallpox patients.
6. Other persons who have a high likelihood of exposure to infectious materials (for example, personnel responsible for hospital waste disposal and disinfection).
7. Personnel involved in contact tracing and vaccination, quarantine/isolation or enforcement, or law enforcement interviews of suspected smallpox patients.
8. Persons permitted to enter any facilities designated for the evaluation, treatment, or isolation of confirmed or suspected smallpox patients (only essential personnel should be allowed to enter such facilities).
9. Persons present in a facility or conveyance with a smallpox case if fine-particle aerosol transmission was likely during the time the case was present (for example, a hemorrhagic smallpox case and/or a case with active coughing).

Additional groups with indirect contact would be considered for voluntary vaccination by the director of the CDC, as follows:

1. Public health personnel in the area involved in critical surveillance and epidemiological data analysis and reporting

2. Logistics, resource, and emergency management personnel
3. Law enforcement, fire, and other personnel involved in other nondirect patient care response support activities, such as crowd control, security, law enforcement, and firefighting and rescue operations.

The Smallpox Response Plan and Guidelines is a draft document; the CDC acknowledges that it will require updates due to changes in resources. Furthermore, the immunological landscape of the United States has changed since the 1970s.

References

Anderson, Roy M., and Robert M. May. 1991. *Infectious Diseases of Humans: Dynamics and Control.* Oxford University Press. 1991.

Bailey, N. T. J. 1953. "The Total Size of a Stochastic Epidemic." *Biometrika* 40:177.

Berry, Brian J. L., L. Douglas Kiel, and Euell Elliott, eds. 2002. "Adaptive Agents, Intelligence, and Emergent Human Organization: Capturing Complexity through Agent-Based Modeling." Arthur Sackler Colloquia of the National Academy of Sciences. *Proceedings of the National Academy of Sciences* 99, supp. 3.

Burke, Donald S. 1998. "Evolvability of Emerging Viruses." In *Pathology of Emerging Infections* 2, ed. Ann Marie Nelson and C. Robert Horsburgh Jr. Washington, DC: ASM Press.

Burke, Donald S., Kenneth A. De Jong, John J. Grefenstette, Connie Loggia Ramsey, and Annie Wu. 1998. "Putting More Genetics into Genetic Algorithms." *Evolutionary Computation* 6:387–410.

Burke, Donald S., Joshua M. Epstein, Derek A. T. Cummings, Jon I. Parker, Kenneth Cline, Ramesh M. Singa, and Shubha Chakravarty. 2004. "Individual-Based Computational Modeling of Smallpox Epidemic Control Strategies." Johns Hopkins University, Bloomberg School of Public Health and Brookings, January.

Epstein, Joshua M. 1997. *Nonlinear Dynamics, Mathematical Biology, and Social Science.* Reading, MA: Addison-Wesley.

Epstein, Joshua M., and Robert L. Axtell. 1996. *Growing Artificial Societies: Social Science from the Bottom Up.* Washington, DC: Brookings Institution Press; Cambridge: MIT Press.

Fenner, Frank, *et al.* 1988. *Smallpox and Its Eradication.* Geneva: World Health Organization.

Grefenstette, John J., Donald S. Burke, Kenneth A. De Jong, Connie L. Ramsey, and Annie S. Wu. 1997. "An Evolutionary Computation Model of Emerging Virus Diseases." NCARAI Technical Report AIC-97-030.

Halloran, M. Elizabeth, Ira M. Longini, Azhar Nizam, and Yang Yang. 2002. "Containing Bioterrorist Smallpox." *Science* 298:1428–32.

Henderson, Donald A., Thomas V. Inglesby, John G. Bartlett, Michael S. Ascher, Edward Eitzen, Peter B. Jahrling, Jerome Hauer, Marcelle Layton, Joseph McDade, Michael T. Osterholm, Tara O'Toole, Gerald Parker, Trish Perl, Philip K. Russell, and Kevin Tonat. 1999. "Smallpox as a Biological Weapon." *Journal of the American Medical Association* 281:2127–37.

Jackson, Ronald J., Alistair J. Ramsay, Carina D. Christensen, Sandra Beaton, Diana F. Hall, and Ian A. Ramshaw. 2001. "Expression of Mouse Interleukin-4 by a Recombinant Ectromelia Virus Suppresses Cytolytic Lymphocyte Responses and Overcomes Genetic Resistance to Mousepox." *Journal of Virology* 75:1205–10.

Kaplan, Edward H., David L. Craft, and Lawrence M. Wein. 2002. "Emergency Response to a Smallpox Attack: The Case for Mass Vaccination." *Proceedings of the National Academy of Sciences* 99:10935–40.

Knuth, Donald E. 1998. *The Art of Computer Programming*. 3rd ed. Vol. 2, *Seminumerical Algorithms*. Reading, MA: Addison-Wesley.

Koopman, Jim. 2002. "Controlling Smallpox." *Science* 298:1342.

Koopman, J. S., S. E. Chick, C. P. Simon, C. S. Riolo, and G. Jacquez. 2002. "Stochastic Effects on Endemic Infection Levels of Disseminating versus Local Contacts." *Mathematical Biosciences* 180:49–71.

Lane, J. M., F. L. Ruben, J. M. Neff, and J. D. Millar. 1969. "Complications of Smallpox Vaccination, 1968: National Surveillance in the United States." *New England Journal of Medicine* 281:1201–8.

Mack, T. M. 1972. "Smallpox in Europe, 1950–1971." *Journal of Infectious Diseases* 125:161–69.

O'Toole, Tara, Michael Mair, and Thomas V. Inglesby. 2002. "Shining Light on 'Dark Winter.'" *Clinical Infectious Diseases* 34:972–83.

Sharma, D. P., Alistair J. Ramsay, D. J. Maguire, M. S. Rolph, and Ian A. Ramshaw. 1996. "Interleukin-4 Mediates Downregulation of Antiviral Cytokine Expression and Cytotoxic T-Lymphocyte Responses and Exacerbates Vaccinia Virus Infection in Vivo." *Journal of Virology* 70:7103–7.

Steinbruner, John, Elisa D. Harris, Nancy Gallagher, and Stacy Gunther. 2002. Controlling Dangerous Pathogens: A Prototype Protective Oversight System. University of Maryland, September, unpublished.

Whittle, P. 1955. "The Outcome of a Stochastic Epidemic—A Note on Bailey's Paper." *Biometrika* 42:116.

GENERATING OPTIMAL ORGANIZATIONS

SOMEONE ONCE ASKED the intriguing question, "What is Beethoven's Ninth?" Surely, it is not merely the printed orchestral score of Beethoven's Ninth, since the Ninth Symphony is a beautiful piece of music, while the score is a silent pile of paper with ink marks all over it. By the same token, there are as many audible realizations of that single printed score as there are conductors and orchestras (each with their idiosyncratic tempi, dynamics, phrasings, and other expressive particularities), open air amphitheaters, and intimate concert halls (each with their individual acoustics). It seems to me that "Beethoven's Ninth" can only denote *the complete set of possible realizations generable under (encoded in) the score.*[1] If the cardinality of this set were one, we'd need only one recorded performance of Beethoven's Ninth. Yet we have new ones each year. The score encodes a seemingly infinite generative capacity.

The agents you are about to meet in the adaptive organization model have fixed behavioral rules and fixed numerical parameters. One can think of this agent microspecification as a fixed genome analogous to a printed score. The agents face dynamic environments to which they must adapt in some way. Even with a fixed genome, different environments produce radically different adaptive histories. In certain environments, the agents endogenously generate hierarchies; in others, they forgo hierarchy and engage in internal trade. While the fixed genome is analogous to the printed symphonic score, the dynamic environment is analogous to the pressures exerted by conductor, orchestra, concert hall, and so on, each of which generates a different performance.

In this chapter, we are going to pose a question whose musical analogue would be strange. It would be: What is the *optimal score* for the Berlin Philharmonic to perform? We will introduce a notion of fitness that will allow us to rank performances. We will fix a dynamic environment (fix the orchestra, and so on). And, by combinatorial optimization, we will determine the optimal genome (the optimal score) in that environment (for that orchestra). And then we'll "listen" to it (will watch the optimal history of organizational adaptation as a movie)!

---

[1] So, in a sense, it remains, and ever will remain, a work in progress. In this sense, all symphonies are unfinished!

Verticality

The models presented thus far unfold on various spaces—two-dimensional lattices, one-dimensional rings, environmental landscapes, towns, social networks. But they are all "flat." There is no *hierarchical* aspect; no agent really has "authority" over any other. There are no superiors or subordinates. The principal way in which the adaptive organization model differs from the rest is precisely that agents generate and dissolve hierarchies locally.

At the most abstract level, the organization's problem is long-range resource allocation, and it must discover when (very expensive but highly efficient) "top down" global reallocation dominates (cheap but sluggish) reallocation through a series of short-range tradelike transactions. In the neoclassical picture, management structure is absent and inputs are adjusted to maximize something (e.g., profit). Here, inputs (labor) are fixed, and it is the management structure that is varied to optimize.

Variable Geometry Firms

For the particular environmental dynamic used in this study, the optimal history of structural adaptation involves oscillations between "flat" trading regimes and hierarchies, in perpetual motion up and down a spatial market as a traveling wave. So the optimal organization does not have a fixed structure. It is a variable-geometry firm.

Chapter 13

GROWING ADAPTIVE ORGANIZATIONS: AN

AGENT-BASED COMPUTATIONAL APPROACH

Joshua M. Epstein*

Introduction

What constitutes an *adaptive organization*? What would constitute *optimal* structural adaptation in a dynamic environment? Can one "grow" optimally adaptive organizations from the bottom up—that is, devise rules of individual behavior that *endogenously generate* optimal structural adaptations? There is, of course, a large literature on the origin of firms, on the size distribution of firms in an economy, and on a host of related topics.[1] However, I am unaware of any explicit model in which *individual agents endogenously generate internal organizational structures that adapt optimally to dynamic environments.* The present chapter develops such a model, using the agent-based technique (Epstein and Axtell 1996).[2] It is important to note that I do not purport to model any existing organization, or to "fit" the model to data of any sort. Rather, the aim, using a highly idealized model, is to illuminate what sorts of individual (micro) agent rules confer adaptiveness on the larger (macro)

* The author is a Senior Fellow in Economic Studies at The Brookings Institution, a Member of the Brookings–Johns Hopkins Center on Social and Economic Dynamics, and a Member of the External Faculty of The Santa Fe Institute.

The author particularly acknowledges software engineer Joshua Miller of the Bios Group for implementing the model in Ascape and for computational analysis. He also thanks Daniel Teitelbaum of the Bios Group for his contributions. For valuable comments and discussions, he thanks Robert Axelrod, Chris Carroll, Joseph Harrington, Myong-Hun Chang, Michael Heaney, William McKelvey, John Miller, Benoit Morel, John Sterman, Anjali Sastry, Nelson Repenning, and Peyton Young. The Cap Gemini Ernst & Young Center for Business Innovation funded this research and the author thanks Christopher Meyer and Eric Mankin for their support and for numerous valuable discussions.

[1] Axtell 1999, 2001; Baum and McKelvey 1999; Carley and Svoboda 1996; Chang and Harrington 2006; Coase 1937; Cohen, March, and Olsen 1972; Lomi and Larsen 1999; Miller 2001; Nickerson and Zenger 2002; Prietula and Carley 1994; Radner 1993; Radner and Van Zandt 1999; Williamson 1975; Young-pa and Durfee 1998.

[2] For an excellent review of the agent-based literature, see Chang and Harrington 2006.

enterprise. There will of course be many simplifying assumptions, but hopefully they will not entirely subvert that broad objective.

Conceptually, we posit that organizations, such as firms in an economy, exist in economic environments (defined carefully below) that may change over time; environments are *dynamic*. Over any period—over any day, let us say—organizations have a *structure*. That is to say, there is, in principle, a graph representing each agent's information (internal and external) and its span of authority (its set of resources and permissible manipulations of them); and there is a specific deployment of resources (e.g., labor and capital). *Ceteris paribus*, the state of the economic environment in a given period, combined with the organization's structure in that period, jointly determine the organization's *performance* in that period: this could mean its total profit, its market share, or other measures. In general, changes in structure induce changes in performance.

Now imagine stipulating a k-period environmental dynamic (we do this concretely below). Then, for every "candidate" k-period history of structural adaptation, there is a corresponding history of organizational performance (e.g., total profit over the k periods). It is therefore perfectly natural to pose the question: *What is the optimal history of structural adaptation in that dynamic environment?*

Given a particular environmental dynamic, would an optimal history of structural adaptation involve periods of extreme hierarchy, separated by relatively "flat" internal trading regimes? This is the sort of general question we wish to explore.

However, we take a *generative* approach. That is, we want a *single fixed set* of operating rules and parameters at the individual agent level that will generate, or "grow," an entire optimal history of structural adaptation "from the bottom up." The autonomous agents, for example, should "grow" hierarchies when they are needed and dissolve them when they are obsolete. The aim is to characterize rigorously what would constitute optimal adaptiveness under various assumptions. While there is no empirical claim particularly, the results appear to call into question the optimality of certain ubiquitous forms, notably, the pyramidal hierarchy, which—at least in this model—turns out to be optimal only in quite restricted cases.

Organization

The chapter proceeds in six parts. Part 1 gives the basic setup of the model. Part 2 explores the spontaneous emergence and dissolution of hierarchy, with internal trade proscribed. In part 3, we explore internal trade alone as a means of reallocating resources within the organization. Having explained the mechanics of hierarchy and trade, I introduce a particular dynamic environment. Against this environment, we examine the effectiveness of pure trade and pure hierarchy as solutions.

Now, without introducing some objective function, no claims can be made about optimality, or about the relative performance, or "fitness," of different modes of organization. Accordingly, in part 4, we introduce a very general objective function for the organization. By setting a parameter (k), it specializes to profit maximization $(k=0)$, to market-share maximization $(k=1)$, and to hybrids of the two $(0 < k < 1)$. Then, in part 5, for various choices of objective function (e.g., profit maximizing), we sweep the entire parameter space of the model for the optimal parameters. For those parameters, we then show the optimal history of structural adaptation. Finally, in part 6 various extensions are proposed.

Part 1. Basic Model

I have used the term *environment* repeatedly and without definition. Of course, an organization's economic environment could include everything from its technological opportunities to pollution regulations to the prime interest rate. While my model can be generalized (see extensions), the environment is represented simply as a dynamic pattern (a flux) of "opportunities," depicted as a flow of red dots moving left to right toward the "market" of the enterprise. This "market" consists of 32 contiguous cells, depicted as a vertical array. The market is manned by the enterprise's labor force of (at most) 32 workers, depicted as solid blue squares. Each of these workers is completely myopic, and controls just the cell he occupies. If an incident red dot runs into a blue square, the red is considered to be intercepted by the blue worker; that opportunity (red dot) has been "taken" by the enterprise. For an enterprise to suffer no red penetrations, workers (blue squares) would have to be positioned to intercept every incident red dot (see figure 13.1).

An initial enterprise and an incoming red opportunity flux (the environment) are shown in figure 13.1. An initial condition for the enterprise always consists of some distribution of blue workers (level-0 managers) and an initial level-1 management layer (comprised of 16 managers). Each level-1 manager controls a two-cell market segment (the cell positioned four spaces to the left and the cell one space north of that). If, at any point, there are workers in cells controlled by a manager, that manager is depicted as a solid dot. If there are no workers under the manager's control the manager is depicted as a hollow dot. So, in figure 13.1, seven solid managers actually control labor; the hollow rest are monitoring their sector of the market for penetrations. The environment is essentially a block of red dots in the south of the market. They are marching toward the enterprise's space.

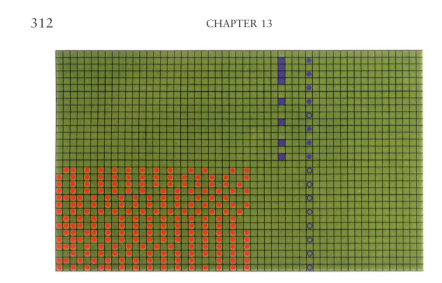

Figure 13.1. Initial enterprise and environment.

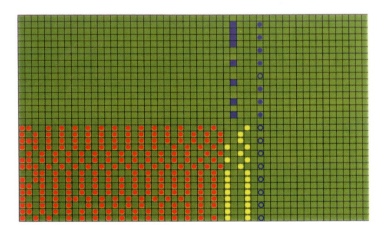

Figure 13.2. Penetrators.

Workers cannot move themselves. The task of management is to allocate them effectively. Clearly, if no reallocation of labor occurs, the red dots will penetrate the blue enterprise's front line. These opportunities are missed. When a red penetrates, the penetrator is colored yellow, as shown in figure 13.2.

Spans of Control

Given certain objective functions for the enterprise, it may prove efficient to generate a hierarchy (the mechanics of this are presented below).

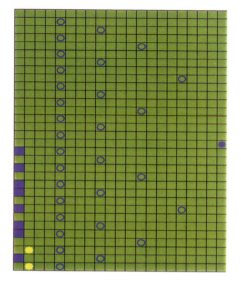

Figure 13.3. The maximum hierarchy.

Figure 13.3 depicts the maximal hierarchy. In a hierarchy of any height, each level-n manager can shift workers about the 2^n cells the manager controls. As noted above, each level-1 manager controls a 2-cell market segment. Each level-2 manager controls a 4-cell segment, each level-3 an 8-cell segment, and so on up to the level-5 CEO, who can shift labor anywhere across the entire 32-cell market of the enterprise. This CEO agent supersedes all subordinates, and so the CEO is the only solid agent. The hollow agents beneath need *not* be interpreted as idle. Presumably, they would transmit the CEO's orders down the line until labor is reallocated. The same point applies to hollow agents in local hierarchies of any height below. In this exposition, the labor force is fixed.

Labor Allocation Rule

The labor allocation rule is syntactically identical for all managers, though of course their spans of control are not identical. The rule, however is

> L: *Within your span of control, identify all labor not currently intercepting (call that List 1) and identify all sites subject to imminent (next period) attack; call that List 2. Choose a random laborer from List 1 and move him to a random site from List 2. Repeat until no sites are threatened or List 1 is exhausted, whichever occurs first.*

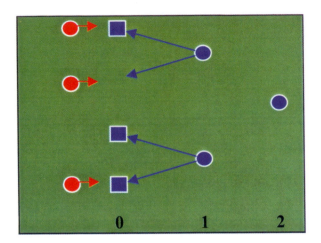

Figure 13.4. The essential problem.

Basically, the rule amounts to identifying everybody who's leaning on his rake, and throwing him at an outstanding problem. Now, the essential problem arises as follows.

As illustrated in figure 13.4, labor is numerically adequate, but it is misallocated. There is a gap in blue coverage at cell 3 (counting up from the bottom), while there is an unopposed blue defender at cell 2. Clearly, to block the penetration, the free defender should be shifted up one site.[3] However, neither level-1 manager is empowered to do that, since each controls only the two sites below him (shown by arrows). There are two pure approaches to the problem:

1. *Hierarchy.* A level-2 manager (controlling all four sites) is created (as shown) to shift the free resource up to its efficient location.
2. *Internal Trade.* The level-1 manager who is facing penetration announces, or "posts," his demand for labor to all other managers in the same level. Those with an excess supply of labor may choose to respond with an internal transfer.

Next I describe in detail the mechanics by which hierarchies endogenously emerge and dissolve, and then the mechanics of internal trade.

---

[3] Terms such as *defender* and *penetrator* suggest a military interpretation of the model that is possible but not necessary, as discussed further below.

MECHANICS OF HIERARCHY WITH TRADE PROSCRIBED

Recall that a level-n manager controls a market segment of 2^n cells. Over this segment, the manager allocates labor according to the labor allocation rule (L) stated above. The manager has two penetration thresholds, *Tmin* and *Tmax*. (To exercise the model, these are initially set by the user. Later we compute their optimal values.) In this paper, we will assume that these values differ by management level but are common across a given level. Every manager has a finite memory (e.g., the last 10 periods), m. The manager computes the average penetrations over this memory (i.e., computes the total number of penetrations of his or her market segment over the last m periods and divides by m); call that result P.

> Upward Hierarchy Rule: If $P \geq Tmax$, then (subject to upward inertia) a manager of level $n + 1$ is created. Otherwise, the downward hierarchy rule applies (see below).

This formal mechanism admits at least three different interpretations:

1. In a military interpretation, it means a superior officer from a higher level of authority is "called in" to allocate across threatened battle sectors. Brigade commanders would "call in" division commanders for support.
2. In the economic version published as a 2003 Santa Fe Institute working paper (Epstein 2003), a higher manager is procured from without (e.g., on the spot market for managerial talent).
3. In a different corporate interpretation, one could imagine that the n-level manager who is reporting excess penetration is promoted internally[4] to a higher ($n + 1$ level) span of control, and *a new subordinate is hired* from without to take the promoted manager's old position at level n.

This last interpretation is more consistent with the observed pattern of managers creating divisions *below* them rather than above. Normally firms expand by adding subordinate units, not by adding superior ones.[5]

However, the mathematics are identical.[6] This, of course, is quite common in science. Both the classical gravitational and electrostatic attractive forces are inverse square laws, for example. Rather than write three versions of commentary for every run presented, the second of the above interpretations will usually be adopted below. But that choice has

---

[4] Or a randomly-selected n-level manager is promoted, for instance.

[5] I am grateful to Joseph Harrington for informing me of this regularity.

[6] The cost accounting is identical between interpretations 2 and 3, which are of prime concern to us here. Clearly, the military case differs in that regard. Salaries and other costs are treated later in this chapter.

more to do with expository efficiency than with a rigid commitment to
that interpretation. Indeed, there are clearly contexts in which others are
more plausible.

In the preceding statement of the rule, *upward inertia* equals the
number of successive iterations for which $P \geq Tmax$, and captures the
reluctance with which managers call for superiors (in interpretations 1
and 2) or engage in literal self-promotion and the hiring of subordinates
(in interpretation 3). Assuming this inertia is not prohibitive, control
now passes to the higher-level manager, responsible for a (twice as large)
market segment of width 2^{n+1}. Within this new span of control, the
new manager applies the labor allocation rule. As a result of these
reallocations of labor, the number of penetrations of this (bigger) segment
may be very low. This manager has her own *Tmin* and *Tmax*.

> Downward Hierarchy Rule: If $P < Tmin$, then (subject to downward
> inertia) the manager is deleted from the structure, and control reverts
> to her subordinates. Otherwise, the upward hierarchy rule applies (see
> above).

Here, downward inertia equals the number of iterations for which
$P < Tmin$, and captures the reluctance of managers to cede control (or
disband subunits as the case may be) when they are no longer needed.
For expository simplicity, we will assume throughout that the *Tmax* of
a subordinate equals the *Tmin* of a superior,[7] and further, that there is a
single value of upward and downward inertia per management level.

That completes the hierarchy specification. Given certain initial labor
allocations and environmental dynamics, these simple rules are sufficient
to generate hierarchies, and to dissolve them, endogenously. Unlike the
standard neoclassical picture, in which management structure is absent
and inputs are varied to maximize profit, in this model inputs are fixed,
and it is the management structure that is varied to optimize. Before
demonstrating that, I present the alternative approach: internal trade.

MECHANICS OF INTERNAL TRADE

Just as before, over the segment they control, managers are computing P (average penetrations over memory) and comparing it to *Tmax*.
Suppose that $P \geq Tmax$. Rather than invoke the upward hierarchy rule,
the manager can "post," to all other managers in the same level, the
manager's excess demand for labor: $P - Tmax$. In turn, those managers
will, with some probability, transfer to the posting agent their excess
labor supply (technically, the minimum of that and the poster's demand).

---

[7]This common value is denoted t in all the tables below.

What is this excess supply? It is total labor under the manager's control minus labor currently intercepting reds, all net of the manager's own excess demand (the manager's own $P - Tmax$). In summary:

> Demand Rule: If $P \geq Tmax$, then (with probability 1 minus demand-inertia) post excess demand, $P - Tmax$.

> Supply Rule: With probability 1 minus supply-inertia, transfer to the posting agent min[the posting agent's excess demand, your excess supply].

With all of this in place, then, imagine that you are the demand-posting agent. In random order, each other manager in your layer is queried. In accordance with the supply rule, each allocates to you some amount of labor (possibly zero). If, in the course of this process, your demand is met, then you stop. If you exhaust all other managers in your layer and still $P \geq Tmax$, then you invoke the upward hierarchy rule. The fact that trade is attempted first commits me implicitly to the view of hierarchy as a kind of (internal) market failure.

Here, horizontal inertia—in demand and supply—would encompass all transaction activities, negotiations, contracts, hoarding, and so on. In a more elaborate version of the model, a literal internal labor market could, of course, be introduced.[8] But, for our purposes, we will optimize on all values of vertical and horizontal inertia, and leave aside the important question of how those values might be induced in practice—via price or other mechanisms. As we shall see, nonzero inertia will prove optimal given certain objective functions in some environments. The entire set of parameters and the management layers to which they apply are displayed in table 13.1. The numerical values employed in the various runs presented as snapshots in the text (and as full movies on the CD) are given in table 13.A.1 of the appendix.

The first four management layers possess penetration thresholds, and upward and horizontal (demand and supply) inertias, while the top layer does not. The lowest level of management has no downward inertia, while the top manager does. The entire matrix of parameters can be thought of as a kind of genome for the organization, shown as G at the bottom of table 13.1.

GÖDEL NUMBER

As a strictly mathematical matter, one can compress the entire genome into a single integer by the expedient of Gödel numbering. Each element

---

[8] According to Coase (1937), however, "the distinguishing mark of the firm is the supersession of the price mechanism."

TABLE 13.1
The Organization's Genome

		Relevant Management Layers				
Hierarchy:						
Penetration Threshold	t	1	2	3	4	—
Upward Inertia	u	1	2	3	4	—
Downward Inertia	d	—	2	3	4	5
Trade:						
Demand Inertia	D	1	2	3	4	—
Supply Inertia	S	1	2	3	4	—

$G = \{t1, t2, t3, t4, u1, u2, u3, u4, d2, d3, d4, d5, D1, D2, D3, D4, S1, S2, S3, S4\}$

of the genome is (or can be converted to) an integer g_i. If L is the length of the genome (here 20) then the product of the first L primes, each raised to g_i, yields a unique integer, G, given by

$$G = \prod_{i=1}^{L} p_i^{g_i}$$

This Gödel number encodes the entire adaptive repertoire of the organization, as we shall see. With this apparatus in place, then, let us put the model through some basic paces, before introducing the objective function required to discuss optimality in any sense.

PART 2. HIERARCHICAL SOLUTIONS WITH INTERNAL TRADE
CLAMPED OFF

To begin, we will ban all internal trade and study the emergence of hierarchy only. As an environment for the first model runs, we posit a heavy opportunity flux (or "attack") in the south. The resources of the organization ("the defense"), however, are deployed largely in the north. This is depicted in frame 1 of run 1. This gross misallocation quickly leads to the southern market segments being overrun, as in frame 2 of run 1. To generate the hierarchy, we set penetration thresholds and upward inertia levels to very low levels. To maintain the hierarchy, we set downward inertias to high levels. Heuristically, we can think of low values as zeroes and high values as ones. In that case, the genome of interest is shown in table 13.2. The actual numerical parameter values used for all runs are given in the appendix (and could, of course, be normalized to fall in the unit interval).

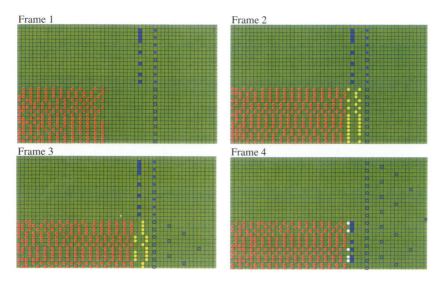

Run 1. Emergence of permanent hierarchy.

TABLE 13.2
The Genome for Immediate and Permanent Hierarchy

| | | Relevant Management Layers | | | | |
		1	2	3	4	5
Hierarchy:						
Penetration Threshold	t	0	0	0	0	—
Upward Inertia	u	0	0	0	0	—
Downward Inertia	d	—	#	#	#	1
Trade:						
Demand Inertia	D	1	1	1	1	—
Supply Inertia	S	0	0	0	0	—

$G = \{0, 0, 0, 0, 0, 0, 0, 0, \#, \#, \#, 1, 1, 1, 1, 1, 0, 0, 0, 0\}$

Hierarchy emerges quickly as agents in the successive layers record excess penetrations, promptly calling into play ever higher levels of management, as in frame 3 of run 1. However, it is not until the very top level of management (level 5) is called into being that an agent (the CEO) has sufficient vision to notice the global north-south misallocation and correct it, as shown in frame 4 of run 1. The entire adaptive history is shown animated as movie 1 (of chapter 13) on the CD.

This adaptation reminds one of a computer company that is focused exclusively on mainframes (the northern market segment), while the opportunity of PCs approaches (the red flux in the south). But it takes a visionary CEO to "see" the strategic error and shift the resources of the firm to the south, exploiting the opportunity.

Of course, having solved the problem, the hierarchy is no longer needed; labor is in place. High-level vision is superfluous. But bureaucracies have immense inertia and once constructed are hard to dissolve. This phenomenon is generated by the model's high level-5 value of downward inertia.

In summary, one recipe for a large persistent hierarchy is as follows: Start with a strategic misallocation of resources. Set penetration thresholds and upward inertias to low values. This grows the hierarchy. High downward inertia then blocks its dissolution. Below, we introduce costs and objective functions for the firm. And we will see that, given certain objectives, the ability to grow a hierarchy is highly adaptive, while the inability to dissolve it is not.

A more adaptive performance is recorded in run 2. Here, everything is as before (the genome is exactly as in table 13.2), except that downward inertias are zero in all management layers. In this case, hierarchy is again spontaneously generated, and it then solves the strategic problem, exactly as in run 1. But, having done so, it dissolves, leaving a "lean" structure overseeing a well-allocated workforce. The corporate culture here is, "When you need help, ask for it (low upward inertia). When you are no longer needed, bow out (low downward inertia)." The upward trajectory is as shown in frames 1–4 of run 1, but the hierarchy dissolves, leaving the structure shown in figure 13.5.

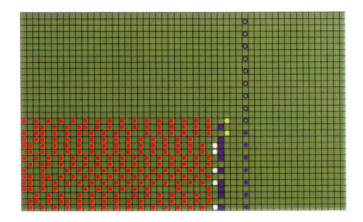

Figure 13.5. Emergence and dissolution of hierarchy: final state.

The full animation of run 2 is movie 2 of the CD. Again, it should be emphasized that the agents are doing all of this from the bottom up without central direction. Each *fixed set* of agent thresholds and inertias generates an entire history of structural adaptation (and labor reallocations) to a dynamic environment. In run 2, the history involves the construction of hierarchy, a strategic reallocation of the organization's resources, and the dissolution of hierarchy, all "self-organized," if you will.

In the runs presented thus far, the agents generate the maximum possible hierarchy (five levels), which then persists entirely (run 1) or dissolves entirely (run 2). In a different environment, fixed settings identical to those of run 1 (permanent hierarchy) generate a very different history of structural adaptation.

In run 3 (see movie 3 for animation), the firm's resources are initially concentrated in the middle third of the market, while the opportunity flux is advancing in the northern and the southern thirds.

Hierarchy arises, as before, to recognize and correct a strategic misallocation. But rather than the single maximum five-level hierarchy, the firm consists of two stable local hierarchies of medium height, one in the north and one in the south.

OBSERVATIONS ON THE PURE HIERARCHY RUNS

The structural adaptations presented thus far unfold without central direction, "from the bottom up," as a result of the agent rules and parameters in the genome. As just illustrated, any fixed genome will generate different histories of structural adaptation in different dynamic environments. In a sense, the genome "encodes" an entire repertoire of adaptations. Thus far, there are no costs of hierarchy and no objective functions, so no ranking of structures is possible. Before introducing costs and objectives, we explore internal trade as the allocative mechanism, with hierarchy proscribed.

PART 3. INTERNAL TRADE

A pure trade genome is given in table 13.3. The value of unity for level-1 upward inertia clamps out any hierarchy, while the level-1 horizontal (demand and supply) inertia values are set to zero. The same run 1 environment and initial misallocation of labor that produced hierarchy and central (top down) reallocation under genome 1 produces a trading solution here. As shown in run 4, trade results in the same final correction as in run 1. Importantly, however, long-range reallocation by internal

Frame 1

Frame 2

Frame 3

Frame 4

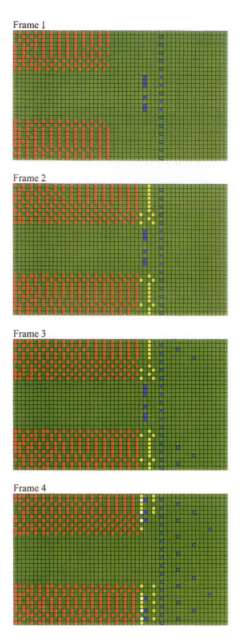

Run 3. Local intermediate hierarchies.

TABLE 13.3
The Genome for Pure Trade

		Relevant Management Layers				
		1	2	3	4	5
Hierarchy:						
Penetration Threshold	t	0	0	0	0	—
Upward Inertia	u	1	#	#	#	—
Downward Inertia	d	—	#	#	#	#
Trade:						
Demand Inertia	D	0	#	#	#	—
Supply Inertia	S	0	#	#	#	—

$G = \{0, 0, 0, 0, 1, \#, \#, \#, \#, \#, \#, \#, 0, \#, \#, \#, 0, \#, \#, \#\}$

trade is generally slower than under in-place hierarchy, since it requires a sequence of low-level, or "local," managerial responses, rather than the single global reallocation possible under extreme hierarchy. (See movie 4 for animation.)

The run 3 split attack that resulted in two local hierarchies can also be handled through trade, as demonstrated in run 5 (with settings as in the previous run). Notice that global reallocations (i.e., allocations across the entire market front) occur here as well. Indeed, stripped of all economic interpretations, this model concerns trade-offs between centralized and decentralized approaches to long-range coordination.

A DYNAMIC ENVIRONMENT

Thus far, the environments have been very straightforward; the flux of incident dots has not changed direction over time. As a final preliminary before introducing the objective function, we explore the performance of pure trade and pure hierarchy in a more complex environmental dynamic. Here the organization is confronted with diagonal patterns of incoming opportunities whose slope alternates periodically; it is a *sawtooth* moving left to right.[9]

Run 6 (see movie 6) demonstrates that trade succeeds in intercepting a number of incoming reds, but the majority penetrate the market. Trade lags behind this dynamic.

---

[9]This may be clearest in the fully animated version on the CD.

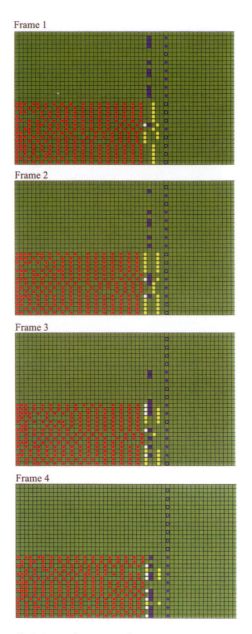

Run 4. Attack handled through pure trade.

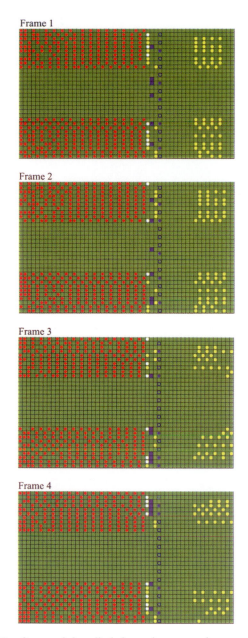

Run 5. The run 3 split attack handled through pure trade.

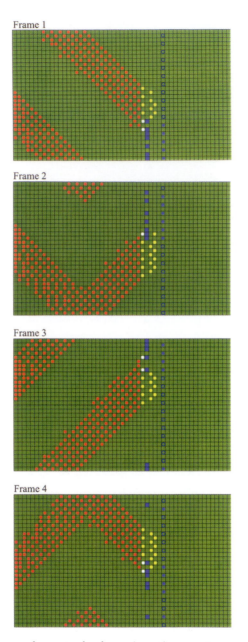

Run 6. Performance of pure trade: dynamic environment.

By contrast, once the full hierarchy is erected, it is vastly more efficient in blocking penetrations, as shown in run 7 (see movie 7).[10] (The black lines indicate top-down directives.)

GENERAL TRADE-OFFS

Now, even without introducing any mathematical objective function, one can clearly see that there are trade-offs between the two pure approaches. The benefit of hierarchy is that it prevents penetrations (protecting market share). However, if salaries increase dramatically with management level, this solution will be expensive. By contrast, pure trade sacrifices market share (allows much penetration) while avoiding the costs of multiple higher levels of management. Given a particular environmental dynamic and cost structure for the firm, then, what levels of penetration and hierarchy actually *maximize* profit, or market share, or combinations of the two? Equivalently, given a particular objective function, what is the optimal (fixed) genome? What history of structural adaptation does it generate?

PART 4. OBJECTIVE FUNCTIONS

All numerical assumptions are provided in table 13.A.2 of the appendix. Over an accounting period t (e.g., a day) define

$R(t)$ = Revenue = The Value of Intercepts

= (The Value Per Intercept)(Total Number of Intercepts).

$W(t)$ = The Wage Bill = $\sum_i$ (Number of Managers in Level i)(Wage at Level i).

Here, we assume that, for managers controlling labor (solid dots), wages increase as the cube of the level. Specifically, the wage at level $i = c(i+1)^3$,

---

[10] For *fixed* memory, trade is slower than hierarchy because long-range reallocation is effected via a series of myopic applications of L (where spans are small) versus a single global application of L, where the CEO's span is global. So, for fixed memory, the relative issue is a sequence of local L-applications versus a single global application. The absolute gap between the two is a function of memory m. The larger is memory, the slower is trade, because for higher m, the moving average over which P is computed is ever more stable, so trade lags the sawtooth. By the time P exceeds the threshold, the environment has moved on. If m is a tiny number, there are huge adjustments to every environmental blip. In principle, one would optimize on m as well as the other parameters.

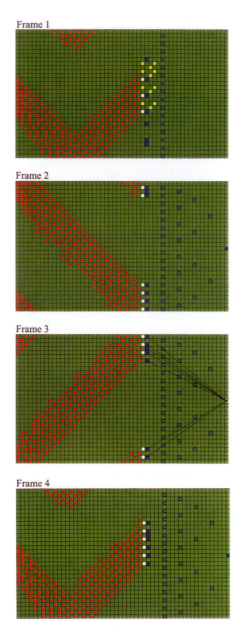

Run 7. Performance of pure hierarchy: dynamic environment.

where c is a multiplier. Managers not controlling labor (hollow dots) are paid ci. Labor is paid unity.

$T(t)$ = The Transaction Costs of Internal Trade

= (Per Trade Cost)(Number of Trades)

$P(t)$ = Penetration Costs

= (The Cost Per Penetration)(Number of Penetrations).

Then, with $0 \le k \le 1$, the general per period objective function $F(t)$ is

$$F(t) = R(t) - W(t) - T(t) - kP(t). \tag{1}$$

If $k = 0$, the firm is a profit-maximizer. It doesn't care about penetrations (i.e., market share). If $k = 1$, the firm is a market-share maximizer, and accounts penetrations as costs. If $k = 0.5$, the firm cares about both market share and profit. Since net returns depend on the genome G, we think of F as parameterized by G, and will adopt the notation $F(t; G)$ to denote the return in period t for genome G. With the entire above apparatus in hand, we can now address the core question.

THE CENTRAL QUESTION

Specify an environmental dynamic over some time horizon T. For any objective function (k-value), each fixed genome G—each vector of thresholds and inertias—will generate some history of structural adaptation, and some corresponding stream of net returns $F(t; G)$. (Again, the G entering into this expression as a parameter could be the Gödel number for the full vector of thresholds and inertias). The problem then is to find that particular G that maximizes cumulative returns over the horizon. In short, the problem is to determine[11]

$$G^* = \arg \max \sum_{t=0}^{T} F(t; G). \tag{2}$$

In a more evolutionary framing, the unit of selection is the genome. Selection pressures are exerted by the specified environmental dynamic over the time horizon T and the specified objective function $F(t; \cdot)$. The genome's fitness is then given by $\sum_{t=0}^{T} F(t; G)$. Then (2) asks for the genome with highest fitness.[12]

---

[11] We are actually less interested in $\sum_{t=0}^{T} F(t; G^*)$.

[12] The computational evidence suggests uniqueness, but is not conclusive.

COMBINATORIAL OPTIMIZATION

Here we have a familiar, if nontrivial, problem in optimization. Our genome is of length 20. Hence if there are four admissible values of each parameter, the space of all genomes is of size 2^{40}. And the fitness landscape associated with it is rugged. There is a vast literature on the general problem, with techniques including various forms of gradient ascent ("hill climbing"), genetic programming, simulated annealing, and so on. We will take a direct approach suitable for the heuristic problem at hand. First, we will shrink the search space and then survey it by brute force. Specifically, we will assume single enterprise-wide values for each of the four inertia variables (up, down, horizontal supply, horizontal demand). This cuts the space roughly by a square root and permits a sweep.

The procedure is as follows: For each of the roughly 2^{20} genomes, we record cumulative returns over the time horizon (500 periods). For $k = 0.0$, $k = 0.5$, and $k = 1.0$, we return that genome (the vector of parameter values) which maximizes the cumulative objective function $\sum_{t=0}^{T} F(t; G)$. Then we apply those values and "watch" the optimal history of structural adaptation. In all of what follows, the dynamic environment is the sawtooth introduced above.

PART 5. RESULTS

Based on earlier runs, we have expectations about the extreme cases of profit maximization ($k = 0$) and market-share maximization ($k = 1$), so they are presented first.

PROFIT MAXIMIZING

For profit maximization, the optimal genome generates the flat pure trade solution. The enterprise accepts substantial penetration (sacrifices substantial market share) but avoids the costs of hierarchy. While the optimal parameters (given in table 13.A.1 of the appendix) differ numerically from those used in run 6, graphically this optimal history of structural adaptation is virtually identical to run 6 and can be viewed as movie 8 of the CD. Note again the way trade lags behind the environment and the high level of yellow penetrators.

MARKET-SHARE MAXIMIZING

For market-share maximization, the optimal genome *immediately* generates the maximum hierarchy. Here, profit is sacrificed (costs are ignored) to avoid penetrations. The optimal upward inertia is literally zero. Since the upward hierarchy rule generates hierarchy when $P \geq 0$, it generates hierarchy *before* any reds are even in view. See movie 9 of the CD. In

Frame 1

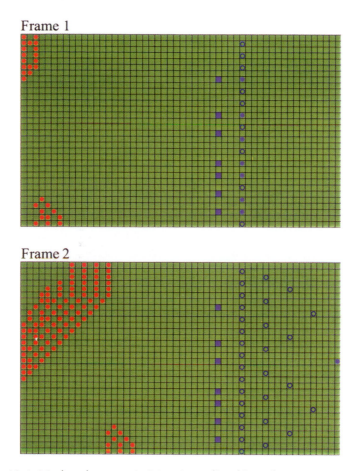

Frame 2

Figure 13.6. Market-share maximizing: immediate hierarchy.

this respect, it differs from run 7. Otherwise, it is exactly as in that run. The first two frames, however, are of interest, and are shown in figure 13.6.

In a sense, this explains the extreme hierarchies observed in military organizations (and perhaps even their peacetime enormity).[13] If the objective is to dexterously allocate forces across a fluid battlefield, "top down command" makes more sense than holding a tank auction. (The same point applies to disaster relief, emergency room operations, etc.).[14]

---

[13] Obviously, this also has a great deal to do with political and industrial interests.

[14] Relatedly, see Weitzman 1974.

These results are in the nature of sanity tests for the model. They are the results we expect, and they were largely anticipated in the earlier runs. What of the hybrid case of $k = 0.5$? Here, the enterprise cares about both profit and market share. If profit maximization is flat (no hierarchy) and market-share maximization is the maximum hierarchy, what would one expect for the hybrid case? Intuition would suggest an intermediate hierarchy.

HYBRID OBJECTIVE

Quite to my surprise, the optimal genome for the $k = 0.5$ case generates no fixed geometry at all. The optimal adaptive organization has a *variable geometry*. This variable geometry firm oscillates between the flat trade regime and a hierarchy of intermediate height. Spatially, the oscillating structure "chases" the sawtooth environment wave up and down the front over time. This "traveling wave" organization has amplitude (maximum hierarchy) 4 and period 10. Run 8 shows this optimal adaptive performance.

The second-best genome looks very much like the winner. But the third-place genome differs in an interesting way. It, too, is an oscillator. But it has lower frequency (30 periods) and higher amplitude (5), building and then dissolving hierarchies of the maximum height possible, as shown in run 9 (movie 11).

Time series of the first- and third-place organizations' oscillations are shown in figure 13.7. Unconstrained maximization of the hybrid ($k = 0.5$) is different in principle from constrained profit maximization. This, too, was studied and generates yet a third, different, oscillating solution.[15]

TOWARD A GENERATIVE DEFINITION OF "DESIGN"

Now consider the question, "What is the winner's 'design'?" Surely, it is not a particular structure, since—given the winner's genome—the structure changes in a dynamic environment. But it is not a particular sequence of structures either, since a different dynamic environment will yield a different adaptive structural history. Rather, the winner's design can only be *the complete adaptive repertoire generable by (encoded in) the enterprise's genome*. And the genome encodes a huge adaptive repertoire, realized differently in different environments, just as different selection pressures induce different realizations of a species' fixed genomic endowment.

---

[15] On dynamic organizational forms, see Nickerson and Zenger 2002; Sastry 1997.

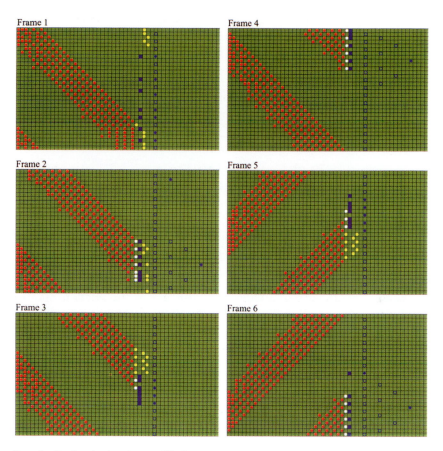

Run 8. Optimal adaptive oscillations.

PART 6. EXTENSIONS AND FUTURE RESEARCH

As an initial demonstration of the approach and of its potential for producing counterintuitive results, perhaps the foregoing exposition is sufficient. But it leaves a great many issues unaddressed. Among the more obvious ones are robustness, mechanism design, imperfect decision-making, multiple firms, and empirical calibration.

ROBUSTNESS

How sensitive are the optimization results to variations in assumptions about manager compensation, transaction costs, and other numerical parameters? Implicitly, the cost of collapsing a hierarchy is zero, when

Frame 1

Frame 2

Frame 3

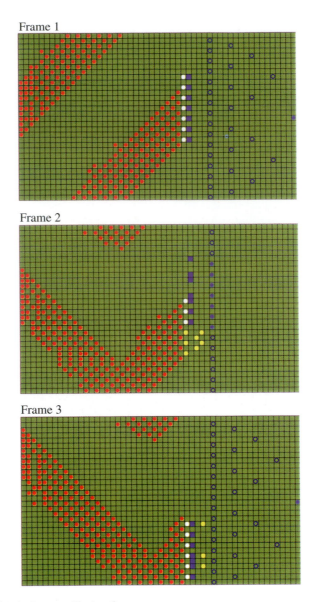

Run 9. Third place oscillating firms.

clearly there are some associated costs. Are there cost levels that would fundamentally alter the relationship between trade and hierarchy?

How sensitive are the optimization results to variations in assumptions about the environmental dynamic? We optimized given a single dynamic

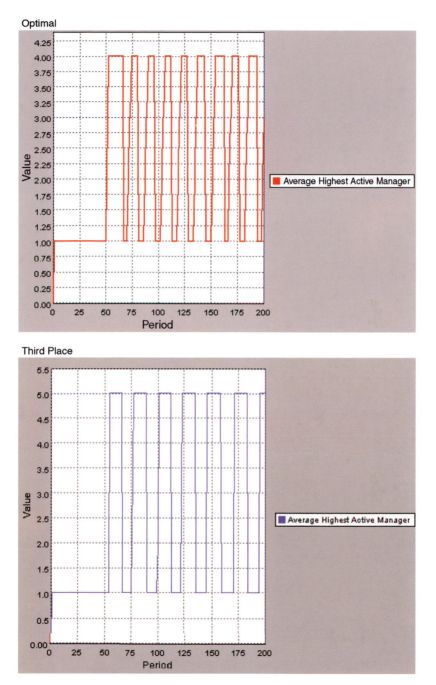

Figure 13.7. Oscillation time series compared.

environment, the sawtooth. But organizations often face different dynamic environments. One could confront organizations with a range of environmental dynamics (e.g., spike, square wave, and sawtooth) and ask for the genome that maximizes the sum or average of returns over those.

In this general connection, the preceding analysis suggests that hierarchy is effective for solving long-range reallocation problems that can suddenly spike in *volatile* environments. But, why is hierarchy ubiquitous in *stable* settings? Behavioral psychology suggests that humans often (*a*) place asymmetric weight on losses (as against gains) and (*b*) confuse probabilities with expected values (overrating risks). One conjecture that might explain the ubiquity of pyramids is that we overrate the loss from unanticipated shocks, and then overrate their probability, thus maintaining a huge overcapacity (hierarchy) to cope (top-down) with rare environmental spikes. In any event, one extension would be to rigorously characterize the circumstances (environment, objective function) under which permanent hierarchy does emerge as optimal. This would be a type of inverse problem.

Having determined (by brute computational force), the optimal propensities to hoard and transfer labor, the economists issue becomes, What mechanism would induce the optimal behavior? What incentive structure would induce *rational* actors to perform optimally? This brings in a huge literature on mechanism design, markets, and games. Modeling approaches doubtless abound. One thought is to explore a variation on classical wage determination. Rather than pay agents their marginal revenue product (MRP), suppose they were paid some convex combination of all the MRPs in their management layer. Rational agents would then have a vested interest in the productivity (intercepts) of others, mitigating the inefficiency of hoarding.

One reason hierarchy allocates effectively in this model is that CEOs are assumed (like everyone else) to follow the labor allocation rule L to perfection, putting unoccupied labor precisely where it is needed. If one were to assume stubborn hidebound CEOs, "sticking to their guns" despite misallocation, hierarchy might fare poorly indeed. Adding noise to the information available at each layer—possibly having it *increase*

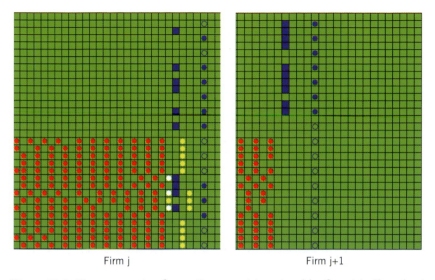

Firm j   Firm j+1

Figure 13.8. Two competing firms. Opportunities missed by firm j (yellow dots) are opportunities (red dots) to firm $j + 1$.

with each layer[16]—might also enrich the tale substantially. It might also prove to be necessary if there were an empirical exercise. Another obvious extension would be to allow CEOs to add labor.

MULTIPLE FIRMS

In the formulation above, firms face an environment, but they do not face other firms. Most firms would probably regard other firms as part of their environment. I introduce multiple firms as follows. Simply imagine firms in a "ring." Opportunities not intercepted by firm j pass through and become opportunities for firm $j + 1$, as illustrated for two firms in figure 13.8. The order in which firms play could be randomized each cycle. New opportunities can be fed in continuously or not.

An illustrative two-firm dynamic case where there are no new opportunities (after 50 cycles) is recorded in run 10 (movie 12). The competitors exhibit different adaptive histories. The right firm remains flat throughout, while the left firm erects the maximum hierarchy (frame 3) and later dissolves it (frame 5), by which point the two firms have essentially divided the market.[17]

---

[16] I thank Ross Hammond for this thought.

[17] Only one horizontal (the fourth from the top) contains blue squares from each firm.

Frame 1

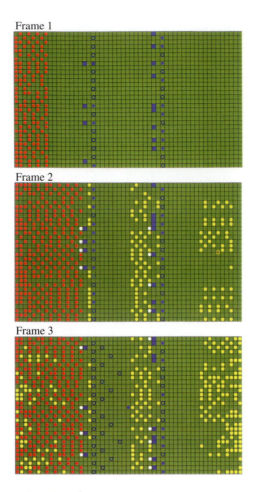

Frame 2

Frame 3

Run 10. Two firms. (*continued on next page*)

A different experiment would be to "peg" one firm to some strategy, and sweep the parameter space for strategies that will defeat it. Friedman (1953) famously argued that the assumption of profit-maximization was warranted on the evolutionary grounds that firms deviating from it would ultimately be selected out. To explore this, one could lock one firm onto profit maximization (our $k = 0$ case) and see whether its competitor (the parameter sweep) could "discover" the strategy of operating at a loss in order to monopolize market share in the short term, driving the first firm out of business, and then "relaxing" into a more profitable strategy having killed off the competition. Of course, it would be natural to coevolve strategies in the general case.

Frame 4

Frame 5

Final frame of movie – steady state

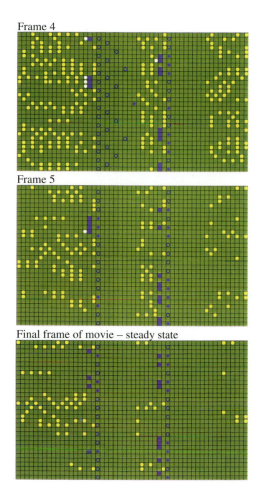

Run 10. (*continued*)

EMPIRICAL CALIBRATION

As noted at the outset, I have made no particular pretense to descriptive realism, and do not purport to have modeled any existing organization (much less a system of competitors), or to "fit" the model to data of any sort. Rather, the aim, using a highly idealized model, has been to illuminate what sorts of individual (micro) agent rules confer adaptiveness on the larger (macro) enterprise. It was an attempt to characterize optimal adaptability in dynamic environments in an idealized fashion. Now, if the variable-geometry firm previously described doesn't look like any existing firm, that could mean simply that existing firms are not adapting optimally, or that the environments they face or objective functions they use are different from the ones I posited. In any event, one obvious direction for future research would be to attempt an empirical calibration of the model to some data on observed organizational adaptation in observed dynamic environments.

Summary

We developed an agent-based model in which hierarchies and internal trading regimes emerge endogenously. And we generated a variety of structural adaptations in a range of contrived dynamic environments. We introduced a rather general objective function for the enterprise and for a variety of special cases (profit maximizing, market-share maximizing, and hybrid), we determined the optimal genome for a particular dynamic test environment. Applying that genome, we then generated, and depicted graphically, the optimal history of structural adaptation to the test environment. This winning history was, to me, quite unexpected, involving oscillating "flat" trading regimes and hierarchies of intermediate height in perpetual motion up and down the spatial market as a traveling wave: a variable geometry firm. A number of extensions and directions for future research were discussed.

APPENDIX: NUMERICAL ASSUMPTIONS AND PARAMETER DEFINITIONS

TABLE 13.A.1
Numerical Assumptions

Run	T1	T2	T3	T4	T5	Up	Down	Supply	Demand	Formation
1	0.7	1.4	2.0	4.0	10	20	1000	0	0	2 (south)
2	0.7	1.4	2.0	4.0	10	20	10	0	0	2
3	0.7	1.4	2.0	4.0	10	20	1000	0	0	3 (split)
4	0.7	MAX	MAX	MAX	MAX	20	10	1	1	3
5	0.7	MAX	MAX	MAX	MAX	20	10	1	1	1 (sawtooth)
6	0.7	MAX	MAX	MAX	MAX	20	10	1	1	1
7	0.7	1.4	2.0	4.0	5.0	20	100	0	0	1
8	0.5	1.0	3.0	4.0	5.0	8.0	0	1	1	1
9	0.0	1.0	1.5	4.0	0.0	8.0	8	0.5	0.5	1
10	0.5	1.0	1.5	2.0	4.0	0.0	4	1	1	1
11	0.0	1.0	3.0	4.0	2.0	0.0	8	1	1	1
12	0.7	1.4	2.0	4.0	10	10	10	1	1	1

Note: The time horizon $T = 500$ in all runs. T1–T5 refer to Thresholds 1–5. Up/Down refer to up and down Inertias (vertical). Supply/Demand refer to probability of supplying/ demanding labor $= 1 -$ inertia (horizontal). Formation is attack formation. MAX is some large number (i.e. 20) chosen to preclude triggering.

TABLE 13.A.2
Parameter Definitions

Parameter	Value	Explanation
Attack formation	1, 2, or 3[a]	Determines the environment (the approach the attackers take)
Cost per penetration	100	Cost to profit function of a penetration
Value per Intercept	200	Revenue per intercept
Cost per trade	5	Cost to profit function of trading labor
Memory	10	The number of periods over which average penetration and intercepts is calculated
k	varies	k-constant in profit function
1–demand inertia	varies	Probability a manager will ask others of its rank for labor
1–supply inertia	varies	Probability a manager will giver others of its rank labor
Prob create attacker	0.7	Probability an attacker is created
Management salary multiplier, c	10	Multiplier used to determine salary
Salary for hollow manager	$c \cdot rank$	Salary controlling no labor
Salary for solid manager	$c \cdot (rank + 1)^3$	Salary controlling labor
Salary for labor	1	Wage to labor
Threshold rank 1	varies	Threshold of managers of rank 1
Threshold rank 2	varies	Threshold of managers of rank 2
Threshold rank 3	varies	Threshold of managers of rank 3
Threshold rank 4	varies	Threshold of managers of rank 4
Threshold rank 5	varies	Threshold of managers of rank 5
Downward inertia	varies	Number of periods a manager's P must be $< T_{min}$
Upward inertia	varies	Number of periods a manager's P must be $\geq T_{max}$
Time horizon, T	500	The number of iterations in the optimizations

Further elements of the Ascape source code not referred to in the text that may be of interest to programmers.

(continued)

TABLE 13.2 (*continued*)

Parameter	Value	Explanation
Max Active Rank	1	Highest ranking manager who is active
Attack Formation Switch Probability	0	Probability of switching directions of certain attacks
Attacker Speed	1	Number of spaces attackers move per iteration
Nearness Line of Sight	TRUE	Reserved
Prob Create Front Line Defender	15	Probability a front-line defender is created at initialization
Radius	1	Reserved
Random Edge Ratio	0	Reserved
Size	(5, 2, 31, 3)	Reserved

[a]Formation 1 is the sawtooth, 2 is opportunity flux only in the south, and 3 is the split attack.

REFERENCES

Axtell, Robert L. 1999. "The Emergence of Firms in a Population of Agents: Local Increasing Returns, Unstable Nash Equilibria, and Power Law Size Distributions." Working Paper 3, Center on Social and Economic Dynamics.

———. 2001. "Zipf Distribution of U.S. Firm Sizes." *Science* 293:1818–20.

Baum, Joel A. C., and Bill McKelvey, eds. 1999. *Variations in Organization Science: In Honor of Donald T. Campbell*. Thousand Oaks, CA: Sage.

Carley, Kathleen M., and David M. Svoboda. 1996. "Modeling Organizational Adaptation as a Simulated Annealing Process." *Sociological Methods and Research* 25(1): 138–68.

Chang, Myong-Hun, and Joseph E. Harrington Jr. 2006. "Agent-Based Models of Organizations." In *Handbook of Computational Economics*, vol. 2, *Agent-Based Computational Economics*, ed. Kenneth L. Judd and Leigh Tesfatsion. Amsterdam: North-Holland.

Coase, Ronald H. 1937. "The Nature of the Firm." *Economica* 4:386–405.

Cohen, Michael D., James G. March, and Johan P. Olsen. 1972. "A Garbage Can Model of Organizational Choice." *Administrative Science Quarterly* 17: 1–25.

Epstein, Joshua M. 2003. "Growing Adaptive Organizations: An Agent-Based Computational Approach." Working paper, Santa Fe Institute, New Mexico, May 29.

Epstein, Joshua M., and Robert L. Axtell. 1996. *Growing Artificial Societies: Social Science from the Bottom Up*. Washington, DC: Brookings Institution Press; Cambridge: MIT Press.

Ethiraj, Sendil K., and Daniel Levinthal. 2002. "Search for Architecture in Complex Worlds: An Evolutionary Perspective on Modularity and the Emergence of Dominant Designs." Wharton School, University of Pennsylvania, pdf copy, November.

Friedman, Milton. 1953. *The Methodology of Positive Economics*. Chicago: University of Chicago Press.

Lomi, Alessandro, and Erik R. Larsen. 1999. "Evolutionary Models of Local Interaction: A Computational Perspective." In *Variations in Organization Science: In Honor of Donald T. Campbell*, ed. Joel A. C. Baum and Bill McKelvey. Thousand Oaks, CA: Sage.

Miller, John H. 2001. "Evolving Information Processing Organizations." In *Dynamics of Organizations: Computational Modeling and Organization Theories*, ed. Alessandro Lomi and Erik R. Larsen. Menlo Park, CA: AAAI Press and MIT Press.

Nickerson, Jack A., and Todd R. Zenger. 2002. "Being Efficiently Fickle: A Dynamic Theory of Organizational Choice." *Organization Science* 13:547–66.

Prietula, Michael J., and Kathleen M. Carley. 1994. "Computational Organization Theory: Autonomous Agents and Emergent Behavior." *Journal of Organizational Computation* 41:41–83.

Radner, Roy. 1993. "The Organization of Decentralized Information Processing." *Econometrica* 61:1109–46.

Radner, Roy, and Timothy Van Zandt. 1999. "Real-Time Decentralized Information Processing and Returns to Scale." *Economic Theory* 17:545–75.

Sastry, M. Anjali. 1997. "Problems and Paradoxes in a Model of Punctuated Organizational Change." *Administrative Science Quarterly* 42:237–75.

Weitzman, Martin L. 1974. "Prices vs. Quantities." *Review of Economic Studies* 41:477–91.

Williamson, Oliver E. 1975. *Markets and Hierarchies: Analysis and Antitrust Implications*. New York: Free Press.

Young-pa, So, and Edmund H. Durfee. 1998. "Designing Organizations for Computational Agents." In *Simulating Organizations: Computational Models of Institutions and Groups*, ed. Michael J. Prietula, Kathleen M. Carley, and Les Gasser. Menlo Park, CA: AAAI Press and MIT Press.

CODA

No one who is still growing intellectually ever feels that he has said all he can in a book. I suppose, therefore, that I should take consolation in the sense of incompleteness I feel in arbitrarily closing the discussion at this point; it is a sign of life.

In a nutshell, I have tried to advance an argument for generative social science and to demonstrate its principal scientific instrument—the agent-based model—in a wide range of applications. If that argument is not now persuasive (and for some it will not be), its repetition at this juncture will not make it so.

What might make it so is further, and better, work. While many of the chapters suggest model extensions and specific areas for further research, I feel that (at least) four overarching areas are undeveloped in this book (and to varying degrees, in the literature as a whole). They are: formalization, networks, psychology, and scaling.

Formalization

As I argued strenuously in the first two chapters, every realization of an agent-based model is a strict deduction, a theorem. As an epistemological matter, it is important to insist that the activity is therefore deductive in nature. At the same time, I observed that we typically quantify over relatively small sets, and thus, these computational theorems are seldom very general. Call me old-fashioned, but I would like to see more work developing an explicit formalism in which to represent agent models. This is not to deny the practical adequacy of statistical approaches or the Russellian beauty of agent models, both of which are very real. But there is unexplored territory, particularly in the area of recursive functions and related fields. This is *terra incognita* to most social scientists, and it is mathematically craggy terrain. But there may just "be gold in them thar hills," particularly in establishing further results on Incompleteness and Computational Complexity in social science.

Endogenous Networks: The Mind of Society[1]

Another underdeveloped area in this book is social networks—how they happen and why they matter. Above, explicit networks are really used

---

[1] The notion of social networks and markets as distributed computational devices is discussed in the opening Generative chapter.

only in the Retirement model, and implicitly in the Smallpox model. How does social network structure affect contagion dynamics—of diseases, of violence, of norms, of technologies, of prices? Much analytical work on such topics treats peer effects in fixed exogenous networks. Some kid contracts a bad habit (smoking, to pick a topic we are working on at our Center[2]) from peers in his network. But, the kid may in fact take up the bad habit in order to gain entry into the network. He gains utility from membership. The network is itself endogenous and dynamic, its structure an "emergent property," if I may be so bold. Agent-based models are well suited to explore this important phenomenon.

Individual Psychology: The Society of Mind[3]

Of course, to psychologists and novelists, the ultimate "emergent phenomenon" is the individual, the agent itself. Another underdeveloped area in this book (if not in the literature as a whole) is individual psychology and learning. As just noted, this is not unrelated to network formation, as individuals may derive utility from network membership, and as the burgeoning literature on happiness shows, may derive disutility from it as well (through invidious comparison to particular reference groups).[4] More generally, the agents in this book are, by design, extremely simple, indeed psychologically impoverished. That is defensible on a variety of methodological grounds, which have been discussed. That said, individuals of any depth and interest are themselves societies—resultants of competing drives, some social, some innate. Want a challenge? Grow Raskolnikov. *That* would be something.

Synthesis

Taking the last two points together, one can imagine agent models with psychologically richer agents coevolving in and generating endogenous social networks. I suspect that such models, while perhaps challenging to analyze, might be worth the effort.

Moore's Law Is Double-Edged

Another question, understudied in this book, is how agent models scale up. With the exception of the Anasazi and perhaps Smallpox models,

---

[2] The Brookings-Johns Hopkins Center on Social and Economic Dynamics.
[3] Marvin Minsky's colorful phase. See *The Society of Mind* (New York: Simon and Schuster, 1985).
[4] See Carol Graham and Stefano Pettinato, *Happiness and Hardship: Opportunity and Insecurity in New Market Economies* (Washington, DC: Brookings Institution Press, 2002).

these are all "Toy" models. I'm not sure it's very interesting, but I know it's very important to understand how model behavior changes when scale (e.g., the sheer number of agents) is dramatically increased. The spectacular growth in computing power will facilitate this. It will also provide strong temptation, when designing models, to put a lot *in*. But that's not the trick. The trick is to get a lot *out*, while putting in as little as possible.

Hence, Einstein affirms "the grand aim of all science, which is to cover the greatest number of empirical facts by logical deduction from the smallest possible number of hypotheses or axioms."[5] As Einstein also knew, prevailing approaches are not abandoned simply because anomalies and criticisms mount. Scientists need a viable alternative. Hopefully, this book helps to provide one.

---

[5] Albert Einstein, *Ideas and Opinions* (New York: Bonanza Books, 1954), 274. I thank Samuel David Epstein for bringing this passage to my attention.